CURRENT TOPICS IN

DEVELOPMENTAL BIOLOGY

VOLUME 17

NEURAL DEVELOPMENT
PART III
Neuronal Specificity, Plasticity, and Patterns

CONTRIBUTORS

SALVATORE CARBONETTO

RONALD L. MEYER

JACK DIAMOND

KENNETH J. MULLER

REHA S. ERZURUMLU

R. K. MURPHEY

SCOTT E. FRASER

RONALD W. OPPENHEIM

H. V. B. HIRSCH

MU-MING POO

HERBERT P. KILLACKEY

GUNTHER S. STENT

DAVID A. WEISBLAT

CURRENT TOPICS IN
DEVELOPMENTAL BIOLOGY

EDITED BY

A. A. MOSCONA

CUMMINGS LIFE SCIENCE CENTER
THE UNIVERSITY OF CHICAGO
CHICAGO, ILLINOIS

ALBERTO MONROY

STAZIONE ZOOLOGICA
NAPLES, ITALY

VOLUME 17

NEURAL DEVELOPMENT
PART III
Neuronal Specificity, Plasticity, and Patterns

VOLUME EDITOR

R. KEVIN HUNT

THOMAS C. JENKINS DEPARTMENT OF BIOPHYSICS
THE JOHNS HOPKINS UNIVERSITY
BALTIMORE, MARYLAND

1982

ACADEMIC PRESS
A Subsidiary of Harcourt Brace Jovanovich, Publishers
New York   London
Paris   San Diego   San Francisco   São Paulo   Sydney   Tokyo   Toronto

ACADEMIC PRESS, INC.
111 Fifth Avenue, New York, New York 10003

United Kingdom Edition published by
ACADEMIC PRESS, INC. (LONDON) LTD.
24/28 Oval Road, London NW1 7DX

LIBRARY OF CONGRESS CATALOG CARD NUMBER: 66–28604

ISBN 0–12–153117–1

PRINTED IN THE UNITED STATES OF AMERICA

82 83 84 85   9 8 7 6 5 4 3 2 1

CONTENTS

CHAPTER 7. From Cat to Cricket: The Genesis of Response
            Selectivity of Interneurons
            R. K. MURPHEY AND H. V. B. HIRSCH

CHAPTER 8. The Neuroembryological Study of Behavior: Progress,
            Problems, Perspectives
            RONALD W. OPPENHEIM

CONTRIBUTORS

Numbers in parentheses indicate the pages on which the authors' contributions begin.

SALVATORE CARBONETTO, *Department of Pharmacology, State University of New York, Upstate Medical Center, Syracuse, New York 13210* (33)

JACK DIAMOND, *Department of Neurosciences, McMaster University, Hamilton, Ontario L8N 3Z5, Canada* (147)

REHA S. ERZURUMLU, *Department of Psychobiology, University of California, Irvine, California 92717* (207)

SCOTT E. FRASER, *Department of Physiology and Biophysics, University of California, Irvine, California 92717* (77)

H. V. B. HIRSCH, *Neurobiology Research Center, State University of New York, Albany, New York 12222* (241)

HERBERT P. KILLACKEY, *Department of Psychobiology, University of California, Irvine, California 92717* (207)

RONALD L. MEYER, *Developmental Biology Center, University of California, Irvine, California 92717* (101)

KENNETH J. MULLER, *Department of Embryology, Carnegie Institution of Washington, Baltimore, Maryland 21210* (33)

R. K. MURPHEY, *Neurobiology Research Center, State University of New York, Albany, New York 12222* (241)

RONALD W. OPPENHEIM, *Neuroembryology Laboratory, Division of Mental Health Research, Dorothea Dix Hospital, Raleigh, North Carolina 27611* (257)

MU-MING POO, *Department of Physiology and Biophysics, University of California, Irvine, California 92717* (77)

GUNTHER S. STENT, *Department of Molecular Biology, University of California, Berkeley, California 94720* (1)

DAVID A. WEISBLAT, *Department of Molecular Biology, University of California, Berkeley, California 94720* (1)

ix

PREFACE

The study of neurogenesis continues its expansive growth, and its ongoing need for increased dialogue between neuroscientists and cell and developmental biologists. Our collection of essays on neural development has expanded accordingly. The present collection of essays on neural specificity, plasticity, and patterns represents the third volume in this treatise.

Nearly four decades have passed since Roger Sperry redefined the problem of developing neural patterns and connections, synthesizing ideas from Cajal and Harrison and Holtfreter, in terms of cytochemical differentiation and intercellular affinity. Sperry's Theory of Neuronal Specificity is widely appreciated for the lively debates it provoked and the rich array of experiments it stimulated, on the development and regeneration of nerve patterns and interneuronal connections. The Theory, of course, was much more than that: a passing of the torch from physiological psychology to contemporary cell biology, a redefinition of prefunctional versus functional determinants in terms of the differentiated states and molecular identities of the nervous system's maturing cells. The power of Sperry's ideas is perhaps best measured in the difficulty we find today in trying to envision the "nerve energies" and "functional compatibilities" that were previously thought to forge functional nerve circuits out of initially equipotential neural networks. As the essays in this volume indicate, the problems of cellular specificity and pattern remain areas of vigorous research, of frustration and progress, and of enduring interest.

R. Kevin Hunt

CHAPTER 1

CELL LINEAGE ANALYSIS BY INTRACELLULAR INJECTION OF TRACER SUBSTANCES

David A. Weisblat and Gunther S. Stent

DEPARTMENT OF MOLECULAR BIOLOGY
UNIVERSITY OF CALIFORNIA
BERKELEY, CALIFORNIA

I. Introduction

Knowledge of the lines of descent of cells that eventually compose the mature organism is useful for understanding the mechanisms that govern embryonic development. The first cell lineage analyses were carried out in 1878 by C. O. Whitman. Working with the glossiphoniid leech *Hemiclepsis* and following the cell cleavage pattern by direct visual observation, Whitman showed that a developmental fate can be assigned to each blastomere of the early embryo. Glossiphoniid leeches are highly suitable for such studies because their eggs are large (with a diameter of approximately 1 mm) and undergo stereotyped cleavages that produce a blastula containing large identifiable cells. These embryos can be observed, manipulated, and cultured to maturity in simple media. Moreover, the development from egg to adult is direct, without larval structures (Schleip, 1936). Cell lineage analyses were later extended to the embryos of other species, not only by direct observation but also with other techniques such as selective ablation, application of extracellular marker particles, and, most importantly, induction of genetic mosaics (Wilson, 1892; Sturtevant, 1929; Tarkowski, 1961; Mintz, 1965; Stern, 1968; Garcia-Bellido and Merriam, 1969; Le Douarin, 1973; Sulston and Horvitz, 1977; Deppe *et al.*, 1978).

1

In this chapter we will review a new method that we devised for determining developmental cell lineages (Weisblat *et al.*, 1978, 1980a,b). In this method an intracellular tracer is injected through a micropipet into identified cells of early embryos. After injection of the tracer, embryonic development is allowed to progress to a later stage, at which time the distribution pattern of the tracer within the embryonic tissues is visualized. In this way one can determine the developmental origin of cells whose small size, envelopment by morphologically similar cells, or remote descent from an identifiable progenitor prevents determination of their lineage by direct observation. The success of the tracer method requires that three conditions be satisfied: (1) After injection of the tracer, embryonic development must continue normally. (2) The injected tracer must remain intact and not be diluted too much in the developing embryo. (3) The tracer must not pass through junctions linking embryonic cells.

Two different types of tracers have been used for cell lineage analysis in the development of the glossiphoniid leech *Helobdella triserialis*: the enzyme horseradish peroxidase and the fluorescent dye rhodamine coupled to a synthetic dodecapeptide. In this chapter we will present an account of the results obtained through use of these two tracers, in particular with respect to the developmental origins of the components of the leech central nervous system.

II. Early Embryonic Development of Glossiphoniid Leeches

The early embryonic development of glossiphoniid leeches has been divided into eight stages (Fig. 1) (Fernandez, 1980; Weisblat *et al.*, 1980a). The times given pertain to the development of *H. triserialis* at 25°C.

Stage 1. Uncleaved egg (0–4.5 hours). The egg of *H. triserialis* is about 500 μm in diameter and is encased in a transparent vitelline membrane. Most of the egg cytoplasm consists of pink yolk.

Stage 2. Two cells (4.5–6.5 hours). The egg cleaves, resulting in cells AB and CD.

Stage 3. Four cells (6.5–8 hours). Cell CD cleaves to produce cells C and D; then cell AB cleaves to produce cells A and B.

Stage 4. This stage has been subdivided as follows: *Stage 4a*. Micromere quartet. Each of the four cells A, B, C, and D cleaves to produce a macromere and a much smaller micromere. Additional rounds of micromere formation directly from the macromeres have been reported for other glossiphoniid embryos (Schleip, 1936). It has not yet been established whether such additional rounds occur in *Helobdella*. *Stage*

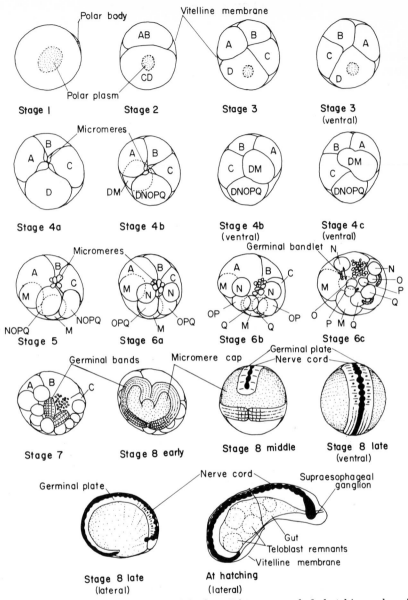

FIG. 1. Schematic representations of the first eight stages and of a hatching embryo in the development of *Helobdella triserialis*. Unless designated otherwise, all drawings are dorsal views, with left and right sides of the embryo lying to the left and right of the drawing, respectively. The two lateral views are shown venter upward, with the head lying to the right. All stages are drawn to the same scale; the diameter of the uncleaved egg (Stage 1) is about 500 μm. In the drawing of Stage 6c, two alternative positions of cell N*l* are shown: the more common one in solid and the less common one in dotted outline, connected by a double-headed arrow.

4b. Macromere quintet. The D macromere cleaves to produce cells DNOPQ and DM. Since Whitman, cell DNOPQ is regarded as the source of ectoderm, and DM as the source of mesoderm. *Stage 4c.* Mesoteloblast formation. Cell DM cleaves to yield the right and left mesoteloblasts M*r* and M*l*.

Stage 5. Ectoteloblast precursor (14–17 hours). Cell DNOPQ cleaves symmetrically to yield the right and left ectoteloblast precursors NOPQ*r* and NOPQ*l*.

Stage 6. Teloblast completion (17–30 hours). During this stage, bilaterally homologous cells cleave synchronously. This stage has been subdivided as follows: *Stage 6a.* N teloblast formation. Cell NOPQ cleaves, yielding the smaller ectoteloblast N and the larger cell OPQ. *Stage 6b.* Q teloblast formation. Cell OPQ cleaves, yielding the smaller ectoteloblast Q and the larger cell OP. *Stage 6c.* O and P teloblast formation. Cell OP cleaves, yielding the final ectoteloblasts O and P. Thus, at the conclusion of Stage 6, one bilateral pair of mesoteloblasts (M*l* and M*r*) and four bilateral pairs of ectoteloblasts (N*l* and N*r*, O*l* and Or, P*l* and P*r*, Q*l* and Q*r*) have been formed. Of these, the M teloblasts are the largest and the O and P teloblasts are the smallest, with the N and Q teloblasts being of an equal, intermediate size. Further cleavages of ectoteloblasts or their precursors, resulting in the formation of additional micromeres during Stages 5 and 6, have been reported for other glossiphoniid embryos (Müller, 1932; Fernandez, 1980). It is likely that such cleavages also occur in *Helobdella*.

Stage 7. Germinal band formation (30–78 hours). Each teloblast initiates a series of unequal divisions that produces a column of small *stem cells*, called a *germinal bandlet*. These stem cells are designated by a lower case letter corresponding to their teloblast of origin. On each side the germinal bandlets merge to form the *germinal band*, as shown in Fig. 3a and b. The left and right germinal bands then move rostrally on the surface of the embryo as more stem cells are added to each bandlet. In the germinal bands, the m bandlet arising from the M mesoteloblast lies under the n, o, p, and q bandlets arising from the N, O, P, and Q ectoteloblasts. The ectodermal bandlets lie in the order of their alphabetic designations, with the q bandlet pair most medial. The germinal bands grow in crescent shape around the *micromere cap* (Mann, 1962). This growth entails a circumferential migration over the surface of the embryo, attended by an expansion of the micromere cap. Right and left germinal bands meet at the future rostral end of the animal.

Stage 8. Coalescence of the germinal bands (78–122 hours). The right and left germinal bands coalesce zipper-like at their lateral edges in a rostrocaudal progression along the future ventral midline to form the *germinal plate*. The germinal bands continue their circumferential migration, meanwhile elongating as the result of further stem cell production by teloblasts and by stem cell divisions. After their circumferential migration, the right and left n bandlets lie most medially in the germinal plate, directly apposed across the ventral midline. The o, p, and q bandlets lie progressively more laterally. As the germinal bands move circumferentially over the surface of the embryo, the area behind them is covered by a layer of cells, some of which, as is to be shown below, are left behind by the migrating bands. Even before the germinal bands have completely coalesced, the embryo begins peristaltic movements along its longitudinal axis.

The definition of later developmental stages of *Helobdella* embryos awaits more detailed observations. The germinal plate expands by cell division and migration in a rostrocaudal progression, covering the surface of the embryo until right and left edges finally meet on the dorsal midline. *Helobdella* embryos hatch from their vitelline membrane at about the sixth day. The bulk of the yolk of the five teloblast pairs is not passed on to their stem cell progeny. It is eventually enveloped by endoderm and, together with the yolk of A, B, and C macromeres, constitutes the content of the gut of the young leech. According to Whitman (1878), the endoderm arises from the A, B, and C macromeres.

III. Development of the Leech Nervous System

By the end of Stage 8, the *Helobdella* embryo shows definite signs of segmentation at the rostral end of the germinal plate, with the appearance of primordial segmental ganglia in register with longitudinally iterated blocks of mesodermal tissue. Segmentation and ganglionic development progress rostrocaudally. By the time of hatching, the rostral ganglia of the embryonic nerve cord already possess many features of the adult ganglia.

The adult nervous system of *Helobdella* is similar to that of other leeches, such as *Hirudo* (Nicholls and Van Essen, 1974). At the front end of the nerve cord is the supraesophageal ganglion, which is linked by a pair of circumesophageal connectives to the subesophageal ganglion. The subesophageal ganglion, in turn, is followed by 21 segmental ganglia linked by the nerve tracts of the interganglionic connectives. The caudal ganglion forms the rear end of the nerve cord. A pair of segmental nerves exits from the left and from the right side of each segmental ganglion. Each segmental ganglion contains on the order of

(a)   **(b)**

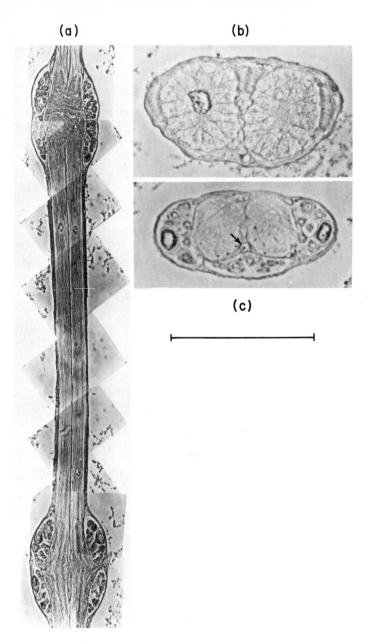

(c)

FIG. 2. Hematoxylin-phloxine-safran-stained sections of a paraffin-embedded adult specimen of *Helobdella*. (a) Horizontal section through the nerve cord at a dorsal level, showing two ganglia linked by the connective. Within the ganglia, nerve cell bodies of

330 neurons (E. Macagno, personal communication), distributed over six distinct *cell packets*, separated by sheets of connective tissue. There are two *lateral* packets on each side, one pair anterior and the other pair posterior, extending over both the ventral and the dorsal aspects of the ganglion, and two *ventromedial* packets, one anterior and the other posterior. The cell packets constitute a cortex of the ganglion. Within the ganglion lies a neuropil of fine processes emanating from neuronal cell bodies both in its own cell packets and in other ganglia of the nerve cord. The roughly bilaterally symmetric right and left halves of the ganglion will be referred to here as *hemiganglia*. In addition to neuronal cell bodies, each hemiganglion also contains a small number of glial cells, as do the interganglionic connectives. *Hirudo* has one bilateral pair of giant glial cell bodies per segment, located midway between the segmental ganglia in the core of the intersegmental connectives. A second pair is located on the ventral aspect of the neuropil of the segmental ganglion. In addition, there is one giant glial cell in each of the six segmental cell packets (Coggeshall and Fawcett, 1964). The disposition, and possibly number, of glial cells in the interganglionic connectives of *Helobdella* differs from that in *Hirudo*, however. *Helobdella* has *two* bilateral pairs of giant glial cell nuclei in its interganglionic connectives, of which one pair is located in the core at each end of these connectives (Fig. 2a and b). On the basis of horseradish peroxidase injections of homologous cells in the related leech *Haementeria*, it seems likely that these nuclei belong to two pairs of glial cells rather than to one binucleate cell pair (K. J. Müller, personal communication). Like *Hirudo*, *Helobdella* has a pair of nerve tract glial cell bodies on the ventral aspect of the ganglionic neuropil (Fig. 2c). Furthermore, as in *Hirudo* (Coggeshall and Fawcett, 1964), the sheath of the interganglionic connective of *Helobdella* contains longitudinally oriented muscle cells (Fig. 2b).

---

the anterior and posterior lateral packets surrounded by the connective tissue sheet are visible. (At this dorsal level, the ventromedial cell packets cannot be seen.) Near either end of the connective, the nuclei of a pair of giant glial cells can be seen (one of these nuclei, lying slightly out of the plane of the section, is only faintly visible). (b) Transverse section through the connective, showing radially oriented glial processes and, in the center of the left connective, the nucleus of one glial cell. Two pairs of longitudinally oriented muscle cells, one pair dorsomedial and the other lateral, can be seen in the sheath of the connective. The arrows point to the muscle cells of the right connective. (c) Transverse section through a ganglion. The arrow points to a glial cell body visible on the midline of the ventral edge of the neuropil, lying in the nerve tract just above the ventromedial cell packet. The second glial cell of this pair lies out of the plane of the section. [For comparison of these sections with corresponding findings in *Hirudo*, cf. Fig. 2 of Coggeshall and Fawcett (1964).] Bar: (a) 270 μm; (b) 50 μm; (c) 100 μm.

At hatching, the rostral segmental ganglia of the embryonic nerve cord have already moved apart; neuropil, connectives, and segmental nerves are present; and the number of nerve cell bodies in the ganglion of the embryo is about the same as in that of the adult. But neither the nerve cell packets nor their glial cells are as yet clearly discernible, and in the connectives, giant glial cell bodies are still missing from the position at which they are found in the adult.

Because of the medial disposition of the n bandlets within the germinal plate, Whitman (1878, 1887) and later workers (e.g., Bergh, 1891) inferred that the ventral nerve cord derives from the N teloblast pair. But as for the supraesophageal ganglion, Whitman (1887) asserted that it has its origin in other than the germinal bands, whereas Müller (1932) claimed that it does arise from the germinal bands. It is hardly surprising that conclusions regarding the developmental origin of tissues of late leech embryos based on direct observations are controversial; after Stage 6, it is difficult to follow the fate of stem cells because they and their descendant cells are small, numerous, and often hidden from view. Therefore, we reexamined the question of the developmental origin of the leech CNS by means of the intracellular tracer method.

IV. Horseradish Peroxidase Method

To identify the progeny of a particular blastomere, a cell of an early leech embryo is penetrated with a micropipet filled with a 2–5% solution of horseradish peroxidase (HRP; Sigma Chemical Co. type VI or IX) in 0.2 M KCl, and a small volume of the solution is forced into the cell by pressure. For injection, the embryo is placed on the tip of a hollow plastic tube, oriented with the target cell facing upward, and immobilized by gentle suction through the tube. The HRP-injected embryo is then allowed to develop to some later stage. At this point it is fixed and stained for HRP by standard histochemical techniques (Müller and McMahan, 1976). The stained embryos can be examined under the microscope, either as whole mounts or, after embedding them in plastic (Epon), as serial thick sections. The examination of sections is aided by counterstaining them with toluidine blue, which causes the yellow HRP-stained cells (that do not take on toluidine blue) to contrast with the blue background of the other cells.

Figures 3 and 4 present the results of some experiments using this technique. The embryos shown in Figs. 3a and 4a are preparations in which cell DNOPQ was injected with HRP at Stage 4b; the injected preparations were fixed and stained after 4 more days of development, by which time the germinal plate was beginning to form. As can be seen, on both sides of the embryo the four ectoteloblasts N, O, P, and Q and the four superficial stem cell columns of the germinal bands were stained, whereas the mesoteloblast M and the stem cells issuing from it were not stained. These results confirm the cell lineage scheme of Fig. 1, according to which the four bilateral ectoteloblast pairs are derived from a single progenitor (DNOPQ) that forms a line of descent different from that founded by the progenitor (DM) of the bilateral mesoteloblast pair. Moreover, these results demonstrate that the germinal bands are formed by the merger of columns of stem cells derived from the teloblasts (Schleip, 1936).

In the embryos shown in Fig. 4b, the right or left NOPQ cell was injected with HRP at Stage 5; the embryos were fixed and stained after allowing 6 more days of development. By then, formation of the germinal plate had proceeded to the caudal end and the embryos had begun peristaltic movements associated with hatching from the egg membrane. Only the right or left half of the embryo was stained. This result shows that each NOPQ cell is the progenitor of only the ipsilateral half of the germinal plate, i.e., that the bilateral symmetry of the ectoderm is established by the division of cell DNOPQ into the sister pair NOPQr and NOPQl, and that there is little cell migration across the ventral midline.

The embryos shown in Figs. 3c and 4c and d are ones whose left ectoteloblast Nl had been injected with HRP at Stage 6a before production of its stem cell column had begun. Development was then allowed to proceed for 6 more days, by which time formation of the ventral nerve cord was essentially complete. Evidently, the left half of the ventral nerve cord was stained. This finding supports Whitman's (1887) inferences, in that, first, progeny of a given teloblast are consigned to discrete structures during normal development rather than randomly distributed throughout the germinal plate, and, second, each N teloblast is a precursor of its ipsilateral half nervous system. The embryo shown in Fig. 4e is a preparation similar to the preparations shown in Figs. 3c and 4c and d, except that in this case the Nl ectoteloblast had been injected with HRP at Stage 7, after n stem cell production had already begun. Here the caudal but not the rostral part of the left half of the ventral nerve cord was stained. This shows that the caudal part of the hemilateral nervous system develops from the younger stem cells in the column produced by the N ectoteloblast. The sharpness of the boundary between stained caudal and unstained rostral portions of the cord and the sharpness of the boundary along the midline between stained left and unstained right hemiganglia again suggests that, at least up to this stage of development of the nerve cord,

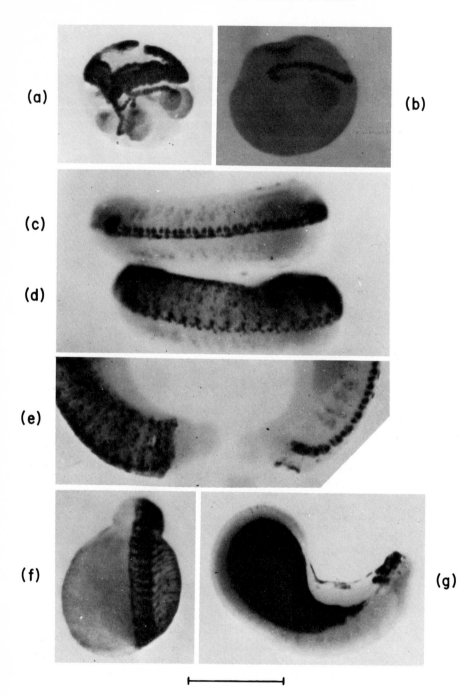

there is little migration of cell bodies either longitudinally or across the midline of the embryo. Nevertheless, some cells do seem to migrate circumferentially; in Fig. 4c, regularly spaced spots of stained tissue appear outside of, and ipsilateral to, the stained hemiganglia. These spots may be primordial sensillae, the peripheral segmental sensory structures that lie circumferentially in the central annulus of the segment (Mann, 1962). Thus, neurons might establish the sensillae by migration from the developing nerve cord. It should be noted, however, that other possibilities, including the one that these spots represent nonneural N teloblast progeny, have not been excluded.

The foregoing results demonstrate that, in satisfaction of the first two conditions for the success of the HRP tracer technique, embryonic development does continue normally after enzyme injection and HRP remains catalytically active in the developing embryo. To demonstrate that this technique satisfies the third condition of cellular localization of HRP, the results of a control experiment are presented in Fig. 4f and g. In this experiment individual NOPQ cells of several embryos were injected with a mixture of HRP and the fluorescent dye, Lucifer Yellow (Stewart, 1978), at Stage 6a. Without allowing further development, the injected embryos, as well as some uninjected embryos, were first

---

FIG. 3. Photomicrographs of HRP-injected *Helobdella* embryos, viewed in whole mount, from the dorsal aspect in (a) and (b) (with future head at top), from the ventral aspect in (c), (d), and (f; [with head at left in (c) and (d) and on top in (f)], and from the lateral aspect in (e) and (g) [with head in center in (e) and on right in (g)]. Bar: 250 μm. (a) Cell DNOPQ was injected at Stage 4b. Fixation and staining occurred at late Stage 7. The n, o, p, and q bandlets and the ectodermal (superficial) part of the germinal bands are stained bilaterally. The ectoteloblasts are lightly stained. This result demonstrates that cell DNOPQ is the progenitor of left and right ectoteloblasts. (b) Cell N*r* was injected at early Stage 7. Fixation and staining occurred at early Stage 8. The right n germinal bandlet is stained. The N*r* teloblast is lightly stained. The initially rearward projection of the n bandlet and its later frontward reversal at the junction with the other bandlets is manifest. (c) Cell N*l* was injected at Stage 6a. Fixation and staining occurred 6 days later. All left (apparent right) hemiganglia of the ventral nerve cord are partially stained. (d) Cell OPQ*l* was injected at Stage 6a. Fixation and staining occurred 6 days later. Left (apparent right) body wall is stained, as are some sectors of left hemiganglia. These stained sectors appear to correspond to those sectors that are unstained in the embryo shown in (c). (e) Right: lateral view of the embryo shown in (c). (cell N*l* injected). Left: lateral view of an embryo similar to that shown in (d) (cell OPQ*r* injected). Neither embryo shows staining in the supraesophageal ganglion. (f) Cell M*l* was injected at Stage 4c. Fixation and staining occurred 6 days later. The left (apparent right) germinal plate is broadly stained. (g) Injection of an A, B, or C macromere at Stage 3. (The exact identity of the injected macromere was not determined.) Fixation and staining occurred 6 days later. Parts of the supraesophageal ganglion and body wall are stained. The large area of intense staining in the center of the embryo derives from the incorporation of the injected macromere into the primordial gut.

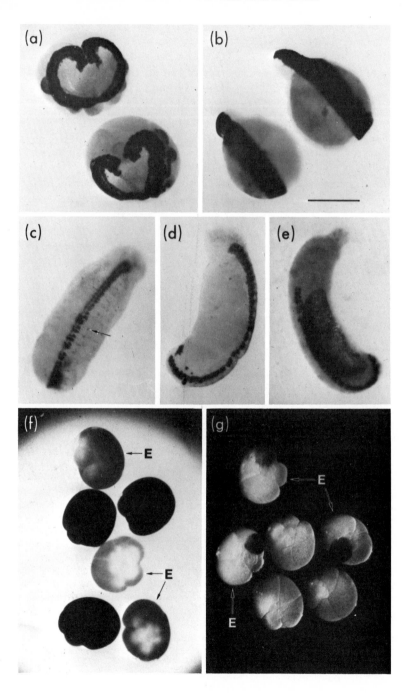

photographed under fluorescence, then fixed, stained for HRP, and re-photographed under standard illumination. As shown in Fig. 4f, injection of a single NOPQ cell caused the whole embryo to fluoresce. Thus, we infer that at this stage all cells of the leech embryo are linked by junctions that permit intercellular passage of Lucifer Yellow (MW about 500). (We have also observed tight electrical coupling between blastomeres using electrophysiological techniques.) Figure 4g shows, by contrast, that the larger HRP molecules (MW about 40,000) remain confined to the injected cell. Hence, the HRP tracer does not move between blastomeres of separate lineages, despite an extensive network of intercellular junctions.

V. Origin of the Segmental Ganglia

As the whole-mount preparations presented in Figs. 3c and 4c and d show, each N teloblast is indeed the progenitor of a substantial part of the ipsilateral half of the nervous system. HRP stain can be seen on the same side as the injected N teloblast in segmentally repeated structures of the appropriate size, shape, and location for hemiganglia of the ventral nerve cord. It is to be noted, however, that the hemiganglia are not stained uniformly; within each hemiganglion the HRP stain appears as a thin longitudinal strip next to the ventral midline and two transverse strips extending laterally from it.

The presence of unstained tissue between the two stained transverse strips in each ganglion following HRP injection of the ipsilateral N teloblast raises the possibility that some nerve cell bodies within the

---

FIG. 4. Photomicrographs of HRP-injected leech embryos. (a) Two embryos whose cell DNOPQ had been injected with HRP at Stage 4b and that were fixed and stained 4 days later. Individual columns of stained stem cells and ectoteloblasts can be seen. (b) Two embryos in each of which an NOPQ cell had been injected with HRP at Stage 5 and that were fixed and stained 5 days later. In each specimen, the germinal plate is stained hemilaterally. (c) Ventral aspect of an embryo whose N*l* ectoteloblast was injected with HRP at Stage 6a and that was fixed and stained 6 days later. The left hemiganglia (apparent right) are stained, as are the putative sensillae (arrow) on the left side. (d) Lateral aspect of the embryo shown in (c). (e) Lateral aspect of embryo whose N*l* ectoteloblast had been injected at Stage 7 after some stem cells had already been produced and that was fixed and stained 6 days later. Only the caudal but not the rostral hemiganglia are stained. (In this photograph, the presumptive gut appears dark as a result of suboptimal dehydration of the embryo.) (f) Fluorescence micrograph of three Stage 6a embryos taken within 30 minutes after cell NOPQ*l* or NO/PQ*r* had been injected with a mixture of HRP and Lucifer Yellow, and of three uninjected control embryos. All cells of the three injected embryos (E) are brightly fluorescent. (g) The same injected (E) and control embryos as in panel (f), after fixation and staining for HRP, photographed under white light. Only the injected cells show the HRP stain. Bar: 0.25 mm in (a) through (e); 0.5 mm in (f) and (g).

(a)   (b)

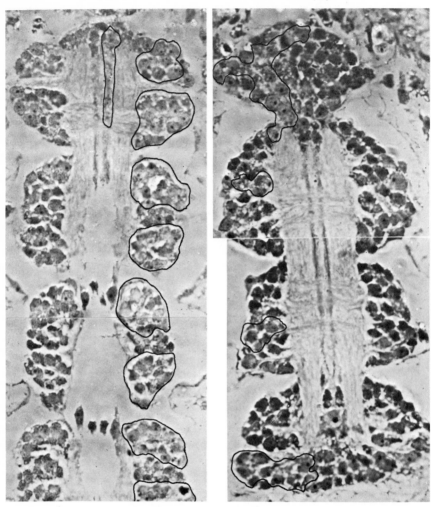

FIG. 5. Sections of HRP-injected embryos. In this and the following two figures, the sections were counterstained with toluidine blue and photographed under phase contrast optics. The outlines indicate areas where HRP stain is visible (in yellow against a blue background) in the original sections. All stained cell bodies lie ipsilateral to the injected cell. (a) Cell N was injected at Stage 6a; fixation and staining occurred 6 days later. This is a slightly oblique horizontal section showing four segmental ganglia through which the section passes at progressively more dorsal levels, from anterior (top) to posterior. First ganglion from top: at this level, most of the cell bodies visible in the hemiganglion are stained. Some unstained cells can be seen at the anterior edge of the hemiganglion

hemiganglion are not derived from that cell. In addition, because the stain does not extend across the ventral midline, it follows from symmetry that any unstained cell bodies within the stained hemiganglion are not derived from the uninjected *contralateral* N teloblast.

The existence of unstained cell bodies within stained hemiganglia is confirmed by serial sections of such preparations (Fig. 5a). The two transverse strips of stain visible in whole mount correspond to anterior and posterior groups of cells in the lateral part of the ganglion. Moreover, the thin longitudinal strips of stain visible in whole mount correspond to stained cell bodies in the ventromedial part of the ganglion. Unstained cell bodies are seen primarily in the central region of the ganglion.

Two additional groups of cells of the nerve cord remain unstained following HRP injection of N teloblasts. The first group consists of two readily identifiable pairs of cells at the anterior edge of the dorsal aspect of the nerve tract of each embryonic ganglion. The nature of these cells—hereafter referred to as "nerve tract" cells—will be discussed later in this chapter. The second group of unstained cells consists of the supraesophageal ganglion (Fig. 3e). It might be inferred that these unstained cells are not derived from the N teloblasts, but such negative evidence is not incontrovertible. It is possible that the unstained cells do derive from the injected teloblast but contain a lower intracellular concentration of HRP than the stained members of the clone because they (1) are products of more generations of cell growth and division, (2) have increased their own cell volume to a greater extent (e.g., by growth of more extensive cell processes), or (3) have developed specific metabolic features that engender a more rapid destruction or elimination of HRP. Moreover, in the case of the supraesophageal ganglion, the additional possibility exists that by the time the N teloblast was injected, its first stem cells had already been produced. In that case, the supraesophageal ganglion might have de-

---

and at the exit of the segmental nerve root. Second through fourth ganglion: at these more dorsal levels, the majority of the cells are still stained, but a greater number of unstained cells is visible between the anterior and posterior stained areas. Third and fourth ganglion: at this most dorsal level, four unstained nerve tract cells are visible at the anterior edge of the ganglion. (b) Cell OPQ was injected at Stage 6a; fixation and staining occurred 6 days later. Tangential horizontal section showing four segmental ganglia. First ganglion from top: at this ventral level, stained neurons can be seen in the rostral, caudal, and lateral edges of the hemiganglion. Unstained neurons can be seen at the medial edge. Second and third ganglion: at this more dorsal level, stained neurons are seen where unstained neurons are visible in the second through fourth ganglion of (a). Fourth ganglion: at this intermediate level, stained neurons are visible in the posterior half of the hemiganglion. Bar: 50 μm.

rived from the eldest n stem cells, which did not contain HRP. Thus, in order to identify positively the progenitors of the unstained groups of CNS cells, HRP injections of other early embryonic cells were undertaken. In one such set of experiments, the OPQ cell (precursor of the O, P, and Q teloblasts) was injected at Stage 6a and development was allowed to proceed for 6 more days before fixation and HRP staining of the embryo. As was to be expected on the basis of the traditional view that the O, P, and Q teloblasts give rise to nonnervous ectoderm (Whitman, 1887; Bergh, 1891; Bürger, 1902; Schleip, 1936), stained tissues are found lateral to the nerve cord and ipsilateral to the HRP-injected OPQ cell (Fig. 3d). However, there also appear to be some segmentally repeated patches of stain whose position near the ventral midline suggests that they are located within the ventral nerve cord.

Comparison of the whole-mount preparations of Fig. 3c and d suggests that these patches of stain are just those groups of cells in each segmental ganglion that fail to stain following injection of the ipsilateral N teloblast. This suggestion is confirmed by sections of OPQ cell-injected specimens (Fig. 5b), which show that the patches of stain correspond to groups of cell bodies located within the hemiganglia that fail to stain following injection of the N teloblast (Fig. 5a).

It can be concluded, therefore, that the N teloblast pair is not the exclusive precursor of the cells of the leech nerve cord and that some of these cells arise from one or more of the teloblasts derived from the OPQ cell pair. Because embryos (not presented here), in which either cell OP or cell Q had been injected with HRP at Stage 6b also show medially located patches of stain, we conclude that the O and/or P, as well as the Q teloblast, contribute to the nerve cord.

The two other groups of cells of the central nervous system that fail to stain after HRP injection of the N teloblast, i.e., the supraesophageal ganglion (Fig. 3e) and the nerve tract cells, also fail to stain after HRP injection of the OPQ teloblast precursor. As for the supraesophageal ganglion, we next tested the possibility that it is derived from the eldest stem cells of the n bandlet, conceivably produced prior to injection of an N teloblast at Stage 6a. For this purpose, HRP was injected into ectoteloblast precursor cell NOPQ of Stage 5 embryos and subsequent development was allowed to proceed for 5 more days. Under this protocol, the N teloblast would certainly receive the tracer enzyme before it begins stem cell production. However, after injection of cell NOPQ, the supraesophageal ganglion was still unstained (Fig. 6a). Thus, if the supraesophageal ganglion is derived from the N teloblasts, its failure to stain after HRP injection of an N teloblast is not attributable to the precocious presence of n stem cells at the time of injection.

The nerve tract cells also fail to stain following HRP injection of cell NOPQ (Fig. 7a), but almost all of the cells of the segmental hemiganglia ipsilateral to the injected NOPQ cell now stain (Fig. 7a). This finding is consistent with our previous conclusion that, between them, the N and Q teloblasts and the O and/or P teloblasts supply the neurons of the segmental ganglia. Nevertheless, in these specimens there are always a few (less than 10) unstained cell bodies visible in each hemiganglion. Although no positive identification of these exceptional cells has been made, it will be shown later than their failure to stain following injection of cell NOPQ is unlikely to be an experimental artifact.

VI. Origin of the Mesoderm

In order to probe further into the origins of the hitherto unstained cells of the nerve cord, one of the pair of M teloblasts of Stage 4c *Helobdella* embryos was injected with HRP and subsequent development allowed to proceed for 6 more days. As was to be expected on the basis of the traditional view that the M teloblast pair is the source of the leech mesoderm (Whitman, 1878, 1887; Schleip, 1936), wholemounts of such preparations show extensive staining of germinal plate tissues outside the ventral nerve cord (Fig. 3f). Moreover, stain is also manifest beyond the lateral edge of the germinal plate, within the domain of the micromere cap. That stain appears in the form of circumferential fibers extending from the germinal plate to the dorsal midline. Thus, the cells composing these fibers seem to have been left behind by the m bandlets during the circumferential migration of the germinal bands at Stage 8. In addition, horizontal sections of such preparations show that nerve tract cells are now stained, as are a few cell bodies within the ganglion itself and thin cell profiles enveloping it (Fig. 7b). Although these stained intraganglionic cell bodies have not been positively identified, they are probably the same few cells that fail to stain after injection of cell NOPQ. Thus, some cells of the leech nerve cord are evidently derived from the M teloblast.

In order to test the traditional view of the M teloblast as a source of mesoderm (Whitman, 1878, 1887), the developmental fate of the progeny of the stem cells of the m bandlet was examined further. For this purpose, cell DM of Stage 4b embryos was injected with HRP and allowed to develop to completion of Stage 8. Sections of such embryos show stained segmental blocks of tissue, corresponding to primordial somites, beneath the ectoderm of the embryo (Fig. 6b). Hence, the traditional assignment of the M teloblast to the role of mesodermal precursor cell is confirmed. By injecting cell DM—the immediate precursor of

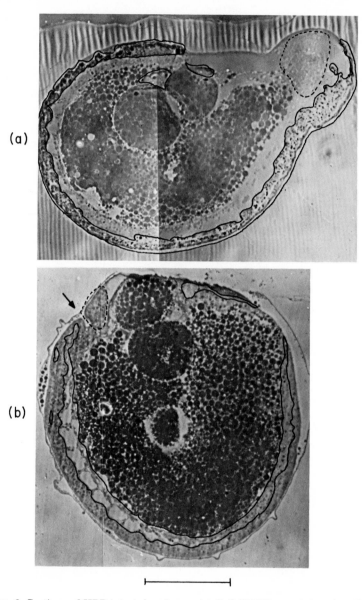

FIG. 6. Sections of HRP-injected embryos. (a) Cell NOPQ was injected at Stage 5; fixation and staining occurred 5 days later. Sagittal section close to the midline, with the head lying to the right. Body wall and nerve cord are heavily stained, except for the head and the supraesophageal ganglion (marked by a dotted outline), which are unstained. The two large cell profiles near the top of the figure are teloblast remnants. The darkly counterstained granules filling the center of the section are yolk platelets from the

the mesoteloblasts—we could be sure that the M teloblasts received HRP before any m stem cells had been produced; under such circumstances, the supraesophageal ganglion still fails to stain (Fig. 6b).

VII. Origin of the Supraesophageal Ganglion

Provided that the failure of the supraesophageal ganglion to stain following injection of HRP into any of the teloblasts or teloblast precursors is not attributable to an artifactual cause, these results support Whitman's contention that the frontmost part of the leech nerve cord is derived from a source other than the germinal bands. A plausible alternative source is the micromere cap. To test this possibility, HRP was injected into an A, B, or C macromere of Stage 3 embryos, i.e., prior to micromere formation. (Because of the small size of the micromeres, the preferable procedure of directly injecting them in Stage 4a *Helobdella* embryos is not feasible with our present techniques.) After 6 days of further development, the embryos were fixed and stained. Whole mounts of such preparations show staining of the frontmost end of the advanced embryo, as well as of the gut containing the remnants of the injected macromere (Fig. 3g). Sections show that the rostral stain is confined to the supraesophageal ganglion and the frontmost body wall (Fig. 7c). In the specimen shown, the stain is confined to a contiguous group of cells on one side bordering the midline. This suggests that the corresponding group of cells on the other side is derived from a different macromere and that these macromere-derived cells do not normally cross the midline. Thus, in accord with Whitman's original inference, the supraesophageal ganglion does have a different embryological origin from the remainder of the nerve cord, namely, the A, B, or C macromere, and presumably, therefore, the micromere cap. (The possible contribution of the D macromere to the supraesophageal ganglion has not yet been determined.)

VIII. Number of Founder Cells of the Ganglion

Inasmuch as the segmental tissues and organs of the leech body arise from the germinal bands, it may be asked how many stem cells found any particular morphologically defined, segmental structure. In the case of the segmental ganglion, it is already clear from the data

---

macromeres and the teloblasts enclosed within the primordial gut. (b) Cell DM was injected at Stage 4b; fixation and staining occurred at the completion of Stage 8. Sagittal section, with future head lying at left. The segmentally iterated blocks of mesodermal tissue are stained. The head region (marked by a dotted outline) lying rostrally to (apparently above) the oral cavity (arrow) is unstained. Three teloblast remnants are visible within the primordial gut. Bar: 100 μm.

(a)   (b)

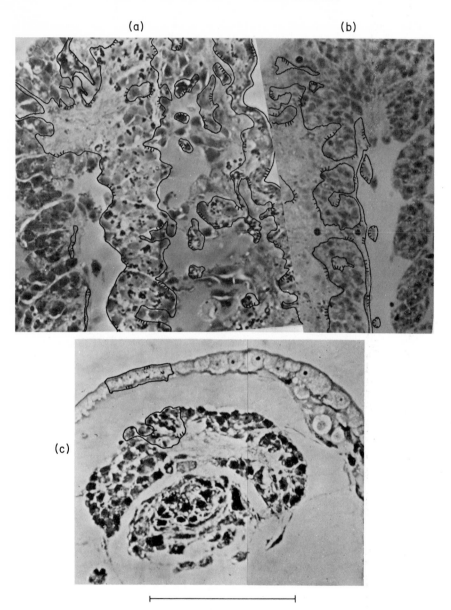

(c)

FIG. 7. Sections of HRP-injected embryos. In this and the following figure, areas where HRP stain is visible in the original sections are indicated by a hachured outline, with the stain lying on the side of the hachure. (a) Cell NOPQ was injected at Stage 5; fixation and staining occurred 5 days later. Horizontal section at a dorsal level, showing three segments. The hemiganglia and the ipsilateral body wall are heavily stained.

presented here that the hemiganglion is founded by more than one stem cell. Because some of the neuronal cell bodies of the hemiganglion arise from stem cells of the n bandlet and others arise from stem cells of the q and o and/or p bandlets, it follows that at least three stem cells must contribute to the founding of each hemiganglion. In addition, because some cells of the segmental ganglion are derived from the M teloblast, a fourth (m bandlet) stem cell must also contribute to the founding of each hemiganglion. The following experiment was carried out to ascertain whether, among the stem cells of just the n bandlet, more than one contributes to the founding of each hemiganglion.

This experiment is made possible by the result shown in Fig. 3e, i.e., if an N teloblast is injected with HRP only after some stem cells have been produced (for instance, at Stage 7), then, upon further development, the caudal, but not the rostral, part of the ventral nerve cord stains. The finding that there is a sharp boundary between stained caudal and unstained rostral portions indicates that a detailed examination of the variation in position of the boundary should resolve the question of the number of n bandlet founder stem cells per hemiganglion. If the number of founder stem cells is not more than one, then the stain boundary should always be *inter*ganglionic, i.e., always fall between a fully stained posterior and a fully unstained anterior hemiganglion. However, if that number is greater than one, the stain boundary should often be *intra*ganglionic, i.e., often fall within a hemiganglion, whose caudal part is stained and whose rostral part is unstained. In fact, the greater the number of founder stem cells per hemiganglion, the higher would be the proportion of intraganglionic to interganglionic stain boundaries found, in a set of embryos whose N teloblast was injected with HRP at Stage 7.

We have found both inter- and intraganglionic stain boundaries in serial sections of different preparations (Fig. 8a and b). In a total of five

---

Stained nerve cell processes extending across the midline are also visible in the ganglion at the top of the figure. The nerve tract cells are unstained. (b) Cell M was injected at Stage 4c; fixation and staining occurred 6 days later. Horizontal section at a dorsal level showing four segments (with the anterior at the bottom). Extensive staining of tissue between nerve cord and body wall is seen. Essentially all of the neurons are unstained. Three stained and three unstained nerve tract cells can be seen. The sheet of connective tissue surrounding ganglionic cell packets is stained, as is one cell body within each of the two middle ganglia. (c) Cell A, B, or C was injected at Stage 3 (the exact identity of the injected macromere, i.e., whether it was A or B or C was not ascertained); fixation and staining occurred 6 days later. Oblique transverse section through the supraesophageal ganglion, shown dorsum up. Stained cells are visible next to the midline in the body wall of the head and supraesophageal ganglion. The esophageal tissue lying below the ganglion is not stained. Bar: (a) 40 μm; (b) and (c) 60 μm.

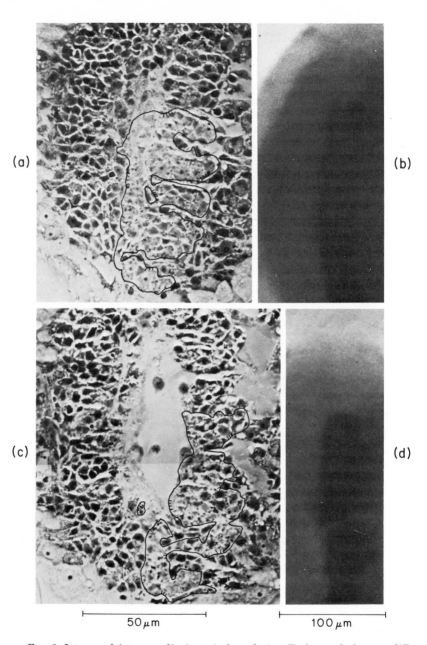

FIG. 8. Inter- and intraganglionic stain boundaries. Each panel shows a different embryo in which cell N was injected with HRP at Stage 7; fixation and staining occurred 3 days later [in (a) and (b)] or 4 days later [in (c) and (d)]. Anterior at the top. (a) and (c). Toluidine-blue counterstained horizontal sections showing three early segmental ganglia

sectioned preparations examined thus far, interganglionic boundaries were found in three cases and intraganglionic boundaries in two cases. Inter- and intraganglionic stain boundaries are manifest also in wholemounts of such preparations (see Fig. 8c and d). In these embryos, which were fixed and stained at the time of hatching (and hence earlier in development than the embryo shown in Fig. 3c), the two transverse strips of stain in each ganglion are morphologically distinct in that one extends further laterally than the other. Comparison of whole-mount views and sections of such embryos shows that the laterally more and laterally less extensive strips correspond to groups of stained cells in the posterior and anterior parts of the hemiganglion, respectively. In some preparations the stain boundary falls rostral to one of the laterally less extensive transverse stained strips, in which case the boundary is interganglionic (Fig. 8c). In other preparations the boundary falls rostral to one of the laterally more extensive transverse stained strips, in which case the boundary is intraganglionic (Fig. 8d). We infer from the occurrence of some intraganglionic boundaries that the number of n bandlet stem cells that found each hemiganglion is greater than one. And because the chance of observing interganglionic boundaries decreases as the number of founder stem cells increases, it can be inferred from the relatively frequent occurrence of such boundaries that this number is unlikely to be much greater than two. If it *is* two, it would appear that one (elder) n stem cell gives rise to neurons of the anterior lateral cell packet and of the ipsilateral half of the anterior ventromedial cell packet. The second (younger) n stem cell would give rise to neurons of the posterior ventromedial packet. However, the present, rather limited data do not rule out the possibility that the ventromedial packets are actually founded by an additional n stem cell.

IX. Rhodamine-Peptide Method

As the foregoing material has shown, injection of HRP tracer into identified cells of early embryos permits the determination of develop-

---

prior to their eventual longitudinal separation. In both (a) and (c), the first ganglion (from the top) does not and the third ganglion does contain stained cells. But in (a) the second ganglion contains stained cells in both anterior and posterior halves of its hemiganglion. By contrast, in (c) the second ganglion contains stained cells only in the posterior half of its hemiganglion. (b) and (d)Whole-mount view of part of the embryonic nerve cord. In both (b) and (d) the caudal part of the nerve cord is stained and the rostral part is unstained. But in (b) the laterally less extensive transverse strip and in (d) the laterally more extensive strip is the first to be stained. In (a) and (b) the stain boundary is *inter*ganglionic, whereas in (c) and (d) it is *intra*ganglionic.

mental cell lineages. However, because the histochemical HRP reaction product is opaque, this method is unsuitable for experiments in which the mitotic state of tracer-labeled cells is to be examined in whole mount using nuclear stains. Moreover, this method cannot be used to test the effectiveness of intracellular Pronase injection as a means of ablating embryonic cells (Weisblat et al., 1979) because HRP is sensitive to proteolytic digestion. Fluorescent dye tracers would overcome both these limitations, but such small molecules (on the order of MW 500) cannot be used directly as cell lineage tracers because, as was seen in Fig. 4f, upon injection they diffuse throughout the entire embryo. A fluorescent dye could be confined to the injected cell and its lineal descendants if attached to a larger carrier molecule, as it has been reported that the molecular weight limit for the permeation of insect salivary gland gap junctions by oligopeptide–fluorescent dye complexes lies between 1200 and 1900 (Simpson et al., 1977). To be suitable for cell lineage tracing, the carrier molecule should be of an appropriate size, have chemical sites to which the dye can be coupled, be reactive with histological fixatives, and resist the action of proteolytic enzymes. It was found that the dodecapeptide (Glu-Ala)$_2$-Lys-Ala-(Glu-Ala)$_2$-Lys-Gly, which has an approximate molecular weight of 1200 and is composed of amino acids in the unnatural D configuration, meets these requirements (Weisblat et al., 1980b). This peptide was synthesized by the Merrifield solid phase method (Stewart and Young, 1969). The synthetic peptide was coupled to rhodamine isothiocyanate, and the product—rhodamine-D-peptide (RDP)—was isolated by column chromatography and lyophilized (Nairn, 1969). In a similar manner, a fluorescein derivative of the peptide (FDP) was prepared.

The utility of RDP as a cell lineage tracer was demonstrated in studies of the cell cleavage pattern within the germinal bandlets of Helobdella embryos (Weisblat et al., 1980b). For this purpose, a teloblast (or teloblast precursor) was injected with RDP. After formation of the germinal bands was underway (i.e., at Stage 7 or 8), the embryo was fixed and treated with the blue-fluorescing, DNA-specific stain Hoechst 33258 (Sedat and Manuelides, 1977). The blue fluorescence of the nuclei and the red fluorescence of the RDP-labeled cytoplasm were then viewed separately through appropriate filters. By focusing through the cleared embryo, the red-fluorescing bandlet can be traced from its point of origin at the teloblast to its final position in the germinal band. Figure 9a shows an embryo whose left N teloblast had been injected with RDP at Stage 6a. The embryo was fixed and stained with Hoechst 33258 at early Stage 8. The numerous fluorescent dots visible in this photograph represent nuclei of diverse embryonic cells, but the nuclei belonging to the n bandlet can be distinguished because they lie within fluorescent (i.e., RDP-labeled) N teloblast progeny. Closer inspection of bandlet nuclei reveals them to be of two types: interphase nuclei with diffuse fluorescence and mitotic nuclei showing brightly fluorescent, condensed chromosomes (Sedat and Manuelides, 1977) (Fig. 9b). The alignment of condensed chromosomes on the metaphase plate indicates the orientation of the spindle axis of cell division. Upon examination of 26 Stage 7 embryos in which an N teloblast had been injected with RDP at Stage 6, no mitotic n stem cells were found at a separation of less than 20 stem cells from the parent teloblast. At that point the bandlet cells are already within the germinal band. The spindle axis of all observed early n stem cell divisions was nearly parallel to the long axis of the bandlet, as shown in Fig. 9b. After that first division, therefore, the bandlet still remains one cell wide. Figure 9c shows a similarly treated embryo whose M teloblasts had been labeled with RDP by injecting their precursor, cell DM, at Stage 4b. In this figure, progeny of m bandlet stem cell mitoses are visible at a distance of about 10 cells from the M teloblast, before that bandlet has joined the ectodermal, n, o, p, and q bandlets in the germinal band. The spindle axis of the initial stem cell division in the m bandlet is perpendicular to the long axis of the bandlet, so that the m bandlet becomes two cells wide after the first stem cell division. Similar observations were made in 17 RDP-labeled m bandlets. Thus, combined use of RDP and Hoechst 33258 reveals that the mesodermal, m stem cells differ from the ectodermal, n stem cells in both the timing and the orientation of their first cleavage: whereas the n stem cells cleave only after entering the germinal band, and with spindle axes parallel to the long axis of the bandlet, the m stem cells cleave prior to entering the germinal band, and with spindle axes perpendicular to the long axis of the bandlet.

Figure 9d presents an embryo in which the left M teloblast had been injected with RDP at Stage 5. In the most rostral sector of the labeled (left) germinal band, segmentation of the mesodermal tissue can be observed, even before its fusion with the right germinal band. RDP-labeled stem cells of the left m bandlet have given rise to clusters of RDP-labeled progeny. The arrangement of the progeny is similar in adjacent clusters, as shown in Fig. 9e. This similarity suggests that the early development of the mesoderm of each body segment proceeds, as does teloblast formation, by stereotyped cell divisions.

RDP injection has also been used in conjunction with cell ablation by Pronase injection (Parnas and Bowling, 1977) in *Helobdella* embryos. As was shown by the HRP tracer method, a topographically coherent fraction of the neurons on one side of each segmental ganglion

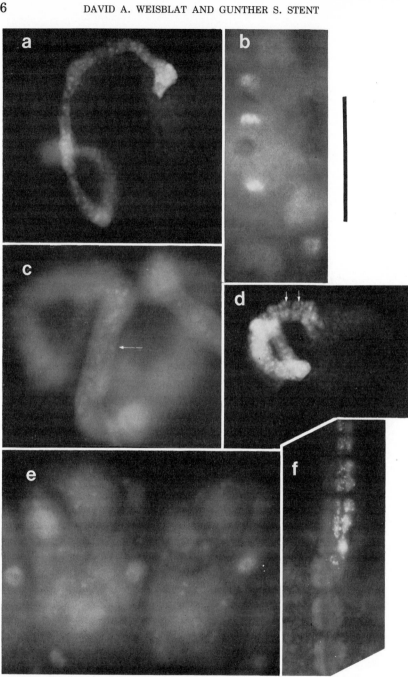

is derived from the ipsilateral N teloblast. Furthermore, injecting Pronase into an N teloblast prior to completion of its stem cell production results in abnormal development of the ipsilateral half of the segmental ganglia (Weisblat *et al.*, 1979). The teloblast may be killed outright by Pronase injection, with complete arrest of stem cell production; but

---

FIG. 9. Photomicrographs of RDP-injected, Hoechst 33258-stained *Helobdella* embryos. (a) Early Stage 8 embryo whose cell N*l* had been injected with RDP at Stage 6a. The fluorescent n bandlet extends caudally from its origin at the N teloblast (left) to its point of entry into the left germinal band (bottom), then rostrally along the lateral edge of the germinal band to the future head (top right). The numerous fluorescent dots are nuclei of cells of the germinal bands and micromere cap. (b) Enlarged view of the n bandlet of an embryo treated similarly to that shown in (a). In the middle of this picture is a telophase cell with the spindle axis parallel to the longitudinal axis of the bandlet. Above and below the dividing cell, interphase nuclei are visible. (c) Early Stage 8 embryo whose M teloblast precursor, cell DM, had been injected with RDP at Stage 4b. The left m bandlet extends from the left M teloblast (out of focus at bottom center) to the point of origin of the germinal band (in focus at top center). Part of the right m bandlet can be seen in the upper right hand portion. Interphase nuclei are visible as pale dots within the m bandlet. The arrow points to the daughter cells of an m stem cell cleavage which took place prior to the entry of the bandlet into the germinal band. The position of the daughter cell nuclei indicates that the spindle axis was perpendicular to the longitudinal axis of the bandlet. Rows of nuclei of unlabeled ectodermal bandlets converge at the point of origin of the germinal band. (d) Late Stage 7 embryo whose cell M*l* had been injected with RDP at Stage 5. The fluorescent m bandlet extends from its origin at the M teloblast (left) to its point of entry into the germinal band (center), then in an arc beneath the ectodermal bandlet to the future head (top center). The arrows indicate two adjacent clusters of m stem cell progeny. (e) Enlarged view of the cell clusters marked by arrows in (d) showing some of the cells in each cluster. Topographically and morphologically corresponding cells can be seen within these clusters. (f) Ventral view of a portion of the nerve cord of a 9-day-old embryo whose cell N*l* had been injected with RDP at Stage 6a and with Pronase at Stage 7. Anterior is at the top; seven ganglia are shown. Ganglia 1, 2, and 3 (from the top) are normal; RDP-labeled cells can be seen in the left (apparent right) sides of these ganglia. Ganglia 6 and 7 are abnormal, in that they contain fewer cells on the left (apparent right) side; the remaining cells are not RDP-labeled. Ganglion 4 (tilted and only partially visible) and ganglion 5 are at the border between the normal, anterior and the abnormal, posterior parts of the nerve cord; ganglion 4 contains some RDP-labeled cells at its left lateral edge, and ganglion 5 at its left anterior midline. The fluorescent background beneath ganglia 6 and 7 results from the fluorescence of RDP and of the dye Fast Green [coinjected with Pronase] in teloblast remnants in the gut. All the embryos shown in this figure were fixed and acid-cleared by the method of Fernandez (1980), except that the fixative included 2.5 μg/ml Hoechst 33258. After fixation (12–24 hours at 4°C) the embryos were mounted between coverslips in 75% glycerol containing 2.5 μg/ml Hoechst 33258. Double-exposure photomicrographs were made under epi-illumination from either a 75-W tungsten-halogen or 50-W mercury light source, first through Zeiss filter set 48 77 14 and then through Zeiss filter set 48 17 02. Bar: 160 μm in (a), (c), and (f); 25 μm in (b) and (e); 250 μm in (d).

it is also possible that the teloblast is merely damaged and continues to produce stem cells in an abnormal manner. To ascertain whether the morphological abnormalities following Pronase injection are due to a disturbance of stem cell production or to its complete arrest, it is necessary to identify the progeny, if any, of the injected teloblast. HRP cannot be used for this purpose, because it is sensitive to proteolytic digestion. Therefore, the following experiment was carried out using RDP, which is Pronase resistant.

At Stage 6a, prior to the onset of a bandlet stem cell production, an N teloblast was injected with RDP. At Stage 7, after stem cell production was underway, the same N teloblast was reinjected with Pronase. Figure 9f shows such an embryo at 9 days. The anterior part of the nerve cord of this embryo consists of morphologically normal (bilaterally symmetric) ganglia, whereas the posterior part consists of abnormal (bilaterally asymmetric) ganglia, deficient in cell number on the side of the ablated N teloblast. Moreover, anterior ganglia contain neurons labeled with RDP, whereas posterior ganglia do not. The boundary between anterior, morphologically normal and posterior, morphologically deficient ganglia coincides with the boundary between anterior, RDP-labeled and posterior, unlabeled ganglia. Thus, it follows that the anterior ganglia of the Pronase-treated embryo contain progeny of n stem cells produced prior to ablation of the N teloblast and that the posterior ganglia are abnormal because they received no cellular contribution from the N teloblast.

X. Summary and Conclusions

The results of the cell lineage analysis by use of the HRP method confirm the previously held view that in glossiphoniid leeches the N and M teloblast pairs are precursors of nervous system and mesoderm, respectively. However, in contrast to the previously held view that the O, P, and Q teloblasts are mainly the source of circular muscles (Bergh, 1891; Bürger, 1902), it has been found that a spatially coherent fraction of the neurons in each ganglion is derived from cells Q and OP. Moreover, the positions within the ganglion of neuronal populations derived from each of these sources (N, OP, and Q) are relatively invariant from segment to segment (cf. Fig. 3c and d) and from specimen to specimen (cf. Fig. 8c and d). This finding suggests that the two cell groups of different developmental lineages might represent functionally distinct classes of neurons, in light of the relative positional invariance of functionally identified neurons in the segmental ganglia of *Hirudo* (Nicholls and Van Essen, 1974) and *Haementeria* (A. Kramer and J. Goldman, personal communication).

The HRP tracer method revealed, furthermore, that the cell bodies contributed by each teloblast are confined to the ipsilateral half of the embryonic ganglion. This finding suggests that the innervation of the contralateral musculature by motor neurons on the dorsal aspect of the adult segmental ganglion in *Hirudo* (Stuart, 1970) and *Haementeria* (A. Kramer and J. Goldman, personal communication) is attributable to a developmental decussation of their axons rather than to an initially uncrossed projection followed by a reciprocal migration of the homologous cell bodies across the midline. In fact, in exceptionally well-stained preparations, stained neuronal processes can be seen in the segmental nerves contralateral to the injected teloblast. [It should be noted that the finding of an exclusively ipsilateral cell origin applies only to *normal* embryonic development; following ablation of one N teloblast, it is possible for the other N teloblast to make a contribution to contralateral hemiganglia (Weisblat *et al.,* 1979).] Moreover, the present experiments show that each N teloblast contributes more than one, probably two, of its stem cells to the foundation of each ipsilateral hemiganglion.

The experiments carried out thus far yield little information regarding the embryonic origin of the glial cells, as most glia are not readily identifiable in even the most advanced embryos for which satisfactory HRP staining has been obtained. For instance, the giant glia of the connective (cf. Fig. 2a and b) are not identifiable until about the thirteenth day of development. However, in transverse sections of N cell-injected embryos fixed and stained at the seventh day of development, HRP stain is seen in cells that appear to correspond to the nerve tract glial cells on the ventral aspect of the ganglionic neuropil (cf. Fig. 2c). This suggests that at least some of the glial cells are derived from the N ectoteloblast.

The mesoteloblast M was also found to contribute cells to the ventral nerve cord. Some of the M-derived cells within the segmental ganglion are likely to correspond to the muscle and sheath cells known to encapsulate the ganglia of *Hirudo* and their nerve cell packets (Coggeshall and Fawcett, 1964). Moreover, it is likely that the M-derived "nerve tract" cells are precursors of the muscle cells of the interganglionic connective as studies with the glossiphoniid leech *Haementeria* have shown such a role for apparently homologous nerve tract cells (A. Kramer, personal communication).

The finding that *only* frontmost body wall and the supraesophageal ganglion stain following HRP injection of a macromere is not in accord with the traditional view that the micromeres arising from cleavage of the macromeres are the main precursors of the micromere cap that covers the surface of the embryo behind the coalescing germinal bands during Stage 8. Under that view it would have been expected that injection of a macromere would result in the presence of stain over large areas of the post-Stage 8 embryo. The observed absence of stain can be explained in at least two ways: In case the micromeres *are* mainly direct descendants of the macromeres, those micromeres that cover most of the embryonic surface might have undergone many more cycles of cell growth and division and hence have diluted to a much greater extent any HRP passed on to them from their parent macromeres than those micromeres that have remained near the site of their origin, i.e., the future head. Alternatively, it is possible that the micromere cap is a mosaic of cells of diverse lines of descent, of which only a minority, and in particular those cells that give rise to the supraesophageal ganglion, are lineal descendents of the A, B, and C macromeres. This latter possibility is supported by the finding (Fig. 3f) that stain appears within the domain referred to as "micromere cap" following injection of the mesoteloblast M.

The demonstration that the supraesophageal ganglion is not derived from the germinal bands indicates that it is not to be included in the number of ganglia that arise through segmentation of the germinal plate. Because in the adult leech both rostral and caudal ganglia are fused and modified, it is not immediately obvious how many ganglionic primordia do arise through segmentation. However, detailed anatomical inspection of the adult subesophageal ganglion in the head and of the caudal tail ganglion at the rear indicates that the former corresponds to a fusion of four and the latter to a fusion of seven glanglia, both in *Hirudo* and in glossiphoniid leeches (Harant and Grassé, 1959). Because the intervening, abdominal part of the ventral nerve cord consists of 21 separate ganglia, a total of 32 ganglia is to be accounted for by germinal plate segmentation. Direct observation of Stage 8 glossiphoniid embryos confirms, furthermore, that 32 distinct ganglionic primordia arise in rostrocaudal progression during segmentation (Fernandez, 1980).

The fluorescent peptides RDP and FDP seem to have considerable promise as cell-lineage tracers and offer some distinct advantages over HRP, in addition to their resistance to proteolysis. For instance, the red-fluorescing RDP and the yellow-fluorescing FDP can be used in combination for double-label experiments; second, their distribution can be observed in living embryos, in contrast to that of HRP, which can be visualized only in fixed preparations. Thus, these fluorescent tracers should make it possible to follow the appearance of successive descendants of an injected early embryonic cell in the same preparation and to know the embryonic origin of nerve and muscle cells identified by intracellular electrophysiological recordings.

ACKNOWLEDGMENT

The authors' work summarized in this article was supported by NIH postdoctoral fellowship grant NS054451, by NIH research grant NS12818, and by NSF grant BN577-19181.

REFERENCES

Bergh, R. S. (1891). *Z. Wiss. Zool.* **52**, 1–17.

Bürger, O. (1902). *Z. Wiss. Zool.* **72**, 525–544.

Coggeshall, R. E., and Fawcett, D. W. (1964). *J. Neurophysiol.* **27**, 220–289.

Deppe, V., Schierenberg, E., Cole, T., Krieg, C., Schmitt, D., Yoder, B., and von Ehrenstein, G. (1978). *Caenorhabditis elegans. Proc. Natl. Acad. Sci. U.S.A.* **75**, 376–380.

Fernandez, J. (1980). *Dev. Biol.* **76**, 245–262.

Garcia-Bellido, A., and Merriam, J. R. (1969). *J. Exp. Zool.* **170**, 61–76.

Harant, H., and Grassé, P.-P. (1959). *In* "Traité de Zoologie" (P.-P. Grassé, ed.), Vol. 5, pp. 509, 513. Masson, Paris.

Le Douarin, N. (1973). *Dev. Biol.* **30**, 217–222.

Macagno, E. R. (1980). *J. Comp. Neurol.* **190**, 283–302.

Mann, K. H. (1962). "Leeches (Hirudinea)." Pergamon, Oxford.

Mintz, B. (1965). *Science* **148**, 1232–1233.

Müller, K. J. (1932). *Z. Wiss. Zool.* **142**, 425–490.

Müller, K. J., and McMahan, U. J. (1976). *Proc. R. Soc. London, Ser. B* **194**, 481–499.

Nairn, R. C. (1969). "Fluorescent Protein Tracing." Williams & Wilkins, Baltimore, Maryland.

Nicholls, J. G., and Van Essen, D. (1974). *Sci. Am.* **230**, 38–48.

Parnas, I., and Bowling, D. (1977). *Nature (London)* **270**, 626–628.

Schleip, W. (1936). *In* "Klassen und Ordnungen des Tierreichs" (H. G. Bronn, ed.), Vol. 4, Div. III, Book 4, Part 2, pp. 1–121. Akad. Verlagsges., Leipzig.

Sedat, J., and Manuelides, L. (1977). *Cold Spring Harbor Symp. Quant. Biol.* **42**, 331–350.

Simpson, I., Rose, B., and Lowenstein, W. R. (1977). *Science* **195**, 294–296.

Stern, C. (1968). "Genetic Mosaics and Other Essays." Harvard Univ. Press, Cambridge, Massachusetts.

Stewart, J. M., and Young, J. D. (1969). "Solid Phase Peptide Synthesis." Freeman, San Francisco, California.

Stewart, W. W. (1978). *Cell* **14**, 741–759.

Stuart, A. E. (1970). *J. Physiol. (London)* **209**, 627–646.

Sturtevant, A. H. (1929). *Z. Wiss. Zool.* **135**, 325–356.

Sulston, J. E., and Horvitz, H. R. (1977). *Dev. Biol.* **56**, 110–156.

Tarkowski, A. K. (1961). *Nature (London)* **190**, 857–860.

Weisblat, D. A., Sawyer, R. T., and Stent, G. S. (1978). *Science* **202**, 1295–1298.

Weisblat, D. A., Blair, S. S., and Stent, G. S. (1979). *Soc. Neurosci. Abstr.* **5**, 184.

Weisblat, D. A., Harper, G., Stent, G. S., and Sawyer, R. T. (1980a). *Dev. Biol.* **76**, 58–78.

Weisblat, D. A., Zackson, S. L., Blair, S. S., and Young, J. D. (1980b). *Science* **209**, 1538–1541.

Whitman, C. O. (1878). *Q. J. Microsc. Sci.* [N.S.] **18**, 215–315.

Whitman, C. O. (1887). *J. Morphol.* **1**, 105–182.

Wilson, E. B. (1892). *Nereis J. Morphol.* **6**, 361–480.

CHAPTER 2

NERVE FIBER GROWTH AND THE CELLULAR RESPONSE TO AXOTOMY

Salvatore Carbonetto

DEPARTMENT OF PHARMACOLOGY
STATE UNIVERSITY OF NEW YORK
UPSTATE MEDICAL CENTER
SYRACUSE, NEW YORK

Kenneth J. Muller

DEPARTMENT OF EMBRYOLOGY
CARNEGIE INSTITUTION OF WASHINGTON
BALTIMORE, MARYLAND

I. Introduction

Nerve cells, with their long axons or nerve fibers, are unusually vulnerable. The skull and spinal vertebrae afford the brain some protection from physical injury, but peripheral axons can be easily damaged. In soft-bodied animals, the problem is greater still. In all ani-

*CURRENT TOPICS IN
DEVELOPMENTAL BIOLOGY, VOL. 17*

mals, the nerve cell body, or soma, represents only a small fraction of the volume of the entire cell; thus, injury that severs an axon leaves the cell with a major job of reconstruction. In this respect, it is remarkable that the nervous system can recover from serious injury, and in many cases grow to restore function (e.g., Eidelberg and Stein, 1974).

In the account that follows, we wish to explore some of the complex events within the nerve cell that are triggered by axotomy, the severing of the axon. We will not attempt to present a complete review of the vast literature on this subject, and, where appropriate, we will direct the reader to timely reviews that treat in depth some of the topics that we touch. Rather, we hope to emphasize the interconnectedness of results obtained in a wide variety of systems, from functioning mammalian central and peripheral nervous systems to embryonic cells regenerating in culture and invertebrate nervous systems. The interactions among cells are of paramount importance, but in the limited scope of this chapter, we cannot discuss many types of intercellular interactions, "trophic" and otherwise, that normally operate as the nervous system attempts to pull itself back together.

We will first examine what changes appear in the axotomized neuron; some may be groundwork for regeneration, but others remain without explanation. Once the transition from injury to regeneration occurs, regenerating neurons acquire many of the specializations of developing neurons. The movement of the growth cone and elongation of the axon are dependent upon the cell's synthetic machinery and require a specialized cytoskeletal, contractile, and transport apparatus whose mechanisms are not well understood. Because some basic mechanisms of neuronal growth might be expected to be shared with nonneuronal motile cells, we will examine growth, metabolism, and cytoskeleton on a more general scale. Nevertheless, axonal transport and certain regulators of growth, such as nerve growth factor (NGF), can only properly be studied in neurons. Growth cones seem to sample their environment as they extend the axon, but the composition of that extracellular environment and the cues it provides are only beginning to be examined in detail. Typically, growth terminates when the fiber reaches its target, whereupon cellular structure and synthetic processes return to normal and function may be restored. Although recovery has been well characterized, how the cell is signaled to start and stop growing remains largely a mystery. In discussing these phenomena, we will draw heavily on examples with which we are directly familiar; we hope that the reader will agree that the overall impression one gains from a wider survey is that diverse neuronal reactions to axotomy share many of their basic mechanisms.

II. Response of the Neuron to Axotomy

A. THE CUT END

When an axon, or nerve fiber, is severed, the open proximal stump seals within minutes, preventing penetration of extracellular tracers such as horseradish peroxidase into the cytosol. Sealing of the membrane can be retarded by chelating the extracellular free Ca^{2+} or by application of phospholipases (Zipser, 1980). Nerve fibers will typically discharge impulses when severed, but prolonged discharge is not characteristic of neurons *in vivo* whose cut axons are bathed in physiological saline (Wall *et al.*, 1974). Firing eventually stops as the damaged and depolarized cell recovers its resting potential; normal function may not return for days (Meiri *et al.*, 1981).

The cut axon normally retracts somewhat, particularly if it has little attachment to its substrate or surrounding tissue (Ramón y Cajál, 1928). Retraction is evident in cultured neurons, where cutting the nerve fiber can cause it to snap back toward the cell body. Mature axons will develop a knob or swelling at the cut end (Ramón y Cajál, 1928; Weiss and Hiscoe, 1948) that only later may grow (see later), whereas axons growing when injured may rapidly generate a new growth cone (Speidel, 1933; Lasek, 1981).

B. MORPHOLOGICAL CHANGES IN THE CELL BODY

The cell body responds to axonal injury with an assortment of morphological and physiological changes. The classic reaction to axotomy, which is now considered to be in some respects extreme, has been termed *chromatolysis* (literally, "loss of color"), although over the years the term has been applied to many morphological changes that accompany axotomy. For a detailed description, the reader is referred to Lieberman's (1971, 1974) reviews of chromatolysis, which are still timely. Nissl (1892) described the reduced coloration of the cytoplasm (specifically of the Nissl substance, or rough endoplasmic reticulum) that stained with basophilic dyes and observed other morphological changes, including enlargement of the soma, nucleus, and nucleolus and a migration of the nucleus to an eccentric position in the cell. Studies using the electron microscope, though limited by sampling size, confirm the light microscopic studies and indicate increases in neurofilaments and in the size and number of lysosomes. The reduced staining of rough endoplasmic reticulum (rough ER) has been attributed to a dispersion of ribosomes within the cytoplasm, thought to result from a shift in the cell's synthetic priorities from material that is normally secreted to the synthesis of structural proteins. The enlargements of nucleolus and cell correlate with increased ribosomal RNA synthesis (Gunning *et al.,* 1977). Some of these changes occur in neurons as diverse as mouse hypoglossal motoneurons (Watson, 1965), frog spinal motoneurons (Edström, 1959), and ganglion cells in the goldfish retina (Murray and Grafstein, 1969), all of which, however, fail to show reduced staining typical of classic chromatolysis. Invertebrate neurons can also exhibit some similar morphological changes with axotomy (Young *et al.,* 1970; Jacklet and Cohen, 1967). In addition to neurons in the brain that undergo chromatolysis (reviewed by Lieberman, 1971), spinal motoneurons (in many mammals), autonomic neurons, and sensory neurons are among the well-studied neurons that become chromatolytic. It is striking that of the centrally and peripherally directed axons of each sensory neuron in the dorsal root ganglia, only severance of the peripherally directed axon—the one that rapidly regenerates—induces a chromatolytic reaction (Carmel and Stein, 1969; Hare and Hinsey, 1940). This suggested to Cragg (1970), among others, that the chromatolytic reaction is more pronounced in cells that regenerate axons more vigorously. [An alternative interpretation is that the peripheral axon is the one normally receiving from end organs "trophic" material that is required to prevent the cell from becoming chromatolytic (Purves, 1976a).] However, because some neurons that regenerate fail to show appreciable chromatolytic changes, any general importance of chromatolysis in regeneration remains unclear.

Although the mechanism that triggers chromatolysis remains uncertain, it seems likely to involve some signal carried from the cut end of the axon to the soma by retrograde axonal transport. The first evidence that such transport, rather than electrical discontinuity, signals to the neuron that its axon has been cut was the finding that axotomy farther from the cell body elicits chromatolytic responses at proportionally later times (Glover, 1967; Lieberman, 1974; Cragg, 1970). Retrograde transport is rapid enough to account for triggering the changes that occur within hours or days in the soma (Kristensson and Olsson, 1974), and application of colchicine, which blocks axonal transport, can mimic axotomy in its effects (Pilar and Landmesser, 1972; Purves, 1976b). Although there is evidence that some effects of axotomy, such as loss of presynaptic inputs to the damaged neuron (see later), may be due to a loss of target-derived substances such as NGF (Nja and Purves, 1978; Hendry, 1975), it is unknown whether chromatolysis in general is simply a response to interruption of transported materials traveling in one direction or another along the axon.

C. Changes in the Neuron's Electrical Characteristics

Vertebrate spinal neurons with peripheral axons are among the most thoroughly studied in their response to axotomy, particularly with respect to electrical properties such as decreased presynaptic input and lowered axonal conduction velocity. As with other changes associated with axotomy, there is some variation in degree both among species and neuron type, and even within the cells of a single motoneuron pool. Nonetheless, the pattern of change is consistent across cell types.

Axotomized motor or autonomic neurons often do not change their electrical resistance or membrane resting voltage measurably (Kuno and Llinas, 1970a; Purves, 1975), but within days of axotomy their synaptic input becomes greatly reduced (Acheson et al., 1942; Kuno and Llinas, 1970b; Purves, 1975; Pilar and Landmesser, 1972). Electrical measurements indicate a selective reduction in somatic and proximal dendritic input (Kuno and Llinas, 1970b; but compare Mendell et al., 1976). Morphological studies confirm that presynaptic terminals withdraw from the axotomized cell at about the same time that synaptic input drops. The pre- and postsynaptic elements often become separated by glial processes (Blinzinger and Kreutzberg, 1968; Purves, 1975). Furthermore, postsynaptic membrane specializations disappear in conjunction with a loss of postsynaptic sensitivity to applied transmitter (Brenner and Martin, 1976). The drop in postsynaptic sensitivity is not simply due to reduced presynaptic input and is strikingly different from the response following presynaptic denervation, in which postsynaptic densities persist and the membrane may become more sensitive to transmitter (Kuffler et al., 1971). As mentioned earlier, in sympathetic neurons the loss of synapses induced by axotomy can be reversed by chronic application of NGF in the vicinity of the cell body (Nja and Purves, 1978). Whereas presynaptic input reappears upon regeneration of the axon, delay of regeneration retards or permanently prevents its return (Goldring et al., 1980; Purves, 1975).

Compared to the proportion of synapses lost, the reduction of presynaptic efficacy is disproportionately small. Two changes in electrical properties of the soma can account for this discrepancy. First, concurrent with the reduction in synaptic input, the dendrites develop all-or-nothing responses (Purves, 1975; Kuno and Llinas, 1970a; but compare to Mendell et al., 1976) that can be triggered by synaptic events as small as 0.5 mV recorded in the cell body. In addition, the excitability of the cell may be increased, for despite no change in resting potential, cat spinal motoneurons show a 30% reduction in "rheobase current" (current required to bring the cell to threshold) (Kuno and Llinas, 1970a). The development of increased excitability resembles certain responses of some invertebrate axons to axotomy. The somata of axotomized (or colchicine-treated) cockroach motoneurons develop an electrical excitability (Pitman et al., 1972), as do single, isolated motoneurons of the leech grown in culture medium (Ready and Nicholls, 1979). For invertebrate neurons, increased excitability may be due to an increased membrane resistance rather than to an increase in channel excitability.

The conduction velocity of regenerating axons is reduced. The reduction correlates with a thinning of the growing axon and evidence that there is a corresponding reduction in the spread of current along it (Czeh et al., 1977; Muller and Carbonetto, 1979). Following regeneration, the axon caliber increases (Ramón y Cajál, 1928; Muller and Carbonetto, 1979; Lasek, 1981), as does the conduction velocity (Kuno et al., 1974). Cutting the peripheral axons of the rapidly conducting population of sensory cells results in chromatolysis together with slowed impulse conduction along the regenerating peripheral and the central axons of the cells. In contrast, cutting the central axon, which fails to evoke chromatolysis, actually increases the conduction velocity of the peripheral axon whereas the cut central axon, which regenerates only a short distance until reaching the central nervous system (CNS), shows no change in conduction velocity. The important point is that regeneration is associated with slowed conduction, and that conduction velocity increases as function returns.

Some of the most detailed studies of sensory and motor neuron responses to axotomy have been performed by Kuno and his collaborators. Many of their results are consistent with the idea that neurons revert to a "predifferentiated" state during regeneration, although this is probably somewhat of an oversimplification (Hall et al., 1978). In addition, in agreement with Purves (1976a), many of their results can be interpreted as evidence for a trophic influence of the target being lost upon axotomy and regained when the nerve regenerates to the target, even if functional innervation is not established. For example, during development, spinal motoneurons innervating slow (soleus) muscle acquire a prolonged afterhyperpolarization following each nerve impulse, whereas the afterhyperpolarization in motoneurons that innervate fast (gastrocnemius) muscles is not lengthened. Axotomy reduces the afterhyperpolarization in soleus motoneurons. The effect is achieved without axotomy, if the soleus muscle is made to atrophy by immobilization at a fixed, short length (Gallego et al.,

1979) or by partially denervating the muscle (Huizar *et al.*, 1977). Similarly, blocking nerve activity with tetrodotoxin can, presumably by its effect on target muscle activity, prompt a decrease in the afterhyperpolarization in soleus motoneurons, a change reversed by daily stimulation of the muscle (Czeh *et al.*, 1978). These results indicate that an electrical property—afterhyperpolarization—gained during development is maintained by trophic influences of the muscle. It remains to be seen how many changes fit this pattern and what the nature of the retrograde signal might be.

The synaptic projections of sensory neurons undergo cyclic changes during sensory regeneration, dropping in efficacy and then returning to normal when the neurons become functionally reconnected with sensory structures in the muscle (Gallego *et al.*, 1980). In contrast, when regeneration of an entire sensory nerve is prevented, the central connections remain diminished in strength. One surprising finding, however, is that those neurons that reach the periphery after 2 months, but that fail to make effective connections, actually increase the effectiveness of their central synapses above normal, suggesting an influence of the periphery nonetheless.

D. Alterations in Axonal Transport

Since Weiss and Hiscoe's (1948) demonstration that ligation of a nerve causes an immediate accumulation of material at the site of the lesion, changes in axonal transport following nerve injury have been under intensive investigation. Because the effects of axotomy on axonal transport of membrane and cytoplasmic proteins from their sites of synthesis and modification in the ER and Golgi to their final destinations have been extensively reviewed (Grafstein and McQuarrie, 1978; Grafstein and Forman, 1980; Schwartz, 1979; Lasek, 1981; Wilson and Stone, 1979), we will only touch on this important feature of axotomy. Simultaneous with morphological and synthetic changes in the cell body, the transport of radioactively labeled material down the axon shifts quantitatively. There is a reduction in the amount of material associated with synaptic transmission that moves at a rate of up to 400 mm/day (fast transport). In several cases, the amount of other material moving at rates of fast (80–400 mm/day) and of slow (0.5–5 mm/day) transport increases, without an increase in the rate of transport. Nonetheless, few new protein species or other labeled materials can be detected on two-dimensional gels following axotomy (discussed later). Primarily fast axonal transport supplies material to the growing tip, because the rate of growth is about the same as that of slow transport. In addition to studies such as those by Kristensson and Olsson

(1974), showing that exogenous materials become transported to the cell body from the site of the lesion, we know that transport continues not only from the cell body, but in both directions within the isolated distal segment. In neurons of *Aplysia*, the sea slug, axotomy of one of two branches of an axon immediately increases the amount of material transported down the intact branch (Goldberg *et al.*, 1976). How material is selected for transport, by exactly what means it travels, and how it is converted into the structural material of the neuron remain unclear (see later).

E. AXONAL AND CELLULAR DEGENERATION

Although some axotomized mammalian neurons degenerate and die, such death is more likely to ensue (1) in very young animals, (2) when the axotomized branch represents a large fraction of a cell's arborization (Fry and Cowan, 1972), or (3) if reconnection with the end or target organ fails to occur (Ramón y Cajál, 1928; Lieberman, 1974; Purves, 1975; Cowan and Wenger, 1967). After axotomy, the distal portion of the axon, detached from its cell body, degenerates. This may occur within a matter of days in mammals (Ramón y Cajál, 1928) and in days or weeks in lower vertebrates and invertebrates (Speidel, 1933; Guthrie, 1962; Letinsky *et al.*, 1976). The degenerating axons are phagocytosed, typically by surrounding Schwann cells, astroglial cells, or "microglial" phagocytes.

Although severed distal axonal stumps usually degenerate, they will survive for months in hibernating mammals (Albuquerque *et al.*, 1978) and many invertebrates (Hoy *et al.*, 1967; Carbonetto and Muller, 1977). The axonal stumps of only certain neurons seem to be long-lived in invertebrates (Bittner and Johnson, 1974). The mechanism for survival of anucleate axons is unclear. One hypothesis is that ensheathing glial cells transfer proteins and other needed macromolecules to the axons (Lasek *et al.*, 1977; Gainer *et al.*, 1977; Meyer and Bittner, 1978). However, the metabolic needs of isolated axons are not known. For example, reduced temperature is thought to be an important element, and several studies indicate that degeneration proceeds more rapidly at temperatures elevated slightly above ambient levels (Nordlander and Singer, 1973; Usherwood *et al.*, 1968; Wilson, 1960; A. Mason and K. Muller, unpublished). Long-lived stumps in invertebrates can have an important role in regeneration as when the regenerating neuron synapses upon its severed axonal stump (Carbonetto and Muller, 1977), which acts as a splice between the injured neuron and its target. It is striking that in the leech a rapid degeneration of certain long-lived stumps seems to be triggered by successful regeneration of a growing axon to synapse with its normal target (Muller and Carbonetto, 1979).

III. Initiation of Nerve Fiber Regeneration

A. TRIGGERS FOR GROWTH

The axotomized nerve fiber sprouts usually near its cut end some time after the onset of chromatolysis in those neurons that undergo an "axon reaction." This is consistent with the idea that growth requires a change in the cell's synthesis and transport of molecules. Possible triggers for sprouting include (1) interruption of trophic agents coming from the end organ, (2) interruption of transport per se, (3) inactivity in the end organ, and (4) removal of an inhibitory effect of the end organ. It is beyond the scope of this chapter to discuss in depth the variety of triggers for growth of mature axons, but even a brief examination may suggest several aspects of axotomy important for axonal growth. Fortunately, there have been several recent reviews in this area (Brown *et al.*, 1981; Purves, 1976a; Diamond *et al.*, 1976).

The peripheral nervous system has provided a wealth of examples of sprouting. As first described by Edds (1950, 1953) and discussed in detail by Brown *et al.* (1981), intact motor axons will sprout at their terminals and send off collaterals at distal nodes of Ranvier upon partial denervation of the muscle that they innervate. Botulinum toxin and α-bungarotoxin bind, respectively, to pre- and postsynaptic receptors at neuromuscular junctions and both trigger sprouting of nerve terminals (Pestronk and Drachman, 1978; Holland and Brown, 1980). It is possible that both ligands act by reducing muscle activity, because sprouting occurs in paralyzed muscle and the ligand-induced sprouting can be blocked by direct stimulation of muscles (Brown *et al.*, 1981). The presence of functionally denervated fibers may then induce sprouting.

Evidence from interneuronal synapses in invertebrates suggests that the loss of the target is not alone sufficient to induce growth by the mature neuron. In addition, the neuron requires another stimulus, such as injury, even if it does not directly affect the axon that has lost its target (Fig. 1) (Muller and Scott, 1980; Scott and Muller, 1980).

Denervation of neurons in autonomic ganglia induces preganglionic and postganglionic collateral sprouting (Courtney and Roper, 1976; Sargent and Dennis, 1977, 1981; Purves, 1976c) and denervation of sensory fields in the skin can cause adjacent intact axons to sprout into the denervated fields (Aguilar *et al.*, 1973). Collateral sprouting into denervated areas also occurs in the CNS in the septal nucleus (Rais-

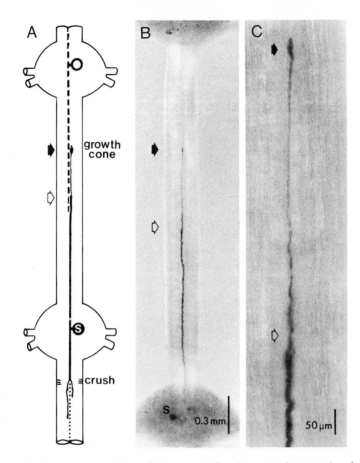

Fig. 1. An intact axon can be made to sprout when its synaptic target is selectively destroyed, provided that the neuron has been otherwise stimulated to grow. An S interneuron in the leech segmental ganglion at top (anterior) was entirely destroyed by intracellular injection with protease [dashed line and open circle in (A)]. A distant axon of the adjacent S cell (S) was then severed by crushing the nerve (crush). The schematic diagram shows that the injured axon regenerated in a normal fashion across the crush; the intact axon of that cell, which had been in synaptic contact with the killed cell near the connective midpoint (open arrow), sprouted a fine process tipped with a growth cone (solid arrow). (B) The sprouting neuron was marked by intracellular injection of horseradish peroxidase (HRP) 25 days after its posterior axon (not visible) had been severed. (C) The sprouted axon is seen at higher magnification, showing the growth cone (solid arrow) extending from the original axon tip (open arrow). (From Muller and Scott, 1980.)

man, 1969), tectum (Lund and Lund, 1971; Yoon, 1972), hippocampus (Cotman and Lynch, 1976), spinal cord (Bjorklund and Stenevi, 1979), and red nucleus (Tsukahara et al., 1975), to name a few prominent examples.

Histofluorescence techniques have made monoamine-containing neurons particularly easy to detect in the brain. It has been reported that neurons in the locus ceruleus can sprout into areas not denervated if locus ceruleus axons projecting elsewhere are cut (Pickel et al., 1974). Unfortunately, we do not know if the same neurons that are injured by the cut are those that sprout, so it is unclear whether it is direct injury to one portion of the neuron that prompts it to sprout an intact axon, as has been described in the leech (Muller and Scott, 1980; Scott and Muller, 1980), or whether injured neurons are somehow signaling intact neurons that innervate a different target to sprout, as occurs in the frog spinal cord (Rotshenker and McMahan, 1976).

B. INITIAL OUTGROWTH OF FIBERS

The sprouts which are tipped by growth cones and which emerge from the severed axon at or near its end, grow in all directions, but they extend preferentially in the direction of preexisting fibers (Fig. 2). Several aspects of this preferential growth in vivo suggest that external cues like those described in tissue culture (Letourneau, 1975b) and the presence of the extracellular matrix left behind by degenerating nerve fibers may help direct the nascent fibers. In addition, regenerating nerve fibers may be predisposed to attain their former shape (Solomon, 1980). These factors may be sufficient to enable the growing nerve fiber to reach its target.

The early stages of nerve fiber growth have been studied extensively in cultures of neurons from cell lines and embryonic tissues. When first cultured, neurons from both sources are spherical cells suspended in culture medium. To begin nerve fiber growth, neurons must first attach to the substrate. Attachment occurs within 1 hour and is followed by a much slower process (1–6 hours), during which the neuron develops a wide, flattened region spread over the substrate (Collins, 1978; Kataoka et al., 1980; Wessells et al., 1978). The flattened region, which we call a "somatic veil" to distinguish it from similar parts of growth cones, results from the progressive extension and spreading of active filopodia that become attached to the substrate. As one or more nerve fibers emerge from the somatic veil, their activities become restricted and are ultimately limited to the tip of the growing nerve fiber or growth cone (Collins, 1978) (Fig. 3). The somatic veil retracts where nerve fibers emerge from the cell soma and eventually

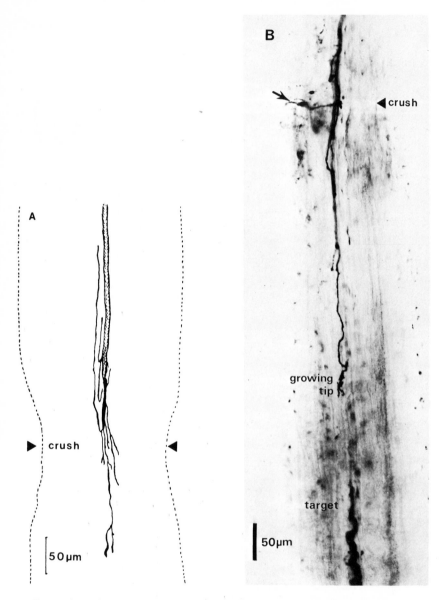

FIG. 2. Axonal sprouting at the site of nerve injury. Individual S interneurons were stained with intracellular injection of HRP 6 (A) and 19 (B) days after their axons were severed by crushing the nerve. The *camera lucida* tracing from wholemount in (A) shows that fibers tipped with growth cones extended in several directions but grew preferentially along the nerve (outlined with dots). In the photomicrograph in (B), the fine axon regenerating along its normal pathway extended almost to the target, whereas other branches at the crush remained short (arrow). (From Muller and Carbonetto, 1979.)

FIG. 3. Scanning electron micrograph of the tip of a regenerating nerve fiber in culture. The fine filopodia (F) are typical of a growth cone. In living cells, these move about to "sample" and sometimes attach to the substrate. Also typical are flattened membranous regions called "veils" (V), which probably attach the growing nerve tip to the substrate. Bar: 5 μm.

disappears, except in unipolar neurons, where it persists; its attachment to the substrate offsets the pull on the soma by the elongating nerve fiber.

The transition of neurons from attached spherical cells to cells extending nerve fibers can be divided into three stages—attachment, spreading, and outgrowth. Attachment of neurons to a substrate is relatively insensitive to metabolic inhibitors and is often described in terms of electrostatic interactions between the substrate and the cell surface. Rather than the net charge on the cell, however, the distribution of charged species on its surface may determine the attachment of cells to substrates or to one another (Wright *et al.*, 1980). Coating culture dishes with fibronectin (a protein found in serum and the extracellular matrix) or collagen will enhance the attachment of cells (Grinnell, 1978), but it is unclear whether there are specific receptors on the cell surface that recognize fibronectin or collagen.

It is unlikely that the attachment of neurons to the culture substrate is directly relevant to the initiation of nerve fiber regeneration *in vivo*. Nevertheless, the attachment of neurons to other cells can be highly specific (Gottlieb and Glaser, 1980) and may be more relevant to nerve fiber initiation (Luduena, 1973) and synaptogenesis (Ruffolo *et al.*, 1978). An intriguing aspect of attachment is that it is required for differentiation of many cells and stimulates protein synthesis in cultured neurons (J. DeGeorge and S. Carbonetto, unpublished observations). Correspondingly, in mitotically active cells, the degree of attachment and spreading is directly correlated with nucleic acid and protein synthesis in the cell (Folkman and Moscona, 1978).

The cytoskeletal reorganization that follows attachment of nonneuronal cells to a substrate is profound (Vasiliev and Gelfand, 1976; Lazarides, 1976) and is likely to be as great for regenerating neurons. Two of the reported cytoskeletal changes in neurons are aggregations of microtubule organizing centers that serve as "seeds" for the growth of microtubles (Spiegelman *et al.*, 1979; Brinkley *et al.*, 1976) and the "unwinding" of bundles of neurofilaments from the cell body into the growing nerve fiber (Sotelo *et al.*, 1979). Presumably, the cytoskeletal reorganization is initiated by interaction of cell surface macromolecules with the culture substrate. Reported changes in adhesivity of neurons following treatment with NGF, cyclic AMP, calcium ionophores, and elevated extracellular K^+ (Schubert *et al.*, 1973) may thus be relevant to the cytoskeletal changes. How cell surface interactions are signaled to the interior of the cell is a question of general biological interest that awaits elucidation.

C. Synthetic Requirements for Nerve Fiber Elongation

Following the damage caused by axotomy and the subsequent increase in ribosomes (Barr and Bertram, 1951), one might expect that nerve fiber regeneration requires protein and RNA synthesis. Consistent with this, nerve fiber regeneration in culture is accompanied by increases in RNA and protein synthesis (Angeletti *et al.*, 1965). Additional data from experiments with inhibitors of protein and RNA synthesis indicate, with some qualifications, that translation but not transcription is required for nerve fiber growth.

In most cases, treatment of neurons in culture with cycloheximide, which blocks protein synthesis, inhibits nerve fiber regeneration (Partlow and Larrabee, 1971; Luduena, 1973; Burnham and Varon, 1974). Those cases in which protein synthesis is not required for nerve fiber regeneration (Seeds *et al.*, 1970) can be understood once it is known that neurons contain presynthesized pools of proteins (Chan and Baxter, 1979) whose incorporation into the nerve fiber does not require additional protein synthesis (Carbonetto and Fambrough, 1979). The presence of such pools in neurons, together with the likelihood that pools are turning over at some finite rate, is a clue that protein synthesis may change little qualitatively following axotomy. When analyzed by two-dimensional gel electrophoresis (Hall et al., 1978; Skene and Willard, 1981; Benowitz et al., 1981), axotomized nerves in vivo reveal only minor changes in the synthesis of new protein species. With this in mind, it is not surprising that nerve fiber regeneration is not inhibited by actinomycin D, which blocks RNA synthesis (Partlow and Larrabee, 1971; Burnham and Varon, 1974; Burstein and Greene, 1978; cf. Mizel and Bamburg, 1976). However, there is also little change in the species of protein synthesized during fiber growth in cultured pheochromocytoma (PC12) cells requiring RNA synthesis to grow nerve fibers (Burstein and Greene, 1978). PC12 cells can differentiate in the presence of NGF from round cells in suspension to generate long nervelike fibers and do so with little change in the types of proteins they synthesize (McGuire et al., 1978; Garrels and Schubert, 1979). Unlike skeletal muscle, where differentiation is mediated by obvious changes in transcription (Patterson and Bishop, 1977) and in spite of extensive morphological changes, regeneration or growth de novo of nerve fibers requires the appearance in the cell of only a few minor protein species. A similar finding during neural development in vivo would emphasize the importance of posttranslational processes during differentiation of the nervous system.

IV. Components of Growing Fibers

A. MORPHOLOGY OF REGENERATING NERVE FIBERS

Elongating nerve fibers have a unique morphology. The main axis cylinder of the fiber, which, when mature, terminates as a synaptic bouton or specialized sensory ending, ends instead as a growth cone. Growth cones consist of a wide basal region from which extend projections called "filopodia," which are approximately 10–30 μm long and 0.2 μm in diameter. In cultured neurons, where direct observation is simple, the movements of the rigid-looking filopodia do not resemble the cytoplasmic streaming in amebae.

The major cytoskeletal structures in neurons are neurofilaments, neurotubules, and microfilaments (for detailed descriptions, see Bunge, 1973; Johnston and Wessells, 1980). This nomenclature is somewhat confusing. Neurotubules (25-nm diameter) are homologous to microtubules in nonneuronal cells. Neurofilaments (10-nm diameter) belong to a class called "intermediate filaments" and are distinct from microfilaments. Microfilaments (6–7 nm in diameter) contain actin and are present in many cell types. Porter and his colleagues have described a cytoplasmic matrix that they call "microtrabeculae" (Wolesewick and Porter, 1979), which may constitute an additional major cytoskeletal structure in neurons (Ellisman and Porter, 1980) while incorporating those previously described.

Differences in appearance of the main axis cylinder and growth cone of nerve fibers are dictated by the underlying cytoskeleton. The growth cone filopodia contain a microfilamentous meshwork and no neurotubules, whereas the main cylinder contains both neurotubules and microfilaments. This segregation of cytoskeletal elements is readily apparent in neurons treated with cytochalasin B or colchicine. Cytochalasin B and colchicine disrupt microfilament and neurotubule assemblies, respectively. Cytochalasin B causes the filopodia of the growth cone to retract rapidly and, at a much slower rate, causes resorption of the growing nerve fiber (Yamada et al., 1970). Colchicine has no initial effect on the growth cone but causes the whole nerve fiber to be slowly resorbed (Daniels, 1972).

Parallel to the axis cylinder within nerve fibers are microfilaments beneath the plasma membrane. Neurotubules and neurofilaments are more apparent in the center of the nerve fiber. Neurotubules, once thought to be unitary and continuous for the length of the fiber, are composed of shorter lengths of tubules varying in number along the fiber (Nadelhaft, 1974; Chalfie and Thomson, 1979). Neurotubules have projections (Fernandez et al., 1971), spaced at regular distances, that are possibly involved in intracellular transport (Grafstein and Forman, 1980). Other organelles within the axis cylinder include smooth membranous reticulum, mitochondria, coated vesicles, cisternae, dense-core vesicles (65–160 nm), and, occasionally, ribosomes.

Neurofilaments, like neurotubules, extend no farther than into the base of the growth cone, which is filled with a meshwork of microfilaments. Most commonly the meshwork extends into the filopodia (Bunge, 1973; Yamada et al., 1971), in some instances forming bundles of microfilaments (Kuczmarski and Rosenbaum, 1979). The bundles of microfilaments are found where filopodia closely adhere to their substrates and are localized with myosin (Letourneau, 1979b, 1981), to form structures that resemble adhesion plaques in fibroblasts (Goldman et al., 1976). In the growth cone, the organization of microfilaments consists either of a meshwork or of bundles, possibly depending on whether filopodia are attached at the time of fixation or on the fixation procedure itself (Kuczmarski and Rosenbaum, 1979; Maupin-Szamier and Pollard, 1978).

Microfilaments are associated with the plasma membrane in a variety of cells (Mooseker, 1976; Small *et al.*, 1978), including neurons (Isenberg and Small, 1978), and the plasma membrane in these cases may serve a function similar to the Z line in skeletal muscle. Consistent with this analogy, after heavy meromyosin treatment of microfilaments in growth cones, the resulting arrowheads that decorate the microfilaments point away from the plasma membrane, suggesting that the microfilaments may pull on the plasma membrane. The direction of polymerization of actin is opposite that of force generation and therefore toward the plasma membrane. However, it is not possible to speculate on the role of these structures unless one knows the disposition in the cell of the interacting proteins. For example, if the microfilaments in growth cones are anchored at an adhesion site and the myosin is attached to a mobile structure, then the myosin could slide toward the plasma membrane over the actin. On the other hand, if both actin and myosin are anchored, an isometric contraction could result.

The base of the growth cone contains an array of organelles that are excluded from filopodia. These include large, dense-core vesicles, smooth membranous reticulum, clear vesicles, lysosomes, mitochondria, and sometimes a few polysomes. This complement of membranous structures undoubtedly reflects the complex dynamics of growth cones whereby retraction and extension of filopodia can occur within seconds. Particular attention has been paid to the mounds of vesicles that are sometimes observed in electron microscopy of growth cones (Pfenninger and Bunge, 1974) and nerve fibers grown in culture (Wessells *et al.*, 1976). Discussion originally centered on whether these vesicles represented a supply of membrane that was a precursor to the plasma membrane (see later). More recently, discussion has shifted to the possibility that the vesicles are a fixation artifact (Hasty and Hay, 1978; Hay and Hasty, 1979; Pfenninger, 1979; Shelton and Mowczko, 1979; Nuttall and Wessells, 1979; Rees and Reese, 1981).

Ideas about the function of organelles in nerve fibers and growth cones arise largely by extrapolation from other cells. The term "cytoskeleton" itself belies our ignorance of the function of the fibrillar structures in neurons. Except for some involvement of microfilaments and neurotubules in axonal transport, little is known of the function of these structures in nerve regeneration. Microfilaments are thought to be the force-generating structures or the "cytomusculature" of neurons. Neurotubules may not be necessary for extension of relatively short nerve processes (Schubert *et al.*, 1973) but could serve as a "scaffolding" that also restricts the regions of outgrowth of the nerve fiber (Bray *et al.*, 1978) and determines the shape of the neuron. In this scheme, neurofilaments, which are more stable polymers than either microfilaments or microtubules, would constitute the superstructure of the neuron. How or if the plasma membrane or extracellular matrix is involved in maintaining or developing the structure of the nerve fiber is unclear. Axoplasm that is extruded from squid giant axons and therefore is devoid of any membrane attachment retains its cylindrical shape. However, the plasma membrane may be important in the genesis of the cytoskeleton. For example, cytoskeletal proteins may insert directly into the plasma membrane (Estridge, 1977) or may interact with membrane proteins. During fiber growth, cytoskeletal proteins may help anchor nerve fibers to external structures.

B. Cell Exterior and Growth Environment: The Extracellular Matrix

Carbohydrate-rich material is ubiquitous throughout the extracellular spaces of the central nervous system. In peripheral nerves, Schwann cells that ensheathe nerve fibers have a basal lamina in which the carbohydrate-rich material is accompanied by collagen fibrils. Such components of the extracellular space are the milieu through which developing neurons migrate (Pratt et al., 1975) and regenerating nerve fibers grow. In peripheral nerves, these extracellular and supporting cellular structures remain after degeneration of axotomized nerve fibers and may guide the regenerating nerve fiber to its target. In addition, experiments of McMahan and his colleagues (Marshall et al., 1977; Sanes et al., 1978; Burden et al., 1979; Edgington et al., 1980) have demonstrated an important regulatory role for the extracellular matrix during the regeneration of nerve–muscle synapses. These workers have shown that regenerating nerve fibers will grow to and innervate "ghost" neuromuscular junctions that consist only of the basal lamina and extracellular matrix left behind by a damaged and retracted muscle. Moreover, regrowth of the damaged muscle is accompanied by formation of a new sarcolemma replete with acetylcholine receptors and other postsynaptic specializations at the region of the old synapse. Thus, constituents of the extracellular matrix can dictate very precisely the location and dimensions of the pre- and postsynaptic structures of a regenerated neuromuscular junction.

Several different structures can be grouped under the title of extracellular matrix. The glycocalyx is a layer of carbohydrate-rich material (Bennett, 1963) covering many types of cells, including regenerating nerve fibers and their growth cones (James and Tresman, 1972). The basal lamina (also called basement membrane) in peripheral nerves includes polysaccharides similar to those in the extracellular matrix of the CNS but also has a layer of collagen and other proteins. In culture, many cells produce an extracellular matrix that is secreted onto the culture substrate and is often referred to as substrate-attached material or microexudate. The molecular components of these various extracellular matrices are still being defined and their precise relationship, especially with regard to neurons in culture, is unclear. Known constituents of the extracellular matrix of neurons include certain glycosaminoglycans, collagen, laminin, and possibly fibronectin (Alitalo *et al.*, 1980; Bunge *et al.*, 1980; Margolis and Margolis, 1979).

Glycosaminoglycans (formerly called acid mucopolysaccharides) are linear carbohydrate polymers usually composed of a disaccharide repeating unit of uronic acid and hexosamine. The disaccharides have either carboxyl or sulfate groups and are polyanions that readily stain with colloidal iron and other cationic stains (Hughes, 1976). Those glycosaminoglycans associated with the nervous system include hyaluronate, chondroitin 4- and 6-sulfate, and heparan sulfate (Margolis and Margolis, 1979). *In vivo*, all of these except for hyaluronate are covalently bound to proteins, forming proteoglycans.

The best-studied proteoglycan is that from cartilage; it consists of a core protein (MW 2×10^5) with multiple side arms of chondroitin sulfate and keratan sulfate. A single proteoglycan subunit may be 0.5 μm in length (Hascall and Heingard, 1975) and linked by hyaluronate (MW up to 10^7) into much larger complexes whose physical properties are well suited to their role as major constituents of the extracellular space. *In vivo*, these extremely large molecular complexes are hydrated and can trap macromolecules and alter the flow of smaller solutes (Hadler, 1980; Ogston, 1970).

Many of the molecular interactions of the negatively charged glycosaminoglycans are electrostatic; however, glycosaminoglycans can also interact highly specifically. For example, the binding of hyaluronate to the core protein of the cartilage proteoglycan is to a well-defined region of the core protein and is not inhibited by other polyanions nor by isomers of hyaluronate (Lindahl and Hook, 1978). Similarly, hyaluronate can bind specifically to cell surface macromolecules in cultured fibroblasts and also to embryonic chick brain cells (Underhill and Toole, 1979). The function of this hyaluronate receptor on the cell surface is unknown. It may anchor the hyaluronate to serve as an assembly site for other extracellular matrix components or possibly it is related to the stimulatory effect of hyaluronate on cell migration (Pratt *et al.*, 1975).

The basal lamina associated with Schwann cells contains glycosaminoglycans and Type I and Type III collagen (Junqueira *et al.*, 1979). In addition, there are probably collagen-associated proteins resembling fibronectin and laminin in the substrate-attached material of cultured neurons (Alitalo et al., 1980; Culp et al., 1980). In primary cultures of neurons, the collagen produced by Schwann cells (Bunge et al., 1980) is assembled along with glycosaminoglycans into a basal lamina (Bunge et al., 1979).

The structure and biosynthesis of collagen has been the subject of extensive study (Bornstein and Sage, 1980), as has the interaction of collagen with cells (Kleinman et al., 1981). One interaction is the well-known ability of collagen to stimulate growth and differentiation of cultured cells, which appears to be mediated by glycoproteins that attach to the collagen. Fibronectin is the best studied of these glycoproteins (Yamada and Olden, 1978), which also include chondronectin in chondrocytes (Hewitt et al., 1980) and laminin in epithelial cells (Chung et al., 1977). As mentioned previously, the adhesion of cultured cells that is mediated by these glycoproteins is a complex process culminating in specialized adhesion sites that contain an extracellular matrix as well as an intracellular scaffold of microfilaments.

Fibronectin contains several separate domains, including one that binds to collagen and another distinct region that binds to the cell surface (Yamada et al., 1980; Ehrismann et al., 1981). Experiments by Rephaeli et al. (1981) demonstrate that this interaction of fibronectin at the cell surface can increase the permeability of cells to calcium and stimulate protein phosphorylation. It is easy to see that these events could be those that signal binding of the cell surface to fibronectin and stimulate the cytoskeletal changes that lead to formation of an adhesion site. A similar sequence of events in regenerating nerve fibers may follow from the attachment of growth cone filopodia and trigger events leading to extension of the nerve fiber.

The structure, composition, assembly, and functional role of the extracellular matrix have been briefly considered in this section in the light of specific interactions between matrix components and the cell surface. These interactions result in complex morphological changes that are important to cell growth and motility. Certainly the experiments of McMahan and his colleagues emphasize the importance of the extracellular matrix for nerve fiber regeneration. Their experiments on the frog neuromuscular junction show that the extracellular matrix is able to stop the growth of the ingrowing nerve fibers and regulate the reconstruction of the pre- and postsynaptic elements. Clearly one next step in our understanding of nerve fiber regeneration is to obtain a description of the components of the extracellular matrix responsible for this control.

C. COMPOSITION OF THE NEURONAL SURFACE

It is widely thought that the specificity of intercellular interactions that occur in the nervous system result from complementary surface macromolecules on the interacting cells (Sperry, 1963). In systems other than the nervous system, there is direct evidence that cell surface macromolecules mediate highly specific associations, as, for example, between gametes (Schmell *et al.*, 1977; Glabe and Vacquier, 1977) and between sponge cells (Humphreys *et al.*, 1977). There is no equivalent evidence that regenerating nerve fibers attain their former pattern of innervation by specific binding of macromolecules on the growth cone with those of the target cell. However, studies on the developing visual system of the chick show that some cell surface macromolecules are present in a gradient over the tectum (Trisler *et al.*, 1981), as predicted by previous physiological and anatomical studies (Fraser and Hunt, 1980). Also, cells from the retina are able to adhere in the expected pattern (dorsal retina to ventral tectum), possibly as a result of this or a similar gradient of surface macromolecules (e.g., Gottlieb and Glaser, 1980). Furthermore, if gradients of soluble rather than membrane-bound molecules guide regenerating nerve fibers to their targets, as is sometimes suggested, they might both stimulate and direct growth, as in the case of NGF action on dorsal root ganglion neurons (Gundersen and Barrett, 1980; Greene and Shooter, 1980).

There are substantial difficulties in determining what molecules on the surfaces of cells might be responsible for specificity of neuronal regeneration. First, the cells constituting the nervous system are heterogeneous. Second, if the molecules in question are proteins, detection may be a problem because there are only trace amounts of plasma membrane proteins in the cell. Third, procedures for isolating surface macromolecules are not straightforward, particularly without a specific ligand that can be used as an affinity reagent. For these and other reasons, our knowledge of the molecular aspects of nerve fiber guidance during regeneration is meager.

One idea being investigated is that unique macromolecular components on the cell surface may be important in neuronal differentiation or function (for several reviews, see Lajtha, 1970; Raff *et al.*, 1979). Thus, some investigators have sought to characterize surface macromolecules on neurons whose localization on growth cones or other parts of the cell might give some clue as to their function in nerve fiber growth (Estridge and Bunge, 1978; Cornbrooks *et al.*, 1980; Wessels *et al.*, 1976; Brown, 1971; Denis-Donini *et al.*, 1978; Carbonetto and Argon, 1980). In general, however, biochemical studies have indicated that there are only subtle changes in the surface proteins of cells growing fibers (McGuire *et al.*, 1978). Antibodies, lectins, and toxins that bind with high affinity to neurons and that can be labeled to high specific activities have been employed to describe further these subtle surface changes in regenerating nerve fibers (Carbonetto and Argon, 1980; Denis-Donini *et al.*, 1978; Letourneau, 1979a; Pfenninger and Maylie-Pfenninger, 1979; Hatten and Sidman, 1977; Gonatas, 1979; Brown, 1971; Wood *et al.*, 1980; Carbonetto and Fambrough, 1979; Greene, 1976; Dvorak *et al.*, 1978). Antibodies, for example, have revealed the existence of surface macromolecules involved in the fasciculation of nerve fibers (Rutishauser *et al.*, 1978). The application of methods for producing monospecific antibodies to very complex antigens (Kohler and Milstein, 1975) promises to facilitate studies of this type. Using these methods, investigators have injected nervous tissue into mice and have selected monospecific antibodies following cell fusion, culturing, and cloning of individual spleen cell hybridomas. The results from a number of laboratories that are applying this technique to the nervous system are very encouraging (Eisenbarth *et al.*, 1979; Momoi *et al.*, 1980; Zipser and McKay, 1981; Trisler *et al.*, 1981).

In mature neurons, the composition of the plasma membrane is varied over the surface of the cell, as not all macromolecules are uniformly distributed within the lipid bilayer. Voltage-sensitive ion channels are concentrated at the nodes of Ranvier (Ritchie and Rogart, 1977; Chiu and Ritchie, 1980) whereas neurotransmitter release sites (Heuser *et al.*, 1979) and receptors (Harris *et al.*, 1971) are concentrated at synapses. In regenerating neurons, there appears to be a gradient of intramembranous particles such that they are abundant at cell bodies and sparse at growth cones (Pfenninger and Bunge, 1974; Pfenninger and Rees, 1976). Furthermore, other cell surface components accumulate along growing nerve fibers or at sites of adhesion to the substrate (Culp *et al.*, 1980). At the nodes of Ranvier of developing and regenerating nerve fibers, the dimensions of the specialized regions of plasma membrane are precisely regulated (Waxman and Foster, 1980). Similar precision of localization occurs upon regeneration of nerve terminals on denervated muscles (Letinsky *et al.*, 1976). How localizations of molecules arise and are sustained in the fluid matrix of the plasma membrane is presently being investigated in a number of cell types (e.g., Bloch and Geiger, 1980). More perplexing is the nonuniform distribution of lipids that may exist in neuronal membranes (DeBaecque *et al.*, 1976; Eisenbarth *et al.*, 1979) or across the lipid bilayer.

D. BIOSYNTHESIS AND TURNOVER OF NEURONAL MEMBRANES

During regeneration of its nerve fiber, an axotomized neuron will undergo an enormous increase in its surface area. This requires that newly synthesized membrane proteins, especially those responsible for transport across the membrane and for ion translocation such as voltage-sensitive and chemically sensitive channels, be inserted into the plasma membrane in the proper orientation. The available evidence suggests that this is accomplished in neurons in a fashion similar to that in other cells (for review, see Holtzman and Mercurio, 1980). Furthermore, the often-cited hypothesis that growing nerve fibers add plasma membrane primarily at the growth cone (Bray, 1970; Koda and Partlow, 1976; Pfenninger and Bunge, 1974) has been questioned. Before examining the problem of where proteins are inserted into the membrane, it is worthwhile to consider their synthesis and turnover.

Most plasma membrane proteins are glycoproteins that have their carbohydrate moieties exposed to the exterior of the cell. The exterior orientation is extremely stable, as the rate of flip-flop of membrane proteins across the membrane is negligible. Membrane proteins are synthesized on membrane-bound ribosomes (rough ER) that, along with other internal membrane compartments such as the Golgi apparatus, serve as a matrix upon which the proteins attain their proper orientation and conformation. Addition of assembled membrane to the surface probably occurs by fusion of vesicles with the plasma membrane, but in neurons one step in this process may include axonal transport. Several aspects of the transport are altered in regenerating nerve fibers (see earlier). Membrane glycoproteins seem to be synthesized by the same cellular organelles that are responsible for the synthesis of soluble proteins secreted by the cell (Palade, 1975). The available evidence indicates that neurons also utilize this pathway during the normal turnover of membrane glycoproteins and during the growth of nerve fibers (Holtzman and Mercurio, 1980).

Whether a protein or glycoprotein is to be "exported" (secreted and membrane proteins) or remain inside the cell appears to be determined early in its synthesis. The most popular hypothesis to explain how and when this decision is made is the "signal hypothesis," which proposes that an initial sequence of nucleic acids in mRNA dictates the attachment of the translating ribosome to the ER and that subsequently the signal peptide is removed. [Wickner (1980) has proposed an alternative or perhaps additional view of the synthesis and processing of membrane proteins. It is unclear which or to what extent either of these pathways is utilized in the cell.] Those proteins without a signal are synthesized on "free" ribosomes and will remain within the cytoplasm. The ribosomes on ER seem to attach to receptors that form a pore in the ER through which the elongating polypeptide chain passes. While still attached to the ribosome and in the course of elongation, the polypeptide chain is glycosylated by enzymes together with lipids that serve as carriers for the carbohydrates (Waechter and Lennarz, 1976; Harford and Waechter, 1979). Transfer of the mannose-rich carbohydrate core to the polypeptides occurs "en bloc." Secreted proteins are completely extruded through the pore and are contained free within the membrane-bound ER. Membrane proteins, however, lie embedded within the membrane, apparently stabilized there by one or more sequences of hydrophobic amino acids.

Synthesis of membrane lipids also occurs within the ER. Some phospholipids and glycolipids in neurons are synthesized and incorporated into axons, but the cell body (Sherbany et al., 1979) seems to be the major site of synthesis (see Holtzman and Mercurio, 1980, for discussion). The spatial overlap of lipid and protein synthesis within the ER may reflect the interdependence of these two processes and perhaps their coordinate regulation during membrane assembly (Grafstein et al., 1975; Sherbany et al., 1979).

Transfer of newly synthesized membrane components from the ER to the Golgi apparatus occurs by budding of vesicles from the ER and fusion with the Golgi. Movement between these two compartments is energy dependent and continues in the absence of protein synthesis (Jamieson and Palade, 1968). Although the transfer is often described as bulk transfer (i.e., involving transport of assembled membrane rather than transfer of individual molecules), it is selective and permits the ER and Golgi to retain their unique protein and lipid compositions. Selective uptake of bulk membrane from the plasma membrane into the cell also occurs (Tsan and Berlin, 1971; Oliver et al., 1974). In the Golgi, glycoproteins and glycolipids are subjected to further processing involving removal of several core carbohydrate residues followed by several cycles of addition and removal of terminal carbohydrates.

Transfer of newly synthesized membrane to the surface of neurons occurs by an extensive transport system that runs the length of the nerve fiber and may extend into dendrites. It is unclear whether membrane is transported on this system in discrete packages or as part of a continuous system of smooth ER (Rambourg and Droz, 1980; Schwartz, 1979).

In mature neurons, the terminal portion of the nerve fiber has specialized regions for addition (and removal) of membrane in association with transmitter release (Heuser et al., 1979), although addition of newly synthesized membrane proteins or lipids is not restricted to the terminal region of the nerve fiber (Bennett et al., 1973). In regenerating nerve fibers, the growth cone was thought to be an organelle specialized for the addition of plasma membrane to growing nerve fibers. This hypothesis was derived from observations of dye particles (Bray, 1970) or red blood cells attached by lectins (Koda and Partlow, 1976) on growth cones of cultured neurons. These relatively large markers, once attached to the growth cone, could be translocated many micrometers toward the cell body (Koda and Partlow, 1976). It was proposed that this retrograde transport in the membrane reflected the flow of macromolecules in the membrane due to a "source" of membrane being added at the growth cone (Bray, 1970; Koda and Partlow, 1976) and a "sink" being internalized at the cell body (Koda and Partlow, 1976). This view was reinforced by freeze–fracture studies of cultured cells (mentioned earlier) that demonstrated mounds of vesicles within growth cones that were thought to be a pool of membrane being added to the surface (Pfenninger and Bunge, 1974).

Other experiments indicate that the retrograde transport of surface markers, especially those bound by lectins, may not reflect the ongoing flow of constituents into the plasma membrane at the growth cone and their uptake at the neuron cell body (Letourneau, 1979a; Carbonetto and Argon, 1980). In other cells, binding of multivalent ligands, such as lectins and antibodies, induce or stabilize associations between plasma membrane proteins and contractile proteins inside the cell (Condeelis, 1979; Mescher et al., 1981). These transmembrane associations, and not the flow of membrane components (Middletown, 1979), are responsible for the movement of macromolecules within the plasma membrane (see Bray, 1978, for discussion). Although direct evidence is lacking that similar mechanisms provide for the retrograde movement of macromolecules in the surface of regenerating neurons, several indirect observations support this proposal.

First, lectin–receptor complexes on growth cones are retrogradely transported in the membrane much like dye particles (Bray, 1980) or lectin-derivatized red blood cells (Koda and Partlow, 1976). This transport is blocked by cytochalasin B, which disrupts microfilaments (Letourneau, 1979a), but it is unaffected by long-term treatment with cycloheximide (Carbonetto and Argon, 1980), which blocks protein synthesis and probably blocks the addition of new plasma membrane proteins (Carbonetto and Fambrough, 1979). Second, following retro-grade transport of the "old" lectin receptor complexes from growth cones, no "new" lectin receptors can be labeled by the lectin. This would not be expected if the transport resulted from the insertion of new membrane from inside the cell (Letourneau, 1979a; Carbonetto and Argon, 1980). Third, the addition of at least one type of membrane protein (viz. α-bungarotoxin receptors) is not localized to growth cones of regenerating nerve fibers (Carbonetto and Fambrough, 1979). Fourth, as mentioned previously, the mounds of vesicles once thought to be the internal supply of plasma membrane in growth cones may be an artifact of fixation.

Little is known about the fate of membrane constituents after they have been added to the plasma membrane of a regenerating nerve fiber. At least some plasma membrane proteins turn over even during vigorous nerve fiber outgrowth (Carbonetto and Fambrough, 1979). The turnover of membrane proteins from the plasma membrane proceeds in part by their internalization and proteolytic degradation, probably within lysosomes. Although there are morphological data to indicate recycling of plasma membrane internalized from nerve terminals (Heuser *et al.*, 1979), it is not known if similar recycling occurs in growing nerve fibers. The rapid extensions and retractions of filopodia on growth cones and the abundance of membranous organelles make the idea attractive.

One way to examine the fate of membrane components has been to follow lectin–receptor complexes on the surface of cultured neurons. Lectin receptors internalized from the plasma membrane transit to a portion of the Golgi apparatus that is thought to be specialized for the production of lysosomes (Gonatas *et al.*, 1977). In regenerating neurons, the internalization proceeds over the entire surface of the cell, and pinocytotic vesicles containing lectin receptors derived from the plasma membrane can be found distributed in the cell body and axo-plasm. The lectin receptors internalized along nerves are transported retrogradely to the cell body (Carbonetto and Argon, 1980).

It is presently unknown whether lectin–receptor complexes of neurons are internalized along the same cellular pathway as other membrane constituents during their normal turnover. Certainly lectins stimulate the internalization of their receptors from the plasma membrane. This phenomenon, sometimes referred to as "down regulation," also occurs following the binding of hormones and growth factors to a variety of cells (Schlessinger, 1980). In neurons, NGF receptors are down regulated or "sequestered" (Levi *et al.*, 1980; Olender and Stach, 1980; for review, see Greene and Shooter, 1980). The physiological action of NGF in supporting nerve fiber outgrowth and chemotaxis of sympathetic and sensory neurons could result from the binding of NGF to its receptors at the cell surface or by NGF–receptor complexes that enter the cell. Thus, peptide hormones and growth factors may exert some of their effects by entering the cell through the same cellular pathways that neurons utilize to metabolize their membrane. This route of internalization may also serve to alter rapidly the composition of the plasma membrane, perhaps locally, and could be important during the interaction of growing nerve fibers with other cells or chemo-attractants.

E. CYTOSKELETON

It may be argued that neurons are best characterized by their remarkable shapes, including extremely long axons and elaborate dendritic arbors. This view is reinforced by findings that cells other than neurons and muscle can be electrically excitable (McCann *et al.*, 1981) and chemically excitable (Kusano *et al.*, 1977). To create their long processes, neurons have developed an extensive and complex cytoskeleton, some components of which are also involved in intracellular transport to and from distant portions of the cell. Unfortunately, less is known about the neuronal cytoskeleton than the cytoskeletons of other types of cells, such as fibroblasts, where the functional relationships of cytoskeletal proteins are being brought into sharp focus. Two-dimensional gel electrophoresis and immunohistochemistry are two of the techniques that are responsible for a virtual explosion of research on cytoskeletal proteins. Future advances are promised by developments in techniques for producing monospecific antibodies of defined specificity mentioned previously (Kohler and Milstein, 1975), for preparing and examining tissues by electron microscopy (Heuser and Kirschner, 1980; Wolosewick and Porter, 1979), and for studying the dynamics of the cytoskeleton by intracellular injection of fluorescently labeled proteins (Taylor and Wang, 1980). There have been several recent reviews of work on neuronal cytoskeletal proteins (Bray and Gilbert, 1981; Lasek and Shelanski, 1981; Wuerker and Kirkpatrick, 1972; Puszkin and Schook, 1979; Johnston and Wessells, 1980; Trifaro, 1978), and there have been many general reviews of cytoskeletal structure and function (e.g., Lazarides, 1980; Stephens and Edds, 1976; Raff, 1979; Goldman *et al.*, 1979; Timasheff and Grisham, 1980; Oliver and Berlin, 1979; Kirschner, 1978; Dustin, 1978).

As mentioned above, the major cytoskeletal structures within neurons are neurotubules, neurofilaments, and microfilaments. In addition to these major structural proteins (tubulin, neurofilament proteins, and actin), neurons also contain other proteins that are involved directly in maintaining and regulating the cytoskeleton or in force generation. These proteins include myosin, tropomyosin, α-actinin, profilin, microtubule-associated proteins, dynein, vimentin (see Bray and Gilbert, 1981, for discussion), as well as calmodulin (Klee et al., 1980), and calcineurin (Klee et al., 1979).

The brain is a rich source of neurotubules, and much of what we know about microtubule assembly in vitro comes from studies of neurotubules. Microtubules are polymers of α- and β-tubulin monomers, each of which is a globular molecule of about 55,000 daltons. Polymerization of tubulin into microtubules requires nucleation sites. The increase in length is the result of net assembly of tubulin at one end of the polymer versus net disassembly at the other end. Under steady-state conditions in vitro, there is a flow of tubulin through microtubules that is called "treadmilling" (Margolis and Wilson, 1978; Bergen and Borisy, 1980). In vivo, treadmilling could serve to transport organelles (Margolis et al., 1978) or molecules in nerve fibers. However, the efficiency of treadmilling is very low in vitro (Bergen and Borisy, 1980) and may not be significant in vivo (cf. Margolis, 1981).

In principle, the regulation of microtubules in the cell can occur at a number of points in their assembly. Nucleation, polymerization, depolymerization, and association of tubulin filaments into microtubules are all potential points of regulation. During growth of nerve fibers, this could occur by alteration of the pool of unpolymerized tubulin, by increased synthesis or decreased degradation, by chemical modification of tubulin, or by interaction of tubulin with various substances from ions to proteins that might promote net assembly. In addition, the formation of nucleating sites can alter the spatial distribution of microtubules within the cell while shifting the assembly–disassembly equilibrium in favor of assembly.

When neuroblastoma nerve fibers grow, the tubulin pool is not increased by protein synthesis (Van de Water and Olmsted, 1980; Seeds and Maccioni, 1978). However, there is new synthesis of a microtubule-associated protein (MAP), which is present in trace amounts in the cell but which may strongly stimulate microtubule assembly (Olmsted, 1981). Similar regulatory proteins may be at work in the developing brain (Weingarten et al., 1975; Mareck et al., 1980). Posttranslational modification of tubulin (Nath and Flavin, 1979), ionic effects on microtubule assembly, and changes in the number and distribution of nucleating sites within the cell (Spiegelman et al., 1979) may also contribute to the regulation of microtubule assembly in growing nerve fibers. The elongation of neurotubules is polar with growth from nucleating sites called "microtubule organizing centers" into the regenerating nerve fiber (Brinkley *et al.*, 1976; Osborn and Weber, 1976; Spiegelman *et al.*, 1979). Organizing centers in neurons are centrosomes or material closely associated with centrosomes. The number of organizing centers in a neuron may affect the initiation of nerve fiber growth and the shape of the neuronal arbor, although there is currently some disagreement about whether there are only 1 or 2 organizing centers per neuron (Brinkley, 1981) or 12 (Spiegelman *et al.*, 1979).

Microfilaments consist primarily of actin; the formation of filaments *in vitro* has been studied extensively with actin isolated from muscle. Filamentous actin (F-actin) under steady-state conditions is in equilibrium with unpolymerized or globular actin (G-actin). Polymerization of G-actin proceeds by association of a few G-actins to form a nucleating site, followed by the assembly of additional monomer. Formation of the nucleating site is the rate-limiting step in filament assembly. There is net assembly at one end of the filament and net disassembly at the opposite end. Assembly–disassembly occurs at each end, but the net flux favors assembly at one end and disassembly at the opposite end. In muscle *in vivo*, the net assembly end is anchored to the Z line; in neurons, the plasma membrane may anchor the assembling filament (Isenberg and Small, 1978). Regulation can occur at a number of steps in the assembly of filaments. For example, the protein profilin that is present in motile cells shifts the G- to F-actin equilibrium in the cell in favor of the G monomer. As a result, there is a substantial pool of unpolymerized actin in motile cells as compared with skeletal muscle. α-Actinin, another regulatory protein, promotes the dissociation of profilin from actin, permitting polymerization of G actin. In the cell, α-actinin is localized at attachment sites of actin filaments, much as at the Z line in muscle and at adhesion plaques in fibroblasts. Other proteins that may regulate filament assembly have cytochalasin-like activity (Isenberg *et al.*, 1980) and inhibit the addition of G actin to nucleating sites. Others act like DNase I to promote the depolymerization of actin. Villin, an endogenous protein in intestinal epithelia, inhibits the association of oligomeric actin into microfilaments (Craig and Powell, 1980). Increases in synthesis of actin during nerve fiber growth have been found in some cells (Fine and Bray, 1971) but not in others (Schmitt, 1976; Schmitt *et al.*, 1977; Rein *et al.*, 1980); it is unlikely to be a principal means of regulating microfilament assembly in regenerating neurons.

Compared to microtubules and microfilaments, our knowledge of the structure and assembly of neurofilaments is primitive. Neurofilament proteins are related to intermediate filaments of other cell types and are more heterogeneous than tubulin or actin. In vertebrates, they are composed of three polypeptides of 200,000, 160,000, and 68,000 daltons (Schlaepfer, 1978; Liem *et al.*, 1978; Thorpe *et al.*, 1979). Neurofilaments from two invertebrates studied so far have only two major polypeptides of about 200,000 and 65,000 daltons (*Loligo*: Roslansky *et al.*, 1980; Lasek *et al.*, 1979) and 160,000 and 150,000 daltons (*Myxicola*: Lasek *et al.*, 1979). Conditions for *in vitro* assembly of neurofilaments have been described (Zackroff and Goldman, 1980).

The experiments of Lasek and his colleagues have presented a framework for understanding several aspects of cytoskeletal assembly (Lasek, 1981). Studying axonal transport in the optic nerve, these workers have shown that tubulin and neurofilament proteins are transported in the orthograde direction at rates of 0.25–1 mm/day (slow axonal transport). These proteins and few others move along the nerve as a single peak, leaving essentially no trailing radioactivity. Actin is transported at a slightly faster rate of 2–4 mm/day. These cytoskeletal proteins are thought to be moving as part of an assembled cytoskeleton from the cell body, where assembly takes place, to the nerve terminal, where it is proteolytically degraded. Neurons do contain proteases that rapidly degrade otherwise stable neurofilaments (Gilbert *et al.*, 1975; Pant *et al.*, 1979; Schlaepfer and Micko, 1979). If these proteases do degrade the cytoskeleton within the cell and act within the growth cone, they might be regulated during nerve fiber growth.

F. Growth Cone Dynamics

Observations of regenerating nerve fibers in culture suggest that growth cones interact with external cues to control the direction and rate of nerve fiber growth and that the cell body exerts limited control. Similar movements of growth cones have also been seen in intact developing and regenerating nervous systems (e.g., Speidel, 1933). Thus, a good starting point to understand how growth cones make local decisions is to observe growing nerve fibers in culture, where visual resolution is best.

Filopodia on neurons (Pomerat *et al.*, 1967), like filopodia on fibroblasts (Albrecht-Buehler, 1976), are in continual motion and appear to "sense" the substrate upon which the regenerating nerve fiber grows (Letourneau, 1975a). In culture, filopodia can elongate 10–30 μm from the nerve fiber in a few minutes and may or may not attach to the substrate. They are also absorbed rapidly into the nerve fiber. A filopodium once attached to a substrate increases in diameter to match that of the main body of the nerve fiber, thereby advancing the growing nerve fiber. The pattern of growth can be altered by various electrical, chemical, and mechanical factors, many of which may act at the growth cone; some of these factors have been reviewed (Jacobson, 1978; Johnston and Wessells, 1980) and suggest mechanisms by which neurons reach their targets.

Some component of directionality in nerve fiber regeneration is intrinsically predetermined. Just as in motile cells (Albrecht-Buehler, 1976), where the movement of daughter cells can be predicted from the movements of the parent, in neurons the movement of growth cones, and hence the shape of the neuronal arbor, can be predicted from the shape of the parent neuron (Solomon, 1979). The pattern is recapitulated during regeneration of resorbed nerve fibers (Solomon, 1980). Intrinsic determination of nerve fiber growth is most clear in the homogeneous environment of the culture dish. However, environments can profoundly influence growth.

Two environmental influences, broadly defined, are materials attached to the substrate or to other cells and growth factors in the medium (see Collins, 1980, for a description of substrate-attached growth factors). Both of these appear to exert an effect on nerve fiber growth by acting directly on growth cones. For example, coating substrates with polylysine or polyornithine will often increase the rate of fiber growth and the amount of branching of neuronal arbors in culture by increasing adhesiveness between growth cones and their substrates (Letourneau, 1975a,b). Moreover, growth cones guide neurons to grow preferentially over their targets when given a choice among different neurons (Bonhoeffer and Huf, 1980). The soluble protein NGF can also direct nerve fiber growth in dorsal root ganglion neurons (Letourneau, 1978), and its effects can be seen when applied locally to growth cones of regenerating neurons (Gunderson and Barrett, 1980). Apparently the continued binding of NGF to the growth cone, which contains only a small fraction of the cells' surface NGF receptors (Carbonetto and Stach, 1982), is sufficient to sustain nerve fiber growth (Campenot, 1977).

Solving two major problems will help us to understand how these agents affect growth cones: (1) How does a filopodium extend? (2) How is the transition in cytoskeletal structure made from the microfilamentous network in the growth cone to the microfilament–microtubule–neurofilament network in the nerve fiber? One prominent example of organelle extension is the acrosome reaction in sperm, which proceeds by the polymerization of actin from presynthesized pools within the cell (Tilney et al., 1978). In another example, during phagocytosis when the cell spreads over a solid particle, the cell surface also extends. It is unknown if either of these processes relate to filopodial extension.

When the filopodium extends and attaches, bundles of microfilaments form from the meshwork in the growth cone (see earlier). The bundles of microfilaments radiate from points of close apposition of the filopodium to its substrate (Letourneau, 1979b) and resemble adhesion plaques in fibroblasts (Goldman *et al.*, 1976). Transmembrane events may trigger the bundling of fibrils (Letourneau, 1979a; Carbonetto and Argon, 1980) in a fashion similar to capping of membrane proteins by multivalent ligands (Schreiner and Unanue, 1976) and the transmembrane association of membrane macromolecules with actin filaments (Condeelis, 1979). In an interesting discussion based upon studies of actin gelation in *Dictyostelium*, Condeelis (1981) describes in some detail how reorganization of the microfilamentous network within the cell might occur. It may be relevant that cell surface receptors on growth cones are particularly susceptible to redistribution triggered by lectins (Letourneau, 1979a; Carbonetto and Argon, 1980). Unlike the effects of forces generated against the lectins, which are free to move, similar contractile forces generated against the substrate might be expected to advance the nerve fiber. Such a contraction is sufficient to move large particles, even though it is likely that few of the receptors in the cell surface are occupied by the immobilized lectins (Koda and Partlow, 1976). Figure 4 presents a summary of how cross-linking of cell surface macromolecules on growth cones by substrate-attached molecules could be involved in nerve fiber elongation.

V. Termination of Growth

During development, neurons stop growing when they reach suitable targets. Although regenerating axons often must grow much farther than developing axons to reach their targets, once there, they too stop growing (Ramón y Cajál, 1928; Müller and Carbonetto, 1979). In a few cases, it has been shown that the regenerating axon can slightly "overshoot" its normal target (Letinsky and McMahan, 1975). Some experiments have shown that in the absence of their normal targets developing neurons may also grow farther than normally and, in some instances, contact unusual or aberrant targets (Schneider, 1973; Laurberg and Hjorth-Simonsen, 1977; Llinas *et al.*, 1973). On the other hand, when the target cell of a regenerating axon has been selectively destroyed without otherwise disrupting the target tissue, regenerating axons either in vertebrates (Marshall *et al.*, 1977; Sanes *et al.*, 1978) or invertebrates (Muller and Scott, 1979; Scott and Muller,

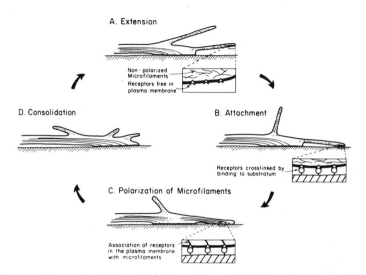

A. Extension

Non-polarized Microfilaments
Receptors free in plasma membrane

D. Consolidation

B. Attachment

Receptors crosslinked by binding to substratum

C. Polarization of Microfilaments

Association of receptors in the plasma membrane with microfilaments

FIG. 4. A model for interactions between growth cone and substrate based upon interactions of lectins with growing nerve fibers. The tip of a growing nerve fiber is diagrammed with two filopodia and an internal cytoskeletal structure consisting largely of microfilaments. (A) Extension. Microfilaments in the growth cone filopodia are in a meshwork, and cell surface macromolecules are freely mobile in the plasma membrane. (B) Attachment. Cell surface macromolecules bind to the substrate. This interaction in effect cross-links the bound macromolecules in a fashion analogous to that of multivalent lectins. (C) Polarization of microfilaments. The cross-linking of membrane macromolecules triggers the association of microfilaments in the cytoplasm with the attached surface macromolecules. This causes a polarization of contractile microfilament proteins into bundles at the attachment site of the filopodium. (D) Consolidation. The attached filopodium may become stabilized by the foregoing events. Furthermore, contraction of the microfilaments against some other cytoskeletal structure or attachment site could generate the tension observed along growing nerve fibers. An increase in filopodial diameter and further assembly of the cellular cytoskeleton would consolidate the attached filopodium as part of the advancing nerve fiber.

1980) that have the "capacity" to grow beyond the region of synapse with the target nonetheless stop growing at the normal location. As mentioned in Section IV,B, frog motoneurons will reinnervate the original end plate region that still contains basal lamina after selective destruction of their target muscle cells. Some unknown property of the basal lamina in that region is sufficient to terminate fiber growth and stimulate formation of a nerve terminal.

Whether the target cell is present or not, once the growing fiber reaches target tissue and stops, the growth cone differentiates into a terminal and the axon increases in caliber (Ramón y Cajál, 1928; Muller and Carbonetto, 1979; Scott and Muller, 1980; Sanes *et al.*, 1978). Because some axons contact a number of targets, it is not surprising that when a "temporary" target is contacted, differentiation and some increase in axon caliber can occur before growth entirely stops (Carbonetto and Muller, 1977). In some cases, nerve fibers may continue to grow, but the new sprouts are continually being destroyed (O'Brien *et al.*, 1978) or contacts at synapses may be constantly displaced and reformed (Bixby and Van Essen, 1979). Certainly, fibers that innervate the taste buds of the tongue never entirely stop growing.

Although physical blocks to growth or the encystment of the growing axon may impede forward progress, they ordinarily do not stop axonal growth. Nonetheless, axons that fail to reach their targets stop growing eventually and are sometimes resorbed by the cell, particularly if other axons of the same cell have reached normal destinations. With the cessation of growth, the cell bodies of chromatolytic neurons whose axons have ended in a neuroma or encystment revert to normal morphology (Cragg, 1970). Significantly, reinjury of the axon can prompt some resumption of chromatolysis (Watson, 1968).

When a growth cone reaches its target, it may stop growing before the cell body has resumed its normal metabolism (Lieberman, 1971) and possibly before an appropriate signal has reached the cell body. In addition, forward progress of the nerve fiber has been shown to be rapidly halted by NGF which directly affects growth cones (Griffin and Letourneau, 1980). Because the growth cone and its filopodia normally contain a well-organized matrix of microfilaments that may be crucial to the forward progress of the growth cone (Kuczmarski and Rosenbaum, 1979; Yamada *et al.*, 1971; see also previous section), it is plausible that forward progress ends when something is released or enters that halts or disrupts the organization of the matrix. For example, an endogenous molecule acting like cytochalasin B may be released, thereby preventing polymerization of the actin microfilament network (Grumet and Lin, 1980). What, then, happens to material for growth that is being shipped to the terminal from the cell soma? Because transport does not immediately stop, it has been hypothesized that proteases selectively degrade material such as neurofilaments, microtubules, and neurofilaments that continue to travel down the axon at approximately the rate of regeneration (Lasek and Black, 1977). Several studies have shown that axoplasm contains proteases that specifically degrade microfilaments and require Ca^{2+} for their activity (e.g., Pant and Gainer, 1980). It has been hypothesized that the activity of such proteases would increase once the growing terminal begins to differentiate to form a synapse and calcium influx increases. It is likely that much of the material continuing to move down the axon would be used to increase the diameter of the axon to normal, just as it might be imagined that the original shrinking in diameter was due to mobilization of components for growth. Once the terminal begins to differentiate, transport of molecules associated with transmitter release resumes and the soma is no longer chromatolytic.

VI. Overview

When a neuron's axon is cut, a signal travels to the soma, probably by retrograde axonal transport, and the soma then responds to axotomy with a burst of synthesis of molecules required to sustain regeneration. This cell body reaction to axotomy is often manifest in preparations stained for light microscopy and has been loosely termed "chromatolysis." The reaction varies greatly in quality and intensity, depending on the location of the lesion and on the age and species of animal. Many of the newly synthesized materials are transported to the nerve fiber terminals at several rates, one of the slower components of which moves at approximately the same rate as the regenerating fiber's growth cone.

During regeneration, the cell body may lose its presynaptic inputs. The growing axon ordinarily shrinks in caliber. At the axon's tip, the growth cone is a strikingly specialized structure sharing many features of motile cells such as fibroblasts, but the growth cone is not the sole site for insertion of membrane constituents. It actively samples and interacts with its environment in a fashion that includes binding exogenous nerve growth factor and, for motoneurons, detecting the specialized properties of the extracellular matrix that mark synapses. The role of the pathway in guiding the growing fiber remains uncertain in most instances, but axons do seem to regenerate preferentially along normal pathways. This preference and the predisposition of regenerating neurons to recapitulate their morphology may be sufficient to guide the nerve fiber to its target.

A key step in the restoration of proper connectivity is the termination of growth. Axons may stop growing before they reach their targets or when they reach them, whereupon the cell ceases to be chromatolytic. If functional contact with the target is restored, the regenerated axon or axons increase to normal caliber and normal presynaptic input may return.

Although the level of description of the process of regeneration is being raised continually, our understanding of molecular signals within and among cells remains limited. The effects of nerve growth factor and the increasingly refined techniques for analysis of axonal transport offer several clues to the nature of some of the signals and mechanisms for growth. With the control of the environment offered by cultures of vertebrate neurons, it has been possible to manipulate growing cells that are subsequently analyzed biochemically. In addition, invertebrate nervous systems that contain identifiable, experimentally accessible cells have shown that single cells can regenerate with nearly perfect accuracy. The overwhelming evidence from the studies has been that diverse systems have many similar intrinsic and extrinsic mechanisms to direct the growing cell, but we are only beginning to glimpse the workings of the machinery and the levers that control it.

ACKNOWLEDGMENTS

We thank Drs. M. J. Anderson, D. Stelzner, and D. Wilson for critical reading of the manuscript. This work was supported in part by NSF grant BNS 80-21860, NIH grant NS15014, and a grant from the Amyotrophic Lateral Sclerosis Society of America.

REFERENCES

Acheson, G. H., Lee, E. S., and Morrison, R. S. (1942). *J. Neurophysiol.* **5,** 269–273.
Aguilar, C. E., Bisby, M. A., Cooper, E., and Diamond, J. (1973). *J. Physiol. (London)* **234,** 449–464.
Albrecht-Buehler, G. (1976). *J. Cell Biol.* **69,** 275–286.
Albuquerque, E. X., Deshpande, S. S., and Guth, L. (1978). *Exp. Neurol.* **62,** 347–373.
Alitalo, K., Kurkinen, M., Vaheri, A., Virtanen, I., Rohde, H., and Timpl, R. (1980). *Nature (London)* **287,** 465–466.
Angeletti, P. V., Gandini-Attardi, D., Toschi, G., Salvi, M., and Levi-Montalcini, R. (1965). *Biochim. Biophys. Acta* **95,** 111–120.
Barr, M. L., and Bertram, E. G. (1951). *J. Anat.* **85,** 171–181.
Bennett, G., DiGiamberardino, L., Koenig, H., and Droz, B. (1973). *Brain Res.* **60,** 129–146.
Bennett, H. S. (1963). *J. Histochem. Cytochem.* **11,** 14–23.
Benowitz, L. I., Shashoua, V. E., and Yoon, M. G. (1981). *J. Neurosci.* **1,** 300–307.
Bergen, L. B., and Borisy, G. G. (1980). *J. Cell Biol.* **84,** 141–150.
Bittner, G. D., and Johnson, A. L. (1974). *J. Comp. Physiol.* **89,** 1–21.
Bixby, J. L., and Van Essen, D. C. (1979). *Nature (London)* **282,** 726–728.
Bjorklund, A., and Stenevi, U. (1979). *Physiol. Rev.* **59,** 62–100.
Blinzinger, K., and Kreutzberg, G. (1968). *Z. Zellforsch. Mikrosk. Anat.* **85,** 145–157.
Bliokh, Z. L., Domnina, L. V., Ivanovna, O. Y., Pletjushkina, O. Y., Svitkina, T. M., Smolyaninov, V. A., Vasilieu, J. M., and Gelfand, I. M. (1980). *Proc. Natl. Acad. Sci. U.S.A.* **77,** 5919–5922.
Bloch, R. J., and Geiger, B. (1980). *Cell* **12,** 25–35.
Bonhoeffer, F., and Huf, J. (1980). *Nature (London)* **288,** 162–164.
Bornstein, P., and Sage, H. (1980). *Annu. Rev. Biochem.* **49,** 957–1004.
Bray, D. (1970). *Proc. Natl. Acad. Sci. U.S.A.* **65,** 905–910.
Bray, D. (1978). *Nature (London)* **273,** 265–266.
Bray, D. (1979). *J. Cell Sci.* **37,** 391–410.
Bray, D., and Gilbert, D. (1981). *Annu. Rev. Neurosci.* **4,** 505–523.
Bray, D., Thomas, C., and Shaw, G. (1978). *Proc. Natl. Acad. Sci. U.S.A.* **75,** 5226–5229.

Brenner, H. R., and Martin, A. R. (1976). *J. Physiol. (London)* **260**, 159–175.

Bretscher, M. S. (1976). *Nature (London)* **260**, 21–22.

Brinkley, B. R., Fuller, G. M., and Highfield, D. P. (1976). *Cold Spring Harbor Conf. Cell Proliferation* 3 [Book A], 435–456.

Brown, J. C. (1971). *Exp. Cell Res.* **69**, 440–443.

Brown, M. C., Holland, R. L., and Hopkins, W. G. (1981). *Annu. Rev. Neurosci.* **4**, 17–42.

Bunge, M. B. (1973). *J. Cell Biol.* **56**, 713–755.

Bunge, M. B., Williams, A., and Wood, P. (1979). *J. Cell Biol.* **83**, 130 (abstr.).

Bunge, M. B., Williams, A. K., Wood, P., Vitlo, J., and Jeffrey, J. J. (1980). *J. Cell Biol.* **84**, 184–202.

Burden, S. J., Sargent, P. B., and McMahan, U. J. (1979). *J. Cell Biol.* **82**, 412–425.

Burnham, P. A., and Varon, S. (1974). *Neurobiology (Copenhagen)* **4**, 57–70.

Burstein, D. E., and Greene, L. A. (1978). *Proc. Natl. Acad. Sci. U.S.A.* **75**, 6059–6063.

Campenot, R. (1977). *Proc. Natl. Acad. Sci. U.S.A.* **74**, 4516–4519.

Ramón y Cajál, S. (1928). "Degeneration and Regeneration of the Nervous System" (translated and edited by R. M. May). Hafner, New York.

Carbonetto, S., and Argon, Y. (1980). *Dev. Biol.* **80**, 364–378.

Carbonetto, S., and Fambrough, D. M. (1979). *J. Cell Biol.* **81**, 555–569.

Carbonetto, S., and Muller, K. J. (1977). *Nature (London)* **267**, 450–452.

Carbonetto, S., and Stach, R. W. (1982). *Dev. Brain Res.* **3**, 463–473.

Carmel, P. W., and Stein, B. M. (1969). *J. Comp. Neurol.* **135**, 145–166.

Chalfie, M., and Thomson, J. N. (1979). *J. Cell Biol.* **82**, 278–289.

Chan, K. Y., and Baxter, C. F. (1979). *Brain Res. 174,* 135–152.

Chang, C. M., and Goldman, R. D. (1973). *J. Cell Biol.* **57**, 867–874.

Chiu, S. Y., and Ritchie, J. M. (1980). *Nature (London)* **284**, 170–171.

Chung, A. E., Freeman, I. L., and Braginski, J. E. (1977). *Biochem. Biophys. Res. Commun.* **79**, 859–868.

Collins, F. (1978). *Dev. Biol.* **65**, 50–57.

Collins, F. (1980). *Dev. Biol.* **79**, 247–252.

Condeelis, J. S. (1979). *J. Cell Biol.* **80**, 751–758.

Condeelis, J. S. (1981). *Neurosci. Res. Program Bull.* **19**, 83–99.

Cornbrooks, C. J., Bunge, R. P., and Gottlieb, D. I. (1980). *J. Neurochem.* **34**, 800–807.

Cotman, C. W., and Lynch, G. S. (1976). *In* "Neuronal Recognition" (S. H. Barondes, ed.), pp. 69–108. Plenum, New York.

Courtney, K., and Roper, S. (1976). *Nature (London)* **259**, 317–319.

Cowan, W. M., and Wenger, E. (1967). *J. Exp. Zool.* **164**, 267–280.

Cragg, B. G. (1970). *Brain Res.* **23**, 1–21.

Craig, S., and Powell, L. D. (1980). *Cell* **22**, 739–746.

Culp, L. A., Ansbacher, R., and Domen, C. (1980). *Biochemistry* **19**, 5899–5907.

Czeh, G., Kudo, N., and Kuno, M. (1977). *J. Physiol. (London)* **270**, 165–180.

Czeh, G., Gallego, R., Kudo, N., and Kuno, M. (1978). *J. Physiol. (London)* **281**, 239–252.

Daniels, M. P. (1972). *J. Cell Biol.* **53**, 164–176.

DeBaecque, C., Johnson, A. B., Naiki, M., Schwarting, G., and Macus, D. M. (1976). *Brain Res.* **114**, 117–122.

DeLeij, J., Kingman, J., and Wilholt, B. (1979). *Biochim. Biophys. Acta* **553**, 224–234.

Denis-Donini, S., Estenoz, M., and Augusti-Tocco, G. (1978). *Cell Differ.* **7**, 193–201.

Diamond, J., Cooper, E., Turner, C., and Macintyre, L. (1976). *Science* **193**, 371–377.

Dustin, P. (1978). "Microtubules." Springer-Verlag, Berlin and New York.

Dvorak, D., Gripps, E., Leah, J., and Kidson, C. (1978). *Life Sci.* **22**, 407–414.

Edds, M. V. (1950). *J. Exp. Zool.* **113**, 517–552.

Edds, M. V. (1953). *Q. Rev. Biol.* **28**, 260–276.

Edington, D. R., Kuffler, D. P., and McMahan, U. J. (1980). *Soc. Neurosci. Abstr.* **6**, 92.
Edström, J.-E. (1959). *J. Neurochem.* **5**, 43–49.
Ehrismann, R., Chiquet, M., and Turner, D. C. (1981). *J. Biol. Chem.* **256**, 4056–4062.
Eidelberg, E., and Stein, D. G. (1974). *Neurosci. Res. Program Bull.* **12**, 189–303.
Eisenbarth, G. S., Walsh, F. S., and Nirenberg, M. (1979). *Proc. Natl. Acad. Sci. U.S.A.* **76**, 4913–4917.
Ellisman, M. H., and Porter, K. R. (1980). *J. Cell Biol.* **87**, 464–479.
Estridge, M. (1977). *Nature (London)* **268**, 60–63.
Estridge, M., and Bunge, R. (1978). *J. Cell Biol.* **79**, 138–155.
Fernandez, H. L., Burton, P. R., and Samson, F. E. (1971). *J. Cell Biol.* **51**, 176–192.
Fine, R. E., and Bray, D. (1971). *Nature (London)* **234**, 115–118.
Folkman, J., and Moscona, A. (1978). *Nature (London)* **273**, 345–349.
Fraser, S., and Hunt, R. K. (1980). *Annu. Rev. Neurosci.* **3**, 319–352.
Frizell, M., McLean, W. G., and Sjöstrand, J. (1975). *Brain Res.* **86**, 67–73.
Fry, F. J., and Cowan, W. M. (1972). *J. Comp. Neurol.* **144**, 1–24.
Gainer, H., Tasaki, I., and Lasek, R. J. (1977). *J. Cell Biol.* **74**, 524–530.
Gallego, R., Kuno, M., Nunez, R., and Snider, W. D. (1979). *J. Physiol. (London)* **291**, 179–189.
Gallego, R., Kuno, M., Nunez, R., and Snider, W. D. (1980). *J. Physiol. (London)* **306**, 205–218.
Garrels, J. E., and Schubert, D. (1979). *J. Biol. Chem.* **254**, 7978–7985.
Geiger, B., Tokuyasu, K. T., Dutton, A. H., and Singer, S. J. (1980). *Proc. Natl. Acad. Sci. U.S.A.* **77**, 4127–4131.
Gilbert, D. S., Newby, B. J., and Anderton, B. H. (1975). *Nature (London)* **256**, 586–589.
Giulian, D., Des Ruisseaux, H., and Cowburn, D. (1980). *J. Biol. Chem.* **255**, 6494–6501.
Glabe, C. G., and Vacquier, V. D. (1977). *Nature (London)* **267**, 836–837.
Glover, R. A. (1967). *Anat. Rec.* **157**, 248.
Goldberg, D. J., Goldman, J. E., and Schwartz, J. H. (1976). *J. Physiol. (London)* **259**, 473–490.
Goldman, R. D., Schloss, J. A., and Starger, J. M. (1976). *Cold Spring Harbor Conf. Cell Proliferation* **3** [Book A], 217–245.
Goldman, R. D., Milsted, A., Schloss, J. A., Starger, J., and Yerna, M. J. (1979). *Annu. Rev. Physiol.* **41**, 703–722.
Goldring, J. M., Kuno, M., Nunez, R., and Snider, W. D. (1980). *J. Physiol. (London)* **309**, 185–198.
Gonatas, N. K. (1979). *J. Histochem. Cytochem.* **27**, 1165–1166.
Gonatas, N. K., Kim, S. V., Stieber, A., and Avrameus, S. (1977). *J. Cell Biol.* **73**, 1–13.
Gottlieb, D., and Glaser, L. (1980). *Annu. Rev. Neurosci.* **3**, 303–318.
Grafstein, B., and Forman, D. S. (1980). *Physiol. Rev.* **60**, 1167–1283.
Grafstein, B., and McQuarrie, I. G. (1978). *In* "Neuronal Plasticity" (C. Cotman, ed.), p. 155. Raven, New York.
Grafstein, B., Miller, J. A., Ledeen, R. W., Haley, J., and Specht, S. C. (1975). *Exp. Neurol.* **46**, 261–281.
Greene, L. A. (1976). *Brain Res.* **111**, 135–145.
Greene, L. A., and Shooter, E. M. (1980). *Annu. Rev. Neurosci.* **3**, 353–402.
Griffin, C. G., and Letourneau, P. C. (1980). *J. Cell Biol.* **86**, 156–161.
Grinnell, F. (1978). *Int. Rev. Cytol.* **53**, 65–144.
Grumet, M., and Lin, S. (1980). *Cell* **21**, 439–444.
Gundersen, R. W., and Barrett, J. N. (1980). *J. Cell Biol.* **87**, 546–554.
Gunning, P. W., Kaye, P. L., and Austin, L. (1977). *J. Neurochem.* **28**, 1245–1248.

Guthrie, D. M. (1962). *J. Insect Physiol.* **8**, 79–92.

Hadler, N. M. (1980). *J. Biol. Chem.* **255**, 3532–3535.

Hall, M. E., Wilson, D. L., and Stone, G. C. (1978). *J. Neurobiol.* **9**, 353–366.

Hanson, H. A. (1973). *Exp. Eye Res.* **16**, 377–388.

Hare, W. K., and Hinsey, J. C. (1940). *J. Comp. Neurol.* **73**, 489–502.

Harford, J. B., and Waechter, C. J. (1979). *Arch. Biochem. Biophys.* **197**, 424–435.

Harris, A. J., Kuffler, S. W., and Dennis, M. J. (1971). *Proc. R. Soc. London, Ser. B* **177**, 541–553.

Hascall, V. C., and Heingard, D. (1975). *In* "Extracellular Matrix Influences on Gene Expression" (H. C. Slavkin and R. Greulich, eds.), p. 423. Academic Press, New York.

Hasty, D. L., and Hay, E. D. (1978). *J. Cell Biol.* **78**, 756–768.

Hatten, M. E., and Sidman, R. L. (1977). *J. Supramol. Struct.* **7**, 267–275.

Hay, E. D., and Hasty, D. L. (1979). *In* "Freeze-Fracture: Methods, Artifacts and Interpretations" (J. E. Rash, ed.), p. 59. Raven, New York.

Hendry, I. A. (1975). *Brain Res.* **94**, 87–97.

Herzog, V., and Farquhar, M. G. (1977). *Proc. Natl. Acad. Sci. U.S.A.* **74**, 5073–5077.

Heuser, J., and Kirschner, M. W. (1980). *J. Cell Biol.* **86**, 212–234.

Heuser, J. E., Reese, T. S., Dennis, M. J., Jan, Y., Jan, L., and Evans, L. (1979). *J. Cell Biol.* **81**, 275–300.

Hewitt, A. T., Kleinman, H. K., Pennypacker, J. P., and Martin, G. R. (1980). *Proc. Natl. Acad. Sci. U.S.A.* **77**, 385–388.

Holland, R. L., and Brown, M. C. (1980). *Science* **207**, 649–651.

Holtzman, E. (1971). *Philos. Trans. R. Soc. London, Ser. B* **261**, 407–421.

Holtzman, E., and Mercurio, A. M. (1980). *Int. Rev. Cytol.* **67**, 1–67.

Hoy, R. R., Bittner, G. D., and Kennedy, D. (1967). *Science* **156**, 251–252.

Hughes, R. C. (1976). "Membrane Glycoproteins." Butterworth, London.

Huizar, P., Kuno, M., Kudo, N., and Miyata, Y. (1977). *J. Physiol. (London)* **265**, 175–191.

Humpheys, S., Humphreys, S., and Sano, J. (1977). *J. Supramol. Struct.* **7**, 339–351.

Isenberg, G., and Small, J. V. (1978). *Cytobiology* **16**, 326–344.

Isenberg, G., Small, J. V., and Kreutzberg, G. W. (1978). *J. Neurocytol.* **7**, 649–661.

Isenberg, G., Aebi, U., and Pollard, T. C. (1980). *Nature (London)* **288**, 455–459.

Jacklet, J. W., and Cohen, M. J. (1967). *Science* **156**, 1640–1643.

Jacobson, M. (1978). "Developmental Neurobiology," p. 157. Plenum, New York.

James, D. W., and Tresman, R. L. (1972). *J. Neurocytol.* **1**, 383–395.

Jamieson, J. D., and Palade, G. E. (1968). *J. Cell Biol.* **39**, 580–588.

Johnston, R. N., and Wessells, N. K. (1980). *Curr. Top. Dev. Biol.* **16**, 165–206.

Junqueira, L. C. U., Montes, G. S., and Kriztan, R. M. (1979). *Cell Tissue Res.* **202**, 453–460.

Kataoka, S., Sandquist, D., Williams, L., and Williams, T. H. (1980). *J. Neurocytol.* **9**, 591–602.

Kaye, P. L., Gunning, P. W., and Austin, L. (1977). *J. Neurochem.* **28**, 1241–1243.

Kelly, P., Cotman, C. W., Gentry, C., and Nicolson, G. L. (1976). *J. Cell Biol.* **71**, 487–496.

Kirschner, M. W. (1978). *Int. Rev. Cytol.* **54**, 1–71.

Klee, C. B., Crouch, T. H., and Krinns, M. H. (1979). *Proc. Natl. Acad. Sci. U.S.A.* **76**, 6270–6273.

Klee, C. B., Crouch, T. H., and Richman, P. G. (1980). *Annu. Rev. Biochem.* **49**, 489–515.

Kleinman, H. K., Klebe, R. J., and Martin, G. R. (1981). *J. Cell Biol.* **88**, 473–485.

Koda, L. Y., and Partlow, L. M. (1976). *J. Neurobiol.* **7**, 157–172.

Kohler, G., and Milstein, C. (1975). *Nature (London)* **256**, 495–497.

Kristensson, K., and Olsson, Y. (1974). *Brain Res.* **79,** 101–109.
Kuczmarski, E. R., and Rosenbaum, J. L. (1979). *J. Cell Biol.* **80,** 356–371.
Kuffler, S. W., Dennis, M. J., and Harris, A. J. (1971). *Proc. Soc. London, Ser. B* **177,** 555–563.
Kuno, M., and Llinas, R. (1970a). *J. Physiol. (London)* **210,** 807–821.
Kuno, M., and Llinas, R. (1970b). *J. Physiol. (London)* **210,** 823–828.
Kuno, M., Turkanis, S. A., and Weakly, J. N. (1971). *J. Physiol. (London)* **213,** 545–556.
Kuno, M., Miyata, Y., and Munoz-Martinez, E. J. (1974). *J. Physiol. (London)* **242,** 273–288.
Kusano, K., Miledi, R., and Stinnakre, J. (1977). *Nature (London)* **270,** 739–741.
Lajtha, A. (1970). "Protein Metabolism of the Nervous System." Plenum, New York.
Lasek, R. J. (1981). *Neurosci. Res. Program Bull.* **19,** 7–32.
Lasek, R. J., and Black, M. M. (1977). *In* "Mechanisms, Regulation and Special Functions of Protein Synthesis in the Brain" (S. Roberts *et al.,* eds.), pp. 161–169. Elsevier, Amsterdam.
Lasek, R. J., and Shelanski, M. L., eds. (1981). "Cytoskeletons and Nervous System Architecture," Neurosci. Res. Program Bull., Vol. 19, No. 1. MIT Press, Cambridge, Massachusetts.
Lasek, R. J., Gainer, H., and Barker, J. L. (1977). *J. Cell Biol.* **74,** 501–523.
Lasek, R. J., Krishnan, N., and Kaiserman-Abramoff, I. R. (1979). *J. Cell Biol.* **82,** 336–346.
Laurberg, S., and Hjorth-Simonsen, A. (1977). *Nature (London)* **269,** 158–160.
Lazarides, E. (1976). *Cold Spring Harbor Conf. Cell Proliferation* **3** [Book A], 347–360.
Lazarides, E. (1980). *Nature (London)* **283,** 249–256.
Letinsky, M. S., Fischbeck, K. H., and McMahan, U. J. (1976). *J. Neurocytol.* **5,** 691–718.
Letourneau, P. C. (1975a). *Dev. Biol.* **44,** 77–91.
Letourneau, P. C. (1975b). *Dev. Biol.* **44,** 92–101.
Letourneau, P. C. (1978). *Dev. Biol.* **66,** 183–196.
Letourneau, P. C. (1979a). *J. Cell Biol.* **80,** 128–140.
Letourneau, P. C. (1979b). *Exp. Cell Res.* **124,** 127–138.
Letourneau, P. C. (1981). *Dev. Biol.* **85,** 113–122.
Levi, A., Schechter, Y., Neufeld, E. J., and Schlessinger, J. (1980). *Proc. Natl. Acad. Sci. U.S.A.* **77,** 3469–3473.
Lieberman, A. R. (1971). *Int. Rev. Neurophysiol.* **14,** 49–124.
Lieberman, A. R. (1974). *In* "Essays on the Nervous System" (R. Bellairs and E. G. Gray, eds.), pp. 71–105. Oxford Univ. Press (Clarendon), London and New York.
Liem, R. K. H., Yen, S.-H., Salomon, G. D., and Shelanski, M. L. (1978). *J. Cell Biol.* **79,** 637–645.
Lindahl, U., and Hook, M. (1978). *Annu. Rev. Biochem.* **47,** 385–418.
Llinas, R., Hillman, D. E., and Precht, W. (1973). *J. Neurobiol.* **4,** 69–94.
Luduena, M. (1973). *Dev. Biol.* **33,** 268–284.
Lund, R. D., and Lund, J. S. (1971). *Science* **171,** 804–807.
McCann, F. V., Pettengill, O. S., Cole, J. J., Russell, J. A. G., and Sorenson, G. D. (1981). *Science* **212,** 1155–1157.
McGuire, J. C., Greene, L. A., and Furano, A. V. (1978). *Cell* **15,** 357–365.
Marchisio, P. C., Weber, K., and Osborn, M. (1980). *In* "Tissue Culture in Neurobiology" (E. Giacobini, A. Vernadakis, and A. Shahar, eds.), pp. 99–109. Raven, New York.
Mareck, A., Fellows, A., Francon, J., and Nunez, J. (1980). *Nature (London)* **284,** 353–355.

Margolis, R. K., and Margolis, R. U., eds. (1979). "Complex Carbohydrates of Nervous Tissue." Plenum, New York.

Margolis, R. L. (1981). *Proc. Natl. Acad. Sci. U.S.A.* **78,** 1586–1590.

Margolis, R. L., and Wilson, L. (1978). *Cell* **13,** 1–8.

Margolis, R. L., Wilson, L., and Kiefer, B. (1978). *Nature (London)* **272,** 450–452.

Margolis, R. L., Rauch, C. T., and Wilson, L. (1980). *Biochemistry* **19,** 5550–5557.

Marshall, L. M., Sanes, J. R., and McMahan, U. J. (1977). *Proc. Natl. Acad. Sci. U.S.A.* **74,** 3073–3077.

Maupin-Szamier, P., and Pollard, T. (1978). *J. Cell Biol.* **77,** 837–852.

Meiri, H., Spira, M. E., and Parnas, I. (1981). *Science* **211,** 709–712.

Mendell, L. M., Munson, J. B., and Scott, J. G. (1976). *J. Physiol. (London)* **255,** 67–79.

Mescher, M., Jose, M. J. L., and Balk, S. P. (1981). *Nature (London)* **289,** 139–144.

Meyer, M. R., and Bittner, G. D. (1978). *Brain Res.* **143,** 213–232.

Middletown, C. A. (1979). *Nature (London)* **282,** 203–205.

Miller, R. A., and Ruddle, F. H. (1974). *J. Cell Biol.* **63,** 295–299.

Mintz, G., and Glaser, L. (1978). *J. Cell Biol.* **79,** 132–137.

Mizel, S. B., and Bamburg, J. R. (1976). *Dev. Biol.* **49,** 20–28.

Momoi, M., Kennett, R. H., and Glick, M. C. (1980). *J. Biol. Chem.* **255,** 11914–11921.

Mooseker, M. S. (1976). *Cold Spring Harbor Conf. Cell Proliferation* **3** [Book B], 631–650.

Muller, K. J., and Carbonetto, S. T. (1979). *J. Comp. Neurol.* **185,** 485–516.

Muller, K. J., and Scott, S. A. (1979). *Science* **206,** 87–89.

Muller, K. J., and Scott, S. A. (1980). *Nature (London)* **283,** 89–90.

Murray, M., and Grafstein, B. (1969). *Exp. Neurol.* **23,** 544–560.

Nadelhaft, I. (1974). *J. Neurocytol.* **3,** 73–86.

Nath, J., and Flavin, M. (1979). *J. Biol. Chem.* **254,** 11505–11510.

Neupat, W., and Schatz, G. (1981). *Trends Biochem. Sci.* **6,** 1–4.

Nissl, F. (1892). *Allg. Z. Psychiatr. Ihre Grenzgeb.* **48,** 197–198.

Nja, A., and Purves, D. (1978). *J. Physiol. (London)* **277,** 53–75.

Nordlander, R. H., and Singer, M. (1973). *J. Exp. Zool.* **184,** 289–302.

Nuttall, R. P., and Wessells, N. K. (1979). *Exp. Cell Res.* **119,** 163–174.

O'Brien, R., Ostberg, A., and Vrbova, G. (1978). *J. Physiol. (London)* **282,** 571–582.

Ogston, A. G. (1970). *In* "Chemistry and Molecular Biology of the Intercellular Matrix," (E. A. Balazs, ed.), Vol. 3, pp. 1231–1240. Academic Press, New York.

Olender, E. J., and Stach, R. W. (1980). *J. Biol. Chem.* **255,** 9338–9343.

Oliver, J. M., and Berlin, R. D. (1979). *Symp. Soc. Exp. Biol.* **33,** 277–298.

Oliver, J. M., Ukena, T. E., and Berlin, R. D. (1974). *Proc. Natl. Acad. Sci. U.S.A.* **71,** 394–398.

Olmsted, J. (1981). *J. Biol. Chem.* (in press).

Osborn, M., and Weber, K. (1976). *Proc. Natl. Acad. Sci. U.S.A.* **73,** 867–871.

Palade, G. (1975). *Science* **189,** 347–358.

Pant, H. C., and Gainer, H. (1980). *J. Neurobiol.* **11,** 1–12.

Pant, H. C., Terakawa, S., and Gainer, H. (1979). *J. Neurochem.* **32,** 99–102.

Partlow, L. M., and Larrabee, M. G. (1971). *J. Neurochem.* **18,** 2101–2118.

Patterson, B. M., and Bishop, J. O. (1977). *Cell* **12,** 751–765.

Pestronk, A., and Drachman, D. B. (1978). *Science* **199,** 1223–1225.

Pfenninger, K. H. (1979). *In* "Freeze-Fracture: Methods, Artifacts and Interpretations" (J. E. Rash, ed.), p. 71. Raven, New York.

Pfenninger, K. H., and Bunge, R. P. (1974). *J. Cell Biol.* **63,** 180–196.

Pfenninger, K. H., and Maylie-Pfenninger, M. F. (1979). *In* "Complex Carbohydrates of Nervous Tissue" (R. V. Margolis and R. K. Margolis, eds.), p. 185. Plenum, New York.

Pfenninger, K. H., and Rees, R. P. (1976). *In* "Neuronal Recognition" (S. Barondes, ed.), pp. 131–178. Plenum, New York.

Pickel, V. M., Segal, M., and Bloom, F. E. (1974). *J. Comp. Neurol.* **155**, 43–60.

Pilar, G., and Landmesser, L. (1972). *Science* **177**, 116–118.

Pitman, R. M., Tweedle, C. D., and Cohen, M. J. (1972). *Science* **178**, 507–509.

Pomerat, C. M., Hewdelman, W. J., Raborn, C. W., and Massey, J. F. (1967). *In* "The Neuron" (H. Hyden, ed.), pp. 119–178. Elsevier, Amsterdam.

Pratt, R. M., Larsen, M. A., and Johnston, M. C. (1975). *Dev. Biol.* **44**, 298–305.

Purves, D. (1975). *J. Physiol. (London)* **252**, 429–463.

Purves, D. (1976a). *Int. Rev. Physiol.* **10**, 125–177.

Purves, D. (1976b). *J. Physiol. (London)* **259**, 159–175.

Purves, D. (1976c). *J. Physiol. (London)* **261**, 453–475.

Puszkin, S., and Schook, W. (1979). *Methods Achiev. Exp. Pathol.* **9**, 87–111.

Raff, E. (1979). *Int. Rev. Cytol.* **59**, 1–96.

Raff, M. C., Fields, K. L., Hakomori, S. I., Mirsky, R., Pruss, R. M., and Winter, J. (1979). *Brain Res.* **174**, 283–308.

Raisman, G. (1969). *Brain Res.* **14**, 25–48.

Rambourg, A., and Droz, B. (1980). *J. Neurochem.* **35**, 16–25.

Ramón y Cajál, S. (1928). "Degeneration and Regeneration of the Nervous System" (translated and edited by R. M. May). Hafner, New York.

Rauvala, H., Carter, W. G., and Kahomori, S. (1981). *J. Cell Biol.* **88**, 127–137.

Ready, D. F., and Nicholls, J. G. (1979). *Nature (London)* **281**, 67–69.

Rees, R. P., and Reese, T. S. (1981). *Neuroscience* **6**, 247–254.

Rein, D., Gruenstein, E., and Lessard, J. (1980). *J. Neurochem.* **34**, 1459–1469.

Rephaeli, A., Spector, M., and Racker, E. (1981). *J. Biol. Chem.* **256**, 6069–6074.

Ritchie, J. M., and Rogart, R. B. (1977). *Proc. Natl. Acad. Sci. U.S.A.* **74**, 211–215.

Roslansky, P. F., Cornell-Bell, A., Rice, R. V., and Adelman, W. J. (1980). *Proc. Natl. Acad. Sci. U.S.A.* **77**, 404–408.

Rotshenker, S., and McMahan, U. J. (1976). *J. Neurocytol.* **5**, 719–730.

Ruffolo, R. R., Eisenbarth, G. S., Thompson, J. M., and Nirenberg, M. (1978). *Proc. Natl. Acad. Sci. U.S.A.* **75**, 2281–2285.

Rutishauser, U., Thiery, J.-P., Brackenbury, R., and Edelman, G. (1978). *J. Cell Biol.* **79**, 371–381.

Sanes, J. R., Marshall, L. M., and McMahan, U. J. (1978). *J. Cell Biol.* **78**, 176–198.

Sanes, J. R., Carlson, S. S., von Wedel, R. J., and Kelly, R. B. (1979). *Nature (London)* **280**, 403–404.

Sargent, P. B., and Dennis, M. J. (1977). *Nature (London)* **268**, 456–458.

Sargent, P. B., and Dennis, M. J. (1981). *Dev. Biol.* **81**, 65–73.

Schlaepfer, W. W. (1978). *J. Cell Biol.* **76**, 50–56.

Schlaepfer, W. W., and Bunge, R. P. (1972). *J. Cell Biol.* **59**, 456–470.

Schlaepfer, W. W., and Micko, S. (1979). *J. Neurochem.* **32**, 211–219.

Schlessinger, J. (1980). *Trends Biochem. Sci.* **5**, 210–214.

Schmell, E., Earles, B. J., Breaux, C., and Lennarz, W. J. (1977). *J. Cell Biol.* **72**, 35–46.

Schmitt, H. (1976). *Brain Res.* **115**, 165–173.

Schmitt, H., Gozes, I., and Littauer, A. X. (1977). *Brain Res.* **121**, 327–342.

Schneider, G. E. (1973). *Brain, Behav. Evol.* **8**, 73–109.

Schreiner, G. F., and Unanue, E. R. (1976). *Adv. Immunol.* **24**, 38–165.

Schubert, D., LaCorbiere, M., Whitlock, C., and Stallcup, W. (1978). *Nature (London)* **273**, 718–722.

Schubert, D. A., Harris, A. J., Heinemann, S., Kidokoro, Y., Patrick, J., and Steinbach, J. H. (1973). *In* "Tissue Culture of the Nervous System" (G. Sato, ed.), pp. 55–86. Plenum, New York.

Schwartz, J. H. (1979). *Annu. Rev. Neurosci.* **2**, 467–504.

Scott, S. A., and Muller, K. J. (1980). *Dev. Biol.* **80**, 345–363.

Seeds, N., Gilman, A. G., Amano, T., and Nirenberg, M. W. (1970). *Proc. Natl. Acad. Sci. U.S.A.* **66**, 160–167.

Seeds, N. W., and Maccioni, R. B. (1978). *J. Cell Biol.* **76**, 547–555.

Shaw, G., and Bray, D. (1977). *Exp. Cell Res.* **104**, 55–62.

Sheffield, J. B., Pressman, D., and Lynch, M. (1980). *Science* **209**, 1043–1045.

Shelton, E., and Mowczko, W. E. (1979). *In* "Freeze-Fracture: Methods, Artifacts and Interpretations" (J. E. Rash, ed.), p. 67. Raven, New York.

Sherbany, A. A., Ambron, R. T., and Schwartz, J. H. (1979). *Science* **203**, 78–80.

Skene, J. H. P., and Willard, M. (1981). *J. Neurosci.* **1**, 419–426.

Small, J. V., Isenberg, G., and Celis, J. (1978). *Nature (London)* **272**, 638–639.

Solomon, F. (1979). *Cell* **16**, 165–169.

Solomon, F. (1980). *Cell* **21**, 333–338.

Sotelo, C., and Triller, A. (1979). *Brain Res.* **175**, 11–36.

Sotelo, J., Toh, B. H., Vildiz, A., Osung, O., and Holbrow, E. J. (1979). *Neuropathol. Appl. Neurobiol.* **5**, 499–505.

Speidel, C. C. (1933). *Am. J. Anat.* **52**, 1–79.

Sperry, R. W. (1963). *Proc. Natl. Acad. Sci. U.S.A.* **50**, 703–710.

Spiegelman, B. M., Lopata, M. A., and Kirschner, M. W. (1979). *Cell* **16**, 253–263.

Stephens, R. E., and Edds, K. T. (1976). *Physiol. Rev.* **56**, 709–777.

Taylor, D. L., and Wang, Y. (1980). *Nature (London)* **284**, 405–410.

Tessler, A., Autilo-Gambetti, L., and Gambetti, P. (1980). *J. Cell Biol.* **87**, 197–203.

Thorpe, R., Delacourte, A., Ayers, M., Bullock, C., and Anderton, B. H. (1979). *Biochem. J.* **181**, 275–284.

Tilney, L. G. (1976). *J. Cell Biol.* **77**, 536–550.

Tilney, L. G., Kiehart, D. P., Sardet, C., and Tilney, M. (1978). *J. Cell Biol.* **77**, 536–550.

Timasheff, S. N., and Grisham, L. M. (1980). *Annu. Rev. Biochem.* **49**, 565–592.

Trifaro, J. M. (1978). *Neuroscience* **3**, 1–24.

Trisler, G. D., Schneider, M. D., and Nirenberg, M. (1981). *Proc. Natl. Acad. Sci. U.S.A.* **78**, 2145–2149.

Tsan, M.-F., and Berlin, R. D. (1971). *J. Exp. Med.* **134**, 1016–1035.

Tsukahara, N., Hultborn, H., Murakami, F., and Fujito, Y. (1975). *J. Neurophysiol.* **38**, 1359–1372.

Underhill, C. B., and Toole, B. P. (1979). *J. Cell Biol.* **82**, 475–484.

Usherwood, P. N. R., Cochrane, D. G., and Rees, D. (1968). *Nature (London)* **218**, 589–591.

Vasiliev, J. M., and Gelfand, I. M. (1976). *Cold Spring Harbor Conf. Cell Proliferation* **3** [Book A], 279–304.

Waechter, C. J., and Lennarz, W. J. (1976). *Annu. Rev. Biochem.* **45**, 95–112.

Wall, P. D., Waxman, S., and Basbaum, A. I. (1974). *Exp. Neurol.* **45**, 576–589.

Watson, W. E. (1965). *J. Physiol. (London)* **180**, 741–753.

Watson, W. E. (1968). *J. Physiol. (London)* **196**, 655–676.

Waxman, S. G., and Foster, R. E. (1980). *Brain Res. Rev.* **2**, 205–234.

Weingarten, M. D., Lockwood, A. H., Hwo, S. Y., and Kirschner, M. W. (1975). *Proc. Natl. Acad. Sci. U.S.A.* **72,** 1858–1862.

Weiss, P., and Hiscoe, H. B. (1948). *J. Exp. Zool.* **107,** 315–395.

Wessells, N. K., Nuttall, R. P., Wrenn, J. T., and Johnson, S. (1976). *Proc. Natl. Acad. Sci. U.S.A.* **73,** 4100–4104.

Wessells, N. K., Johnson, S. R., and Nuttall, R. P. (1978). *Exp. Cell Res.* **117,** 335–345.

Wickner, W. (1980). *Science* **210,** 861–868.

Wilson, D. L., and Stone, G. C. (1979). *Annu. Rev. Biophys. Bioeng.* **8,** 27–45.

Wilson, D. M. (1960). *J. Exp. Biol.* **37,** 57–72.

Wolosewick, J. J., and Porter, K. R. (1979). *J. Cell Biol.* **82,** 114–139.

Wood, J. G., Wallace, R. W., Whitaker, J. N., and Cheung, W. Y. (1980). *J. Cell Biol.* **84,** 66–76.

Wood, P. M. (1976). *Brain Res.* **115,** 361–375.

Wright, T. C., Smith, B., Ware, B. R., and Karnovsky, M. J. (1980). *J. Cell Sci.* **45,** 99–117.

Wuerker, R. B., and Kirkpatrick, J. B. (1972). *Int. Rev. Cytol.* **33,** 45–75.

Yamada, K. M., and Olden, K. (1978). *Nature (London)* **275,** 179–184.

Yamada, K. M., Spooner, B. S., and Wessells, N. K. (1970). *Proc. Natl. Acad. Sci. U.S.A.* **66,** 1206–1212.

Yamada, K. M., Spooner, B. S., and Wessells, N. K. (1971). *J. Cell Biol.* **49,** 614–635.

Yamada, K. M., Olden, K., and Han, L.-H. E. (1980). *In* "The Cell Surface: Mediation of Developmental Processes" (S. Subtelny and N. K. Wessells, eds.), p. 43. Academic Press, New York.

Yoon, M. (1972). *Exp. Neurol.* **35,** 565–577.

Young, D., Ashhurst, D. E., and Cohen, M. J. (1970). *Tissue Cell* **2,** 387–398.

Young, R. W. (1976). *Invest. Ophthalmol.* **15,** 700–725.

Zackroff, R. V., and Goldman, R. D. (1980). *Science* **208,** 1152–1154.

Zipser, B. (1980). *Brain Res.* **182,** 441–445.

Zipser, B., and McKay, R. (1981). *Nature (London)* **289,** 549–554.

CHAPTER 3

DEVELOPMENT, MAINTENANCE, AND MODULATION OF PATTERNED MEMBRANE TOPOGRAPHY: MODELS BASED ON THE ACETYLCHOLINE RECEPTOR

Scott E. Fraser and Mu-ming Poo

DEPARTMENT OF PHYSIOLOGY AND BIOPHYSICS
UNIVERSITY OF CALIFORNIA
IRVINE, CALIFORNIA

I. Introduction

The differentiation of many cells is accompanied by the appearance of a patterned plasma membrane organization. The accumulation of the sodium channels at the nodes of Ranvier, the assembly of active-zone particles in the membrane of the presynaptic motor nerve terminal, and the localization of the acetylcholine receptor in the end plate region of muscle cells are some notable examples. It is the patterned topography of membrane proteins that enables these (and other) excitable cells to perform their specialized functions. However, the work of the last decade has shown that the cell membrane is best viewed as a dynamic mixture of proteins and lipids. Many membrane proteins are clearly capable of long-range lateral migration in the plane of the cell membrane. When a patterned membrane topography is viewed against

77

this conceptual backdrop, several questions immediately emerge. How is membrane topography developed during cellular differentiation? How is this topography maintained in the mature cell and stabilized in the presence of the metabolic turnover of the components that make up the membrane specialization? Can a modulation of membrane topography occur in a mature cell, and if so, how? What are the consequences of modulating the topography?

These questions are forced to the forefront when the neuromuscular junction is considered. The synapse between motor neurons and the muscle fiber is one of the most studied and perhaps most dramatic examples of membrane specialization. In an adult animal, the acetylcholine (ACh) receptors are localized at the synapse, which occupies only about 0.1% of the muscle cell membrane. In contrast, receptors are distributed widely over the surface before and during innervation of the muscle and following denervation of the muscle. A great deal has been learned about the physiological and pharmacological properties of the neuromuscular ACh receptor, as well as the behavior of receptor aggregates, but the mechanisms that induce the localization of the ACh receptors at the synapse and that stabilize this arrangement remain a mystery.

It is the purpose of this chapter to provide a brief overview of the problem of ACh receptor topography at the neuromuscular junction and to discuss some possible mechanisms for the development and stabilization of this topography. We will begin with a discussion of the fluidity of the cell membrane and the speed with which membrane proteins may diffuse laterally in the membrane. This discussion is necessary because many of the proposed mechanisms depend directly or indirectly on the lateral mobility of proteins in the cell membrane. Following a brief review of experimental studies on the development of ACh receptor topography (at the developing synapse and in denervated muscle) will be a discussion of some possible underlying mechanisms. A novel model—that of electromigration of the receptors to the region of the synapse—will also be presented. The strengths and weaknesses of the various models and some experimental approaches to test them will be discussed. In the final section, we will mention some possible functional consequences of the dynamics of the cell membrane as it pertains to the plasticity of synaptic function.

This chapter is not intended to be an exhaustive review of any component of the literature. Instead, it will attempt to present a synthesis of ideas on the mechanisms underlying membrane topography. The perspective of this contribution is derived from the converging viewpoints of the authors—those of membrane biophysics (MMP) and multicellular neuronal plasticity (SEF). A number of excellent reviews on the ACh receptor and the neuromuscular junction have appeared (for example, Fambrough, 1979; Lømo and Jansen, 1980; Dennis, 1981), hence, only a selective overview of the data will be given here. It is our hope that the ideas generated by this collaboration will shed some light on this fascinating problem.

II. Fluidity of the Cell Membrane and the Lateral Mobility of Membrane Proteins

Since the original observation of rapid intermixing of surface antigens on mouse–human heterokaryons (Fry and Edidin, 1970), our understanding of cell membrane organization has evolved rapidly. The initial measurements of the rate of lateral diffusion of membrane proteins yielded values in the range of 1 to 5×10^{-9} cm²/sec (Edidin and Fambrough, 1973; Poo and Cone, 1974; Liebman and Entine, 1974), suggesting free diffusion of these molecules in the plane of the lipid bilayer. Since 1976, however, the results from fluorescence photobleaching recovery experiments have consistently yielded a mobility that is 10 to 10^4 times slower (see review by Cherry, 1979). Moreover, this method indicates that a substantial fraction of the membrane proteins are immobile. The photobleaching technique, when applied to exogenous fluorescent lipids incorporated into the membrane yields a diffusion rate close to that found in the lipid bilayer membrane. Thus, it has been argued that the slower mobility or immobility of membrane proteins reflects a restriction imposed by extramembranous organization on the molecules within the fluid membrane.

Although the notion of restricted mobility may be correct, the evidence that leads to it, especially that from photobleaching experiments, demands closer scrutiny. The fluorescence photobleaching technique requires intense local irradiation of the membrane with laser light. Workers in this field have only recently begun to examine the effect of laser irradiation on the biological integrity of the cell membrane and its components (Sheetz and Koppel, 1979; Wolf et al., 1980). It is still not clear whether the intense laser irradiation of the fluorophores creates a local injury of the membrane. If so, the rate of fluorescence recovery may, instead, reflect a membrane healing process. The simple control of monitoring the membrane potential and/or membrane resistance during the course of photobleaching has not yet been reported. If it turns out that there are no photobleaching-induced artifacts, one still wonders whether the fluorescent ligand used in labeling the protein affects the lateral mobility of the protein being studied. Most cell surfaces are covered with an extensive extracellular glycocalyx. Attach-

ment of an exogenous ligand, even a small one, will inevitably intro-
duce additional viscous drag on the lateral mobility of the "tagged"
molecule. There is no evidence that the drag produced by the extracellular matrix is negligible compared to that of the lipid bilayer.

It is important to know exactly how fast proteins in their native
state may diffuse around in the membrane. Bretscher (1980) has emphasized this point for the case of understanding many biological
mechanisms on the cell surface. In considering the development of cell
surface topography, it is clear that certain proposed mechanisms for
receptor localization depend critically upon the diffusion coefficient.
Some of these models for membrane topography are not feasible if the
diffusion distance covered by the molecule within its lifetime on the
surface is negligible compared to the dimension of the cell. For example, if the lateral diffusion of macromolecules is severely limited, a
patterned topography of membrane proteins on the cell surface cannot
be achieved through random incorporation into the membrane followed
by migration to the designated place by a diffusional mechanism alone.
As discussed later in Section IV, the rate of lateral diffusion of membrane proteins may be of crucial relevance in deciding the plausibility
of various models.

The case of the ACh receptor in the embryonic muscle membrane
exemplifies this point. Edidin and Fambrough (1973) measured the
rate of spreading of fluorescence-labeled Fab antibody directed against
muscle cell surface proteins after local application of the antibody to
cultured muscle cells. The diffusion coefficient of the surface antigens
was calculated to be approximately $1-3 \times 10^{-9}$ cm²/sec (at 37°C). This
would allow the diffusion of a protein from one side of a muscle fiber to
the other in about 1 hour. Using the fluorescence photobleaching recovery method, however, Axelrod *et al.* (1976) found that dispersed ACh
receptors labeled with fluorescent α-bungarotoxin on the cultured rat
myotubes diffuse slowly, with a diffusion coefficient of 5×10^{-11} cm²/sec
(at 22°C). Many proteins in the *Xenopus* myoblast membrane undergo
rapid lateral migration in the presence of an extracellular electric field
(Poo, 1981). This migration leads to an asymmetric distribution of cell
surface molecules. For many proteins, the relaxation of this field-
induced asymmetry can be used to determine a diffusion coefficient, as
has been done for the heterogeneous population of concanavalin A (Con
A)-binding proteins (see Poo *et al.*, 1979). These studies yield an average diffusion coefficient for Con A receptors of 5×10^{-9} cm²/sec (at
21°C) in *Xenopus* myotomal membrane.

As a result of our concern over the generally slow diffusion rates of
membrane proteins reported by the fluorescence photobleaching

method, a new technique has been developed to assay the diffusional mobility of the ACh receptor (Poo, 1982). Electrophysiological techniques were used to assay the distribution of ACh receptors on the embryonic muscle surface (see Orida and Poo, 1978). The initial uniform distribution of ACh receptors was altered by asymmetrically inactivating the receptors with a brief pulse of α-bungarotoxin, applied to one side of the cell with a micropipet. This resulted in a large inactivation of the ACh receptor on the treated side of the cell and a slight inactivation of the receptors on the opposite side. The resulting asymmetry in the ACh sensitivity of the two sides of the cell can be determined by alternately iontophoretically pulsing ACh on the two sides of the cell and measuring the resulting depolarization of the cell. This asymmetry was found to decay rapidly (Fig. 1), returning to a uniform

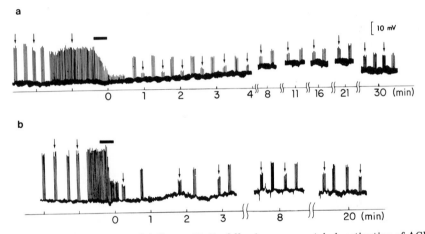

FIG. 1. Redistribution of ACh sensitivity following asymmetric inactivation of ACh receptors. Spherical-shaped *Xenopus* muscle cells in 1.5-day-old cultures were used in these experiments. ACh sensitivity at two spots on opposite sides of the muscle surface is monitored alternatingly by an iontophoretic method (Orida and Poo, 1978). Intracellular recording was performed using a glass microelectrode filled with 3 M potassium acetate (resistance 80–150 MΩ). Microelectrodes for iontophoretic application of ACh were filled with 3 M acetylcholine chloride (Sigma), had resistances from 200 to 350 MΩ, and required braking currents between 1.0 and 2.0 nA to prevent ACh leakage from the pipet and depolarization of the cells. Constant iontophoretic current pulses of amplitude 5 to 10 nA and 1 millisecond duration were delivered through a microelectrode preamplifier (Getting) to the ACh pipet. A third glass micropipet with a tip opening of about 2 μm was filled with α-bungarotoxin (α-BTX, 50 μg/ml) and positioned near the iontophoretic pipet. Asymmetric inactivation of the ACh receptor was done by applying a small positive pressure through the toxin pipet. (a) Alternating recordings of ACh-induced responses on two opposite poles of the spherical muscle cell in a typical experiment. Before preferential inactivation, the responses were nearly identical. After pulse inactivation

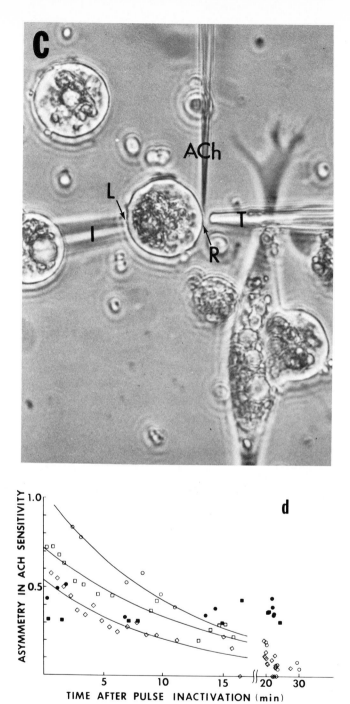

FIG. 1c and d. See pages 81 and 83 for legend.

distribution of ACh sensitivity in 20 to 30 minutes. Pretreating the cells with Con A (100 μg/ml) for 8 minutes prior to toxin treatment resulted in an asymmetry that was stable over long periods of time (Fig. 1b and d). Such Con A treatment is known to render the ACh receptors electrophoretically immobile in these muscle cells (Orida and Poo, 1978). Taken together, these experiments are consistent with the notion that preexisting, functional ACh receptors diffuse rapidly in the embryonic muscle cell membrane. The decay of the sensitivity asymmetry yields a diffusion coefficient of about 2.6×10^{-9} cm^2/sec. In other words, it would take less than 2 hours on the average for an ACh receptor to diffuse across a muscle cell 50 μm wide. Furthermore, the consistent return to equal ACh sensitivity indicates that there is no immobile fraction of ACh receptors on these cells. These results are inconsistent with both the diffusion coefficient (D) and the existence of an immobile fraction of dispersed receptors reported by the photobleaching experiments. The D value reported by those experiments would require more than 4 days for the receptor to cross the same 50-μm cell. In contrast, the lifetime of the extrajunctional ACh receptor in many systems appears to be shorter than 1 day (see Fambrough, 1979; Dennis, 1981). Recent studies of the lateral mobility of extrajunc-

---

with α-BTX, the right side (arrow) of the cell was much less sensitive to ACh than the left side (without arrow). This asymmetric sensitivity decayed with time over a period of about 30 minutes after pulse inactivation. Bar represents the duration of positive-pressure application through the toxin pipet. The membrane potential at the onset of recording was 102 mV, and it varied by less than 15 mV during the 30-minute period. Note the concurrent drop in ACh sensitivity on the left side as the right-side sensitivity increased. (b) Typical recording of preferential inactivation experiments on cells preincubated for 8 minutes in concanavalin A (Con A), a treatment known to prevent electric field-induced migration of ACh receptors (Orida and Poo, 1978). Note the absence of any significant redistribution of ACh sensitivity over a period of 20 minutes after asymmetric receptor inactivation. (c) Phase-contrast micrograph of a representative spherical muscle cell, together with the position of the pipets for intracellular recording (I), iontophoresis (ACh), and toxin application (T). (L) Left; (R) right. Bar: 20 μm. (d) Asymmetry in ACh sensitivity on the two opposite poles of the spherical $Xenopus$ muscle cells plotted against the time after pulse inactivation of ACh receptors at one pole of cells with α-bungarotoxin. Asymmetry is determined by the formula $(V_A - V_B)/(V_A + V_B)$, where V_A and V_B are amplitudes of ACh potential induced by identical ACh pulses on the minimally and maximally inactivated side of the cell, respectively. Data points depicted by open symbols represent three sets of recordings from three different cells of a 1.5-day-old culture. Solid curves represent best least-square fit to these three sets of data with first-order decay kinetics, including points beyond 20 minutes. The characteristic $1/e$ times for the decay were 11.0, 12.8, and 11.8 minutes, respectively, for the cells 23, 20, and 20 μm in radius. These decay rates suggest a diffusion coefficient of the ACh receptors of 4.0, 2.5, and 2.7×10^{-9} cm^2/sec, respectively (Poo, 1982). Filled symbols represent two sets of recordings from cells pretreated with Con A.

tional ACh receptors in intact myotomal fibers of *Xenopus* tadpoles have confirmed that the rapid lateral diffusion of ACh receptors indeed occurs in the developing muscle membrane *in vivo* (Young and Poo, 1982). In short, our evidence suggests that the ACh receptor is capable of rapid long-range diffusion in the embryonic muscle cell membrane and that this movement would permit extensive rearrangement of the ACh receptor topography by diffusional mechanisms alone.

Before leaving our discussion of the lateral mobility of membrane proteins, a word of caution must be voiced. The lateral mobility of membrane proteins has been demonstrated in a number of embryonic, tissue culture, and specialized cell types (e.g., photoreceptors, erythrocytes); however, there is no clear indication that such mobility is a general feature of mature tissues. One can adopt the view we have, that the mobility seen in the few cases studied probably reflects a general feature of membrane organization, but the possibility that the lateral mobility of membrane proteins may be restricted in some mature cell types must be kept in mind. If it turns out that the development of patterned membrane topography is coincident with a loss or reduction in the lateral mobility of membrane constituents, a number of problems arise: (1) How is the turnover of membrane proteins accomplished if they cannot migrate to (from) some endocytotic (exocytotic) site? (2) How are membrane components sorted and recycled at the presynaptic terminal if intermixing and lateral mobility of these components is not allowed? (3) How do receptor–hormone complexes find their way to coated pits for endocytosis or couple with effector molecules in the membrane if they are not laterally mobile? These are but a few of the cases in which the ability of membrane constituents to intermix may play some beneficial role. If lateral mobility is not a general phenomenon, then more complicated mechanisms must step in to perform these tasks.

III. Localization of the ACh Receptor at the Neuromuscular Junction

The best-characterized developing synapse is the neuromuscular junction of vertebrates (see reviews by Fambrough, 1979; Lømo and Jansen, 1980; Dennis, 1981). The developing neuromuscular junction has been studied primarily in the chicken, the rat, or the frog, either *in vivo* or in a culture dish. The distribution of the ACh receptors in the muscle cell membrane can be assayed either by iontophoretically mapping the sensitivity of the cell to ACh or by examining the binding of the derivitized snake toxin α-bungarotoxin (α-BTX). In the iontophoresis method, an intracellular microelectrode is used to measure the membrane depolarization produced when ACh is applied to local regions of the membrane. For a given amount of ACh ejected from the micropipet, a larger depolarization is taken to indicate a higher density of ACh receptors at that particular region. In the second method, radioactive or fluorescence-labeled α-BTX is applied to the muscle cell. Because this toxin binds irreversibly in the time frame of most experiments, the distribution of the toxin observed by autoradiography or fluorescence microscopy is a direct measure of the distribution of the ACh receptors. In the few cases where both techniques have been used in parallel (iontophoresis and α-BTX), the observed distributions have been nearly identical (see Land *et al.*, 1977; Frank and Fischbach, 1979). Many of the experiments on the neuromuscular junction have been performed on denervated or reinnervated cells in an attempt to understand the development of the synapse. Although in most cases denervation effects probably reflect underlying mechanisms similar to those in embryonic synaptogenesis, one must remain cautious in extrapolating the denervation results to the embryonic situation.

A. APPEARANCE AND CLUSTERING OF THE ACh RECEPTOR

From what we now know of the developing neuromuscular junction, at the very least it must be considered a highly dynamic structure. It appears that the receptors are distributed in a random, dispersed manner as they first appear on the embryonic muscle surface. In *Xenopus*, their appearance is extremely rapid; myoblasts go from having no sensitivity to bath-applied ACh to a sizable sensitivity in times on the order of 1 hour (Blackshaw and Warner, 1976). With time in culture, the muscle cells demonstrate a nonuniformity in the distribution of the receptors, forming "hot spots" or clusters of ACh receptors (see Anderson and Cohen, 1977; Land *et al.*, 1977). In *Xenopus*, it appears that such clusters can also form *in vivo* under certain conditions (Chow and Cohen, 1978; Jacob and Lentz, 1979; Chow, 1980). During this period and well into the period of synaptogenesis, the ACh receptor is rapidly synthesized, inserted into the membrane, and then degraded after a few hours (in chick and rat). In some animals, this rapid metabolic turnover of the receptor continues until long after synaptogenesis, and then the turnover rate slows to many days (see review by Fambrough, 1979). The hot spots (clusters) on an uninnervated muscle cell remain stable in position over many hours, moving less than 1 μm in 19 hours (Fambrough and Pagano, 1977), even though the turnover rate of the individual receptors is on the order of a few hours. The clusters are commonly associated with some form of membrane contact, ranging from contact with the substrate of a culture dish to contact by a latex bead (Peng *et al.*, 1981). Thus, it appears that both specific stimuli (contact by a nerve, see next section) and nonspecific stimuli (as above) can serve to localize ACh receptors. These clusters are not dispersed by metabolic inhibitors, except in the case of cultured rat myoblasts (Bloch, 1979). Large clusters can be formed by electrophoresing ACh receptors to one pole of the cell with an extracellular electric field (Orida and Poo, 1978; Poo, 1981a). The stability of these large clusters demonstrates an inherent propensity of the ACh receptor to aggregate.

Some factors found in the media conditioned by a hybrid neuroblastoma cell line have been shown to promote an increased number of ACh receptor clusters on cultured myotubes (Christian *et al.*, 1978). These factors did not increase the total number of ACh receptors but instead appeared to have rearranged the distribution of ACh receptors either by aggregating mobile receptors or by stabilizing labile receptor aggregates. In other studies, extracts from the brain or spinal cord of the rat and chick were found to increase both the total number of ACh receptors and the number of receptor clusters found on uninnervated myotubes in culture (Podleski *et al.*, 1978; Jessel *et al.*, 1979). The factors responsible for the effects in these latter two studies appear to have different molecular weights. It remains a possibility that these factors do not have any direct aggregating effect, but instead produce the increased number of aggregates by elevating the total number of receptors on the cell surface. If the equilibrium for a receptor to be in the free or aggregated state were remain the same, an increased number of receptors might directly produce an increased number of aggregates.

In addition to these diffusible substances, there seems to be a specialization in the subsynaptic basement membrane that can lead to the clustering of receptors. Damaged muscle cells degenerate and are phagocytosed, leaving their basement membranes intact. Myoblasts then reform the fibers within this empty sheath of basement membrane. Even in the absence of innervation, these regenerating fibers demonstrate areas of high receptor density under the "synaptic" region of the basement membrane (Burden *et al.*, 1979). Recent immunological evidence indicates that the subsynaptic basement membrane has at least three different factors that are not found outside of the junctional region (Sanes and Hall, 1979). The distribution of some of these factors closely parallels the distribution of α-BTX binding sites on muscle fibers. Thus, it is likely that factors in the basement membrane help to develop or maintain the localization of the ACh receptor.

B. CULTURE EXPERIMENTS ON SYNAPTOGENESIS

In the controlled environment of a culture dish, nerve–muscle contact and synaptogenesis can be studied in detail. Following the contact of a nerve, the ACh receptors appear to form a cluster at the contact site. This cluster formation proceeds in the presence of transmission blocking agents (curare, α-BTX) but seems to require that the contacting axon be cholinergic (Cohen et al., 1979; Cohen and Weldon, 1980; Kidokoro et al., 1980). Immediately after innervation, clusters are observed both under the axon and distributed over the surface (Betz and Osborne, 1977). With time, the receptors form into the cluster at the contact site and the extrajunctional clusters apparently disappear (Anderson and Cohen, 1977). Convincing evidence has been obtained that supports the idea that the nerve induces the cluster following contact, and that with time, the clusters appear only at regions of neurotransmitter release (Cohen and Fischbach, 1977; Frank and Fischbach, 1979). The presence of clusters at the contact point has led to the speculation that neurons trap the receptors by a surface interaction or by the secretion of some trophic factor (Edwards and Frisch, 1976; Frank and Fischbach, 1979; Axelrod et al., 1981).

The clear message of the bulk of this work is that the cell surface progresses from a state where the receptors are diffusely arranged to one in which the receptors are more localized to the region of the end plate. With time, the distribution becomes increasingly distinct, with ACh receptors located only under the presynaptic nerve terminal.

C. POLYINNERVATION OF MUSCLE FIBERS

Immediately after innervation, many of the muscle cells are polyinnervated, i.e., many motor neurons innervate each muscle cell. With time, this multineuronal innervation is pared down until only a single motor neuron innervates any one muscle fiber, although each motor neuron may innervate many muscle cells (see Lømo and Jansen, 1980; Dennis, 1981). Chronic stimulation of the muscle may speed this process (O'Brien et al., 1978), whereas paralysis of the muscle may slow it partially (Thompson et al., 1979; Srihari and Vrbova, 1978). The elimination of polyinnervation parallels the decline in the number of extrajunctional receptors. Ectopic innervation of the muscle cell is effective at inducing receptor clusters only when extrajunctional receptors are present, either early in development or after denervation (Lømo and Slater, 1978; Weinberg et al., 1981; see review by Lømo and Jansen, 1980). Thus, any complete mechanistic explanation of the development of innervation patterns in muscle cells must explain both the localization of the ACh receptor to the synaptic region and the transition from polyinnervation to monoinnervation.

D. DENERVATION AND DRUG EFFECTS

The effects of denervation and other insults to the intact neuromuscular junction have been explored with the hope of understanding the mechanisms of receptor localization. Following denervation, extrajunctional receptors, which in the normal innervated adult muscle are hardly detectable, appear in high concentration over the entire surface of the muscle cell, a phenomenon termed "denervation supersensitivity" (see review by Lømo and Jansen, 1980). Presynaptic blockage of the propagation of the neuronal action potential by tetrodotoxin (TTX) or local anesthetics causes a similar appearance of extrajunctional receptors, although it is only half as effective as denervation (Lavoie *et al.*, 1976; Warnick *et al.*, 1977; Pestronk *et al.*, 1976a). Similar effects were seen after treatment with botulinum toxin, at a level that allowed reduced miniature end-plate potentials and subthreshold end plate potentials to occur (Thesleff, 1960; Tonge, 1974; Bray and Harris, 1975; Pestronk *et al.*, 1976b). A postsynaptic block of neuromuscular transmission (with curare or α-BTX) also causes a dramatic increase in the number of extrajunctional ACh receptors (Burden, 1977; Berg and Hall, 1975) even though the axons innervating the muscle cells remain active and continue releasing quanta. Direct tetanic stimulation of the denervated muscle cells causes a decrease in the synthesis of ACh receptors and in the number of extrajunctional receptors appearing after denervation (Reiness and Hall, 1977; Jones and Vrbova, 1974; Purves and Sakmann, 1974; Reiness *et al.*, 1977; Linden and Fambrough, 1979). Blockage of muscle cell activity with TTX leads to an increased synthetic rate for the ACh receptor (Shainberg *et al.*, 1976). Thus, the picture that emerges is that blockage of neuromuscular transmission leads to the appearance of extrajunctional ACh receptors and that muscle cell activity is an important factor in the localization and synthesis of the ACh receptors.

The findings of these denervation and drug treatment studies may *at first glance* seem at odds with the cluster formation data given in Section III,B, until one realizes that the experiments are measuring different phenomena. In the studies on the developing synapse, the appearance of clusters at the nerve contact site was the measured phenomenon. The presence of some extrajunctional receptors would be considered insignificant. In fact, the "junctional" clusters only attained an ACh receptor density 25 times that of the extrajunctional receptors. In the denervation experiments, the picture is exactly the opposite. In the adult, the junctional receptors are 10^3 to 10^4 times more dense than extrajunctional receptors. The appearance of extrajunctional receptors was measured, and the residual synaptic clusters were considered less important. Thus, these two classes of experiment report on different facets of the neuromuscular junction: one on the appearance of clusters, the other on the appearance of extrajunctional receptors. Together, these complementary experimental approaches form a large data base on the factors important in the formation and maintenance of the neuromuscular junction.

IV. Mechanisms for the Development and Maintenance of ACh Receptor Localization

The localization of ACh receptors to the neuromuscular junction could be the result of a number of mechanisms acting alone or in combination. The phenomena that need to be explained are (1) the localization of clusters at the site of nerve muscle contact; (2) the disappearance of extrajunctional receptors; (3) the maintenance of the localized receptor distribution at the end plate region; and (4) the competition between synapses that results in the transition from multiinnervated to singly innervated muscle cells. Although each of these behaviors may be explained by the action of a particular mechanism or by a set of mechanisms acting in concert, a single model that can explain them all would be especially attractive. At present, there are a few simple models that have been proposed for the localization. Many of these models assume that the extrajunctional receptor protein is the same gene product as the junctional receptor protein and that the mere location of the receptor in the synapse confers the differences thought to exist between junctional and extrajunctional receptors. These differences include the slightly different isoelectric focusing point, the longer mean open time and lower conductance of extrajunctional receptors/ ionic channels, and some difference in the binding of antisera from myasthenia gravis patients (see Reiness and Hall, 1981; reviews by Fambrough, 1979; Dennis, 1981). All of these could be explained by the receptor's local environment. For example, lipids in the junction may be of different types. Modification of the lipid charge and/or dipole moment can have profound effects on the behavior of excitable channels (Hall, 1981). In addition, bound lipids could easily alter the isoelectric point of the protein, considering that almost one-fifth of the apparent molecular weight of the ACh receptor is due to bound detergent in some preparations (see Edelstein et al., 1975). The most notable difference between junctional and extrajunctional receptors seems to be the longer metabolic half-life of junctional receptors (many days versus many hours; see Fambrough, 1979). The environment may also affect the turnover time for the receptor. It should be mentioned that posttranslational modification of the receptor, once it arrives at the junction, might also account for all these differences (see Changeaux and Danchin, 1976). Although there may be important differences between extrajunctional receptors and those seen at adult synapses, simple mechanisms for localization must be adequately tested before they are discarded in favor of more complicated schemes involving directed insertion of different gene products.

A. Diffusion-Trap Model

A powerful and conceptually simple model for the localization of the ACh receptor is that the nerve contact site acts as a trap for freely diffusing ACh receptors (Edwards and Frisch, 1976). Because the mechanism requires that the receptor be free to wander randomly and passively through the membrane until it encounters the axon. this mechanism depends upon a diffusion of the receptor much faster than that reported by fluorescence photobleaching techniques. The diffusion rate reported by the local inactivation of the ACh receptor with α-BTX is fast enough to allow such mechanisms to work within realistic time periods (see Section II; also Chao *et al.*, 1981; Poo, 1982). This trap could be due to a cell surface interaction or the secretion of some substance that reduces the mobility of the ACh receptors. Consistent with such a scheme are the findings that an ingrowing nerve can cause the accumulation of receptors, even if synaptic transmission is blocked by α-BTX or curare (see review by Cohen *et al.*, 1979). However, it must be remembered that these clusters never develop the dramatic density of ACh receptors seen at the mature untreated synapse. The ability of basement membrane to alter the distribution of receptors indicates that factors in the basement membrane may play a "trapping" role (see Section III,A; also Burden *et al.*, 1979). The finding that such accumulation occurs only if a cholinergic axon contacts the myoblast (see review by Cohen *et al.*, 1979) is also consistent with a role for specific, cell surface interactions or immobilizing factors in the process. Because receptors are trapped as they come toward the synapse from extrajunctional regions, the mechanism is consistent with the finding that newly added receptors are found at the edge of the synaptic cluster (Weinberg *et al.*, 1981).

The finding that poisoning mature synapses leads to the appearance of a high level of extrajunctional ACh receptors indicates that muscle cell activity must play some role in the control of ACh receptor clusters (see Section III, D). For example, treatment with α-BTX leads to the appearance of almost as many extrajunctional receptors as does denervation. Because α-BTX leaves the receptor free to bind to other extracellular proteins (e.g., concanavalin A), we must assume that most of the ACh receptor is still free to interact with the nerve. Furthermore, using TTX to block transmission of action potentials in the nerve, which leaves miniature end-plate potentials and axoplasmic transport untouched, causes extrajunctional receptors to appear at a density about half that seen for a denervated muscle. Botulinum toxin treatment leads to a similar appearance of extrajunctional receptors. These findings might be taken as evidence against the diffusion-trap model but may be the result of the mechanism being overwhelmed by an increased ACh receptor synthesis rate. Because muscle activity has been shown to have a strong effect on the synthetic rate of the ACh receptor (see Fambrough, 1979; Dennis, 1981), it is not unreasonable to propose that poisoning of the synapse leads to a greatly increased rate of synthesis. Because the trap area probably remains somewhat constant, this may lead to a buildup of extrajunctional receptors.

The trap mechanism alone seems hard pressed to explain the reduction in polyinnervation seen during development. Competition for the receptors by trapping alone seems unlikely because the clusters seen at nerve contact sites in culture already extend past the boundaries of the contact (Frank and Fischbach, 1979). The addition of a factor responsible for competition could easily bring the diffusion-trap mechanism into agreement with the data. Moreover, junctional ACh receptor clusters appear to form only at transmitter release sites, not along the entire nerve–muscle contact (Cohen and Fischbach, 1977). Thus, the trap might result from factors released only at transmitter release sites. It remains possible that much of the clustering observed in culture experiments represents a general phenomenon of the receptor aggregating at any favorable site of close contact (as seen with yolk granules in contact with the myoblast) rather than the primary mechanism of synaptogenesis. The major advantage of this model is that it requires so little of the cell: merely random insertion of the receptor, followed by diffusion toward the trap.

B. LOCAL INCORPORATION MODEL

Directed incorporation models for the localization of the ACh receptor are characterized by subsynaptic insertion of the receptor in response to trophic factors exuded from the nerve, local excitation of the muscle membrane, or cell–cell contact. In the absence of a stimulus, the receptors would be inserted randomly in the cell membrane.

Local-excitation-directed insertion of the ACh receptor seems consistent with the normal development of the neuromuscular junction and with the appearance of extrajunctional receptors following the poisoning of a mature synapse. However, local excitation is not consistent with the appearance of subsynaptic clusters in the presence of transmission blocking agents unless this block is less than total (see Cohen, 1972). Trophic factor-directed or contact-directed insertion would be consistent with the subsynaptic clustering in the presence of transmission blocking agents but seems unable to explain the appearance of extrajunctional receptors following the treatment of mature fibers with curare, α-BTX, TTX, or botulinum toxin. All of these agents leave nerve–muscle contact unaltered. TTX prevents the firing of the innervating axon but leaves miniature end-plate potentials intact. Botulinum toxin merely decreases the amplitude of the end-plate potential such that it is subthreshold for a muscle action potential. Curare or α-BTX treatment of muscles leaves the firing activity, axonal transport, and presumably the quantal release of spinal neurons unaltered. We would therefore expect that the hypothetical trophic substance release would continue in the presence of these drugs. The directed insertion site would have to be just at the edge of the existing junctional cluster to explain the finding that recently synthesized receptors are found at the periphery of the cluster (Weinberg *et al.*, 1981). In addition, something must serve to stabilize the synaptic clusters for this mechanism to work. The rapid lateral diffusion seen in *Xenopus* myoblasts (see Section II) makes it clear that receptors do not stay exactly at their site of insertion. The stability of end-plate clusters could be due to the junctional receptors being a different gene product, but the turnover rate of junctional receptors remains high for many days after innervation in some systems (Burden, 1977). If the longer-lived junctional receptors are also thought to be a different gene product, then this line of reasoning would lead to three gene products: (1) diffusible receptors, (2) junctional receptors (no diffusion) with rapid turnover; and (3) junctional receptors with slow turnover. Thus, such arguments can lead to as many genes as there are characteristics for the ACh receptor, making the model more complicated than other models and more difficult to test. Perhaps some additional trapping factor still needs to be postulated to "glue" the receptors together at a subsynaptic site of insertion.

C. SELECTIVE STABILIZATION MODELS

Another class of models, which will be referred to as the selective stabilization models, allows for random insertion of the ACh receptor followed by a stabilization of those receptors at the region of synaptic contact. One of these, put forth by Changeaux and Danchin (1976), relies upon a rapid diffusion of the ACh receptor. Receptors that arrive at an active region of nerve–muscle contact are stabilized by the presence of both a muscle cell-produced factor and a neuronally produced factor. The activity of the muscle cell is proposed to regulate the number of diffusible ACh receptors manufactured in the cell. By limiting the total amount of muscle-produced factor and making synaptogenesis dependent upon this factor, the model is able to account for the reduction in polyneuronal innervation seen during development. Stent (1973) offers another mechanism for the reduction of polyinnervation. He proposes that the transient polarity reversal in the muscle cell membrane potential produced by suprathreshold stimulation of the cell leads to the destruction or removal of all but the ACh receptors at the end plate responsible for the stimulation of the cell. These receptors are protected by the synaptic currents, which serve to reduce the local polarity reversal. Thus, with time, the neuron most able to drive the muscle cell to fire its own action potential will dominate the muscle cell. All other synapses will be eliminated by the destruction of the ACh receptors. Although this model has some attractive features, the observation that a muscle cell can entertain two synapses if they are separated by a distance of more than 3 mm (Kuffler *et al.,* 1977) seems troublesome. The action potential of the muscle fiber propagates over the entire fiber; thus, all of the muscle fiber membrane is subject to the field reversal proposed to destroy receptors.

D. ELECTROMIGRATION MODEL

We would now like to present a different model for the localization of the ACh receptor: the electromigration model. Jaffe has proposed that steady biogenic electric fields have a profound role in development (Jaffe, 1977, 1979; Jaffe and Nuccitelli, 1977). In the electromigration model (Poo and Fraser, 1982), we propose that transient electric fields associated with neuronal activities serve to develop, maintain, and modulate the topography of the membrane components responsible for these activities. For example, the electric fields generated by synaptic currents cause the electrophoretic migration of the ACh receptors in the membrane to the region of synaptic contact. This mechanism has the advantages that it is dependent on nerve and muscle activity to maintain the distribution and that it requires no directed insertion of the receptor.

Any synaptic potential must have a longitudinal component (along the membrane) as well as the typically measured transmembrane component. It is this longitudinal component that is responsible for the spread of the synaptic current along the cell membrane. Although these currents may seem small, it must be remembered that the distances involved are equally small. For example, a synaptic potential that decrements by 1 mV over 5 μm produces a transient longitudinal electric field on the order of 1 V/cm along both the extracellular and cytoplasmic faces of the membrane (especially *in vivo*, where the extracellular space is very restricted). This is well within the range of field strengths required for electrophoretic accumulation of ACh receptors by exogenous dc fields (Orida and Poo, 1978) and the same as the strength that produces accumulation of the Con A receptor in pulsed exogenous electric fields (Fraser and Poo, 1982). Pulsed fields much weaker than this are ineffective at electrophoresing the Con A receptor. Electrophoresis (by exogenous dc fields) of the ACh receptor over the embryonic *Xenopus* myoblast takes on the order of 0.5 hour for electric fields in the range of 1 V/cm. After removal of the field, the receptors do not diffuse back to a uniform distribution, apparently as a result of an inherent ability of the receptors to aggregate. If a second electric field is applied at 90° to the first field, the aggregate moves *en masse* toward the new cathode-facing pole of the cell (see Poo, 1981).

Building upon the experiments on the diffusional and the electrophoretic mobility of the ACh receptor, we can now describe the electromigration model in detail. The basic premise is that the longitudinal electric field associated with synaptic potentials causes the localization of the ACh receptor by electrophoresing the receptors to the subsynaptic membrane. The ability of the ACh receptors to form aggregates among themselves following localization by an electric field (Orida and Poo, 1978) assures that the subsynaptic cluster of receptors will be selective and diffusionally stable. That is, the receptors will not diffuse away from the cleft when the field that brought them there is removed, whereas other proteins will, leaving a cluster enriched for the receptor. The initial localization of the receptors has the characteristics of a positive feedback mechanism. The electrophoresis of the ACh receptors to the subsynaptic membrane will tend to increase the efficacy of the developing synapse. This will lead to larger synaptic potentials and, thus, a greater longitudinal electric field, making the synapse more effective in electrophoresing the remaining extrajunctional receptors. Experimentally, such behavior of the developing synapse has been observed. The initial contact of the nerve causes small, slow miniature end plate potentials, even before the appearance of subsynaptic clusters of the ACh receptors (Kidokoro, 1980). With time, these potentials increase in size, and clusters of ACh receptors are found at the nerve–muscle contact site. The stable size of the mature subsynaptic receptor cluster is the direct result of muscle activity decreasing the synthetic rate of the ACh receptor (see reviews by Fambrough, 1979; Dennis, 1981). By adding this influence to the electrophoretic accumulation proposed earlier, the model demonstrates negative feedback control on the number of ACh receptors on the muscle fiber surface. For a schematic presentation of the components of the model, see Fig. 2. The electromigration model appears to be consistent with both the initial appearance of subsynaptic clusters (Kidokoro *et al.*, 1980; Frank and Fischback, 1979) and the stable distribution of ACh receptors found under mature synapses. It has been reported that subsynaptic ACh receptor clusters form even in the presence of α-BTX or curare (Cohen *et al.*, 1979). Of course, if these synaptic blocks were anything less than complete (see Cohen, 1972), then part of the synaptic field would remain and the model could directly account for the findings. If the block is shown to be absolute, then the electromigration model may have to be modified to account for all published results.

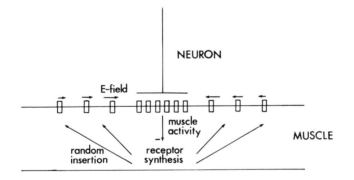

FIG. 2. The electromigration model. ACh receptors are inserted into the membrane at random, where they are free to diffuse (Poo, 1981b). The presence of a nerve terminal that is releasing ACh leads to small synaptic potentials. These potentials have a longitudinal component that causes the electromigration of the receptors to the site of transmitter release. With time, these small potentials localize many of the receptors to the transmitter release site, increasing the efficacy of the synapse and thereby increasing the electromigration of the receptors. The "E-field" vectors in the figure demonstrate schematically the extracellular component of the electric field (the intracellular component may also play a role). Once the receptors are brought together by the electric field, they form diffusionally stable aggregates (Orida and Poo, 1978). When the activity level of the cell reaches a certain level, the synthesis of receptors is inhibited. This provides a negative feedback control of the number of receptors on the surface of the cell, such that sufficient receptors are available to form an effective synapse. The exact number of receptors will depend on the level of activity that suppresses receptor synthesis and on the activity of the innervating neuron.

Following poisoning of an adult synapse or denervation, appearance of extrajunctional receptors is predicted, because the electric fields will be reduced or abolished. As a result, there will be no localization of newly inserted receptors. It is interesting that both poisoning the nerve with TTX and blocking the synapse with botulinum toxin result in the appearance of fewer extrajunctional receptors than found after denervation (see Section III). Tetrodotoxin eliminates all but miniature end-plate potentials and botulinum toxin reduces the size of end-plate potentials until they are subthreshold. Thus, both of these treatments would leave a portion of the electrophoretic fields intact, although reduced in amplitude. The model may also offer an explanation for the reduced miniature end-plate potential amplitude seen in rat diaphragm muscle following a 3-day treatment with curare (Berg and Hall, 1975). Curare would block the synaptic potentials, greatly reducing or abolishing the localization of the ACh receptor. Degradation of the receptor presumably proceeds unaltered during this 3-day period. Thus, a significant fraction of the receptors would be lost from the junction, reducing the efficacy of the synapse. The model therefore appears to be consistent with both denervation and poisoning data.

The electromigration model is also consistent with the reduction in polyinnervation of muscle fibers with time. The receptors will move in response to the net electric field that the individual receptors experience. This field could be the sum of many sources, as is the case for a polyinnervated muscle cell. The largest component of the field will be toward the most active synapse, making the most active synapse most effective in accumulating new receptors. Thus, the synapse most able to excite the myoblast will become even more able with time. Because it is competing with other synapses for a fixed number of receptors, this increase in sensitivity is at the expense of other developing synapses. The turnover of ACh receptors assures that a synapse that is not accumulating new receptors will soon disappear. It is interesting to note that electrically stimulating the muscle fibers directly, which is known in other systems to reduce receptor synthesis (see Fambrough, 1979; Dennis, 1981), appears to speed the reduction of polyinnervation (O'Brien et al., 1978). This stimulation effect may be the result of directly reducing the pool of available extrajunctional receptors, thereby making the competition for the receptors more critical. Because the ACh receptors are proposed to be inserted at random and the synaptic potentials degrade with distance, two synapses may cohabit the fiber if separated by distances great in comparison with the spread of the synaptic currents (Kuffler et al., 1977). Each synapse could merely localize all of the receptors inserted into its "territory" and feel little competition from the other.

Based on the preceding arguments, the electromigration model appears to fit all of the features of normal development and many of the findings in chemically treated synapses. The ability of the model to generate the localization of the receptors, the disappearance of extrajunctional receptors, the maintenance of the distribution, and the reduction from polyinnervation to single innervation, makes it an attractive mechanism for further testing. The ability of exogenous fields to accumulate the receptors is good supporting evidence, but what is really needed is evidence that local, cell-generated fields are capable of localizing the receptors.

In the previous discussion, we have concentrated our attention on the lateral reorganization of membrane proteins by the endogenous electric fields. Clearly, these electric fields could have many other actions. Electromigration of proteins and vesicles in the cytoplasm seems an obvious extension of our arguments. Such electrophoresis could serve to localize some cell machinery or may produce an intracellular concentration gradient that may then serve some useful role. By concentrating on the lateral reorganization of membrane proteins we in no way wished to ignore these possible effects, but instead hoped that the lateral reorganization mechanism would prove to be the most easily tested. Other possible electric field induced effects could represent a hybrid of the models presented above. For example, the electric fields could serve to localize a "receptor glue" that then traps the receptor to the site of the synapse. If some of this "glue" were to become associated with the basal lamina of the muscle cell then the ability of the synaptic lamina to cluster ACh receptors becomes easily understandable (see Section III,A). There seems no compelling reason to insist that only one mechanism is responsible for synaptogenesis, and we should therefore remain receptive to the possibility that a number of mechanisms may act in concert to produce the observed events.

V. Conclusion: Plasticity in the Organization of Excitable Membranes

In the preceding sections we have attempted to review the spectrum of experimental findings obtained from work on the topography of the ACh receptor and a few of the models that may account for the topography of the receptor distribution. This topography may be the result of any of the models mentioned earlier, a combination of these models, or perhaps a mechanism yet to be described. It seems that two of the models—the diffusion-trap model and the electromigration model—can explain a rather wide range of the data and require little of the cell machinery. The cell can merely insert the ACh receptor at random and the ACh receptor is then localized by the physics of the membrane. Thus, a complete description of these models is possible, based upon the properties of the membrane alone. This makes these models easily testable with available methods. The ability of the electromigration model to fit the data on cluster formation and competition between synapses makes it particularly attractive.

As the electromigration model is tested more thoroughly, it may have some important ramifications. As we mentioned at the outset of this chapter, the patterned topography of excitability-inducing proteins is crucial for excitable cells to perform their specialized functions, namely, the transmission of nervous signals. Because the electromigration model appears to agree with the data available at present, we propose that the dynamic nature of the plasma membrane enables the development of a patterned membrane topography through the lateral rearrangement of these proteins. The electric fields associated with the functioning of excitable cells could themselves serve to develop the membrane topography underlying these functions. Aggregates of the ACh receptor are easily electrophoresed by exogenous electric fields, even though they are diffusionally stable (see Poo, 1981). This tempts us to propose that once a stable topography is formed in a mature cell, a modulation of topography may still occur through the frequent occurrence of electrical events associated with nervous activity. If so, the electromigration mechanism may have profound effects on signal processing in the adult nervous system. Modulation of the topography of transmitter receptors at the synaptic membrane by this mechanism could directly produce a long-term facilitation of an active synapse. Modulation of the distribution of calcium channels or active-zone particles at the presynaptic nerve terminal and the sodium channel distribution near an axonal branch point offer more examples in which an alteration in the topography may lead to major use-dependent changes in signal transmission in the nervous system. It is our hope that the speculations and new findings presented in this chapter will increase the interest of investigators in the problem of neuronal plasticity at the level of the plasma membrane.

ACKNOWLEDGMENTS

We are happy to thank Drs. D. Petersen and R. A. Cone for numerous discussions that helped to refine the electromigration model, Drs. M. Bronner-Fraser, I. Chow, and S. H. White for their excellent advice on the manuscript, and M. Tacha for her expert secretarial assistance. The preparation of this chapter was supported by NSF (BNS-8023638 to S. E. Fraser; BNS-8012348 to M-m. Poo).

REFERENCES

Anderson, M. J., and Cohen, M. W. (1977). *J. Physiol. (London)* **268**, 757–773.

Axelrod, D., Ravdin, P., Koppel, D. E., Schlessinger, J., Webb, W. W., Elson, E. L., and Podleski, T. R. (1976). *Proc. Natl. Acad. Sci. U.S.A.* **73**, 4594–4598.

Axelrod, D., Bauer, H. C., Stya, M., and Christian, C. N. (1981). *J. Cell Biol.* **88**, 459–462.

Berg, D. K., and Hall, Z. W. (1975). *J. Physiol. (London)* **244**, 659–676.

Betz, W., and Osborne, M. (1977). *J. Physiol. (London)* **270**, 75–88.

Blackshaw, S. E., and Warner, A. E. (1976). *Nature (London)* **262**, 217–218.

Bloch, R. J. (1979). *J. Cell Biol.* **82**, 626–643.

Bray, J. J., and Harris, A. J. (1975). *J. Physiol. (London)* **253**, 53–77.

Bretscher, M. (1980). *Trends Biochem. Sci.* **5**, 4–5.

Burden, S. (1977). *Dev. Biol.* **61**, 79–85.

Burden, S. J., Sargent, P. B., and McMahan, V. J. (1979). *J. Cell Biol.* **82**, 412–425.

Changeaux, J.-P., and Danchin, A. (1976). *Nature (London)* **264**, 705–712.

Chao, N-m., Young, S. H., and Poo, M-m. (1981). *Biophys. J.* **36**, 139–153.

Cherry, R. J. (1979). *Biochim. Biophys. Acta* **559**, 289–320.

Chow, I. (1980). Doctoral Thesis, McGill University, Montreal.

Chow, I., and Cohen, M. W. (1978). *Soc. Neurosci. Abstr.* **8**, 368.

Christian, C. N., Daniels, M. P., Sugiyama, H., Vogel, Z., Jacques, L., and Nelson, P. G. (1978). *Proc. Natl. Acad. Sci. U.S.A.* **75**, 4011–4015.

Cohen, M. W. (1972). *Brain Res.* **41**, 457–463.

Cohen, M. W., and Weldon, P. R. (1980). *J. Cell Biol.* **86**, 388–401.

Cohen, M. W., Anderson, M. J., Zorychta, E., and Weldon, P. R. (1979). *Prog. Brain Res.* **49**, 335–349.

Cohen, S. A., and Fischbach, G. D. (1977). *Dev. Biol.* **59**, 24–38.

Dennis, M. J. (1981). *Annu. Rev. Neurosci.* **4**, 43–68.

Edelstein, S. J., Beyer, W. B., Eldefrawi, A. J., and Eldefrawi, M. E. (1975). *J. Biol. Chem.* **250**, 6101–6106.

Edidin, M., and Fambrough, D. (1973). *J. Cell Biol.* **57**, 27–37.

Edwards, C., and Frisch, H. L. (1976). *J. Neurobiol.* **7**, 377–381.

Fambrough, D. M. (1979). *Physiol. Rev.* **59**, 165.

Fambrough, D. M., and Pagano, R. (1977). *Year Book—Carnegie Inst. Washington* **76**, 28–29.

Frank, E., and Fischbach, G. D. (1979). *J. Cell Biol.* **83**, 143–158.

Fraser, S. E., and Poo, M-m. (1982). In preparation.

Frye, L. D., and Edidin, M. (1970). *J. Cell Sci.* **7**, 319–335.

Hall, J. E. (1981). *Biophys. J.* **33**, 373–381.

Jacob, M., and Lentz, T. L. (1979). *J. Cell Biol.* **82**, 195–211.

Jaffe, L. F. (1977). *Nature (London)* **265**, 600–602.

Jaffe, L. F. (1979). *In* "Membrane Transduction Mechanisms" (R. A. Cone and J. E. Dowling, eds.), pp. 199–231. Raven, New York.

Jaffe, L. F., and Nuccitelli, R. (1977). *Annu. Rev. Biophys. Bioeng.* **6**, 445–76.

Jessel, T. M., Siegel, R. E., and Fischbach, G. D. (1979). *Proc. Natl. Acad. Sci. U.S.A.* **76**, 539.

Jones, R., and Vrbova, G. (1974). *J. Physiol. (London)* **236**, 517.

Kidokoro, Y., Anderson, M. J., and Gruener, R. (1980). *Dev. Biol.* **78**, 464–483.

Kuffler, D., Thompson, W., and Jansen, J. K. S. (1977). *Brain Res.* **138**, 353–358.

Land, B. R., Podleski, T. R., Salpeter, E. E., and Salpeter, M. M. (1977). *J. Physiol. (London)* **269**, 155–176.

Lavoie, P.-A., Collier, B., and Tenenhouse, A. (1976). *Nature (London)* **260**, 349–350.

Liebman, P., and Entine, G. (1974). *Science* **185**, 457.

Linden, D. C., and Fambrough, D. M. (1979). *Neuroscience* **4**, 527–538.

Lømo, T., and Jansen, J. K. S. (1980). *Curr. Top. Dev. Biol.* **16**, 253–281.

Lømo, T., and Slater, C. R. (1978). *J. Physiol. (London)* **275**, 391–402.

O'Brien, R. A. D., Ostberg, A. J. C., and Vrbova, G. (1978). *J. Physiol. (London)* **282**, 571–582.

Orida, N., and Poo, M-m. (1978). *Nature (London)* **275**, 31–35.

Peng, H. B., Cheng, P. C., and Luther, P. W. (1981). *Nature (London)* **292**, 831–834.

Pestronk, A., Drachman, D. B., and Griffin, J. W. (1976a). *Nature (London)* **260**, 352–353.

Pestronk, A., Drachman, D. B., and Griffin, J. W. (1976b). *Nature (London)* **264**, 787–789.

Podleski, T. R., Axelrod, D., Raudin, P., Greenberg, I., Johnson, M. M., and Salpeter, M. M. (1978). *Proc. Natl. Acad. Sci. U.S.A.* **75**, 2035.

Poo, M-m. (1981). *Annu. Rev. Biophys. Bioeng.* **10**, 245–276.

Poo, M-m. (1982). *Nature (London)* **295**, 332–334.

Poo, M-m., and Cone, R. A. (1974). *Nature (London)* **247**, 438–441.

Poo, M-m., and Fraser, S. (1982). In preparation.

Poo, M-m., Lam, J. W., Orida, N., and Chao, A. W. (1979). *Biophys. J.* **26**, 1–22.

Purves, D., and Sakman, B. (1974). *J. Physiol. (London)* **237**, 157–182.

Reiness, C. G., and Hall, Z. W. (1977). *Nature (London)* **268**, 655–657.

Reiness, C. G., and Hall, Z. W. (1981). *Dev. Biol.* **81**, 324–331.

Reiness, C. G., Hogan, P. G., Marshall, J. M., Hall, Z. W., Griffin, G. E., and Goldberg, A. L. (1977). *In* "Cellular Neurobiology" (Z. W. Hall, R. B. Kelly, and C. F. Fox, eds.), pp. 207–215. Alan R. Liss, New York.

Sanes, J. R., and Hall, Z. W. (1979). *J. Cell Biol.* **83**, 357–370.

Shainberg, A., Cohen, S. A., and Nelson, P. G. (1976). *Pfluegers Arch.* **361**, 255–261.

Sheetz, M. P., and Koppel, D. E. (1979). *Proc. Natl. Acad. Sci. U.S.A.* **76**, 3314–3317.

Srihari, T., and Vrbova, G. (1978). *J. Neurocytol.* **7**, 529–540.

Stent, G. S. (1973). *Proc. Natl. Acad. Sci. U.S.A.* **70**, 997–1001.

Thesleff, S. (1960). *J. Physiol. (London)* **151**, 598–607.

Thompson, W., Kuffler, D. P., and Jansen, J. K. S. (1979). *Neuroscience* **4**, 271–282.

Tonge, D. A. (1974). *J. Physiol. (London)* **241**, 127–139.

Warnick, J. E., Albuquerque, E. X., and Guth, L. (1977). *Exp. Neurol.* **57**, 622–636.

Weinberg, C. B., Reiness, C. G., and Hall, Z. W. (1981). *J. Cell Biol.* **88**, 215–218.

Wolf, D. E., Edidin, M., and Dragsten, P. R. (1980). *Proc. Natl. Acad. Sci. U.S.A.* **77**, 2043–2045.

Young, S. H., and Poo, M.-m. (1982). Submitted for publication.

CHAPTER 4

ORDERING OF RETINOTECTAL CONNECTIONS: A MULTIVARIATE OPERATIONAL ANALYSIS

Ronald L. Meyer

DEVELOPMENTAL BIOLOGY CENTER
UNIVERSITY OF CALIFORNIA
IRVINE, CALIFORNIA

I. An Appraisal of the Field

It is perhaps the morphological simplicity of the retinotectal system that has for so long lured us into a prejudice toward simple explanations of its development. For the purposes of many studies dealing with the formation of retinotectal connections, the system can be considered to comprise only three parts: a sheet of retinal ganglion cells, a sheet of tectal cells, and a cable of ganglion cell fibers that connects to tectum in retinotopic fashion. In the course of several decades of experimentation, a number of competing models have emerged to explain the development of the retinotopography of this projection. Each model has been characteristically simple in conception but yet considered comprehensive enough to account for nearly every observation. Some of the ensuing argument has persisted to the present time. In this chapter it will first be briefly argued that the traditional view of the problem must be considered to be erroneous, that is, that no single simple explanation can account for all the observations. Instead, the task must be to integrate a number of interactive processes into a comprehensive under-

101

standing. Such an integrative approach will be set forward in the rest of the chapter. It represents an attempt to analyze the system in a more empirical and open-ended fashion than that generally associated with past models. In addition, this approach offers an alternative to the current wave of numerical modeling and so skirts thorny theoretical issues that are associated with simulations of poorly described complex systems.

Despite some diversity, most hypotheses about how optic fibers form a retinotopic projection onto tectum have relied on a single dominant process to generate order. These processes largely fall into one of three categories. In the oldest category are those hypotheses that assume that individual retinal fibers or tectal cells are intrinsically identical with each other, that is, fibers and cells possess no cytochemical or other marker denoting position. Growing fibers are presumed to be "ignorant" about retinal position and to be obliged to rely on external guidance and circumstance to attain their proper positions on tectum. Guidance is assumed to come from glial structures or from the shafts of preexisting "pioneer" fibers along which fibers grow. A sequential and patterned differentiation of retinal ganglion cells with subsequent outgrowth of fibers coupled to a corresponding patterned development of tectum is considered adequate to generate the retinotectal projection. Prior to the 1940s, such time–position–mechanical guidance explanations predominated but were generally felt to be incompatible with some later experiments, particularly those using regenerating nerves. It is currently being argued, however, that the data against these hypotheses have been misinterpreted and that even during regeneration considerable evidence for time, position, and mechanical guidance can be found (Horder and Martin, 1978). In other systems such as the optic system of *Daphnia*, these explanations are thought to be strongly supported (Lopresti *et al.*, 1973).

In the early 1940s, Sperry found that when the optic nerve of an adult lower vertebrate was cut and the fibers scrambled, fibers would regenerate, apparently simultaneously, to reform the original retinotopic projection (Sperry, 1944, 1945). He inferred that optic fibers and tectal cells were cytochemically differentiated according to their position and that fibers were directed by a contact chemotaxis to their matching target cells in tectum (Sperry, 1965). Various modifications of this general idea have since appeared in order to accommodate recent "plasticity" experiments. In the latter, fibers were found to grow to abnormal tectal positions following removal of part of retina or tectum. These modified chemoaffinity models postulate that the cytochemical markers are subject to experimental alteration (Yoon, 1972a; Meyer

and Sperry, 1973, 1974) or that the matching process is context dependent (Hunt and Jacobson, 1974a; Prestige and Willshaw, 1975) such as
is expected from a thermodynamic steady state in the adhesive interactions between the available fibers and tectal cells (Meyer, 1974, 1975a).
In common with all of these hypotheses is the general notion that the
elements are "smart," that both fibers and tectal cells possess positional information, and that these positional markers are read out by
an interaction between optic fibers and tectum. Although fiber–tectum
affinity alone has difficulty in accounting for all plasticity results, the
evidence for some form of chemoaffinity interaction has become increasingly strong, and this notion is widely accepted.

The third category of explanation arose as an answer to the plasticity results. In particular, these hypotheses address the capacity of optic
fibers to preserve retinotopography when a whole retina "compresses"
onto a surgically formed half tectum (Gaze and Sharma, 1970) or when
a half retina (or compound retina composed of two mirror-image
halves) expands across a whole tectum (Yoon, 1972b). Led by Gaze,
several workers have questioned whether tectal cells possess any positional information and have instead suggested that only retinal fibers
know their position (Gaze and Keating, 1972; Chung and Cook, 1975;
Hope et al., 1976). It was postulated that fibers form retinotopography
by interacting with each other, not with tectum. The instructional role
of the tectum, if any, was thought to be limited to providing a minimal
polarity cue for the map. Except for one activity model (Willshaw and
von der Malsburg, 1976), Sperry's idea of a specific cytochemical tag for
individual retinal fibers was implicitly retained. The major shortcoming of this approach is in explaining "specificity," wherein fibers target
in on their original tectal sites in the absence of, or even in violation of,
interfiber retinotopography. The proponents have responded by dis-
carding some data for reasons of possible species differences or by ques-
tioning the generality of regeneration experiments, which have provided the bulk of the specificity data. For regeneration, they argue that
tectal position markers do not exist when fibers normally grow into
tectum but are subsequently created by the optic innervation (Chung
and Cook, 1975; Schmidt, 1978). These ideas have generated a great
deal of discussion.

The preceding trichotomy is perhaps drawn with a heavy hand.
Naturally, not every researcher and idea fits tidily into one of the three
categories. Some researchers have offered explanations that combine
ideas from two or more of these categories (Levine and Jacobson, 1974;
Meyer, 1974, 1975a; Fraser and Hunt, 1980a), and certainly most are
not dogmatic in their views. This is particularly true of workers in

mammalian development (see Schneider and Jhaveri, 1974; Lund, 1978). Nevertheless, there has been a strong tendency to view the search for an explanation as a decision between incompatible alternative models. Symptomatic of this are the many arguments and counterarguments in the literature. Evidence accumulated over the last several years leads to two related conclusions about each of the preceding kinds of processes. One is that each almost certainly makes a contribution to the patterning of the projection. The other is that any one of the preceding processes cannot, by itself, account for substantial segments of the data. Because some of the specific experiments will be described in the context of the main discussion, this evidence will only be outlined here.

In support of time–position–mechanical guidance are electron microscopic (EM) observations that developing and regenerating fibers follow glial channels to and through tectum (Turner and Singer, 1974; Murray, 1976), even though some fibers are in the wrong part of tectum and are surrounded by foreign optic fibers (Meyer, 1980). Also, an experimental manipulation of the position and timing of a set of regenerating fibers has been found to produce an anomalous projection (Meyer, 1979a). Numerous electrophysiological recordings from animals in which position and timing had been experimentally disturbed have invariably shown degraded retinotopography, and this has now been demonstrated with anatomical methods (Meyer, 1980). On the other side of the coin, however, are the findings that many regenerating fibers take highly anomalous routes, yet terminate at their appropriate tectal position (Udin, 1978; Meyer, 1980). In normal *Rana* and cat, fibers that terminate next to each other in tectum or dorsal lateral geniculate nucleus are widely dispersed within the optic nerve (Maturana, 1960; Horton *et al.*, 1979).

In strong confirmation of selective fiber–tectum interaction or affinity are a number of experiments in lower vertebrates showing that the "correct" fibers can track down rotated, translocated, or inverted pieces of tectum (Yoon, 1975; Levine and Jacobson, 1974). These experiments have been extended to "virgin" tecta lacking any previous optic innervation (Straznicky, 1978). In addition, a number of other examples of tectal specificity that were previously limited to regenerating systems have been extended to normal development (Crossland *et al.*, 1974; Frost and Schneider, 1979). On the other hand, there have been a number of cases in which fibers have ignored the original position markers of a translocated piece of tectum (Jacobson and Levine, 1975; Sharma, 1975; Rho, 1978). In an intact tectum, fibers can, under certain conditions, ignore markers (Meyer, 1978a) or map with a reversed polarity (Meyer, 1979a).

The situation is similar for fiber–fiber interactions. In some of the preceding tectal translocation experiments, fibers that innervate the transplated tissue showed a strong tendency to align themselves with fibers in the surrounding tectum irregardless of the orientation of the transplant (Jacobson and Levine, 1975). The available evidence suggests that the intrinsic polarity of these transplants is not lost (Rho, 1978). In other experiments, a set of foreign optic fibers were transplanted into a denervated area of an otherwise innervated tectum. These foreign fibers aligned themselves with the host fibers instead of with tectal polarity cues (Meyer, 1979a). However, in most tectal transplantation experiments, including cases in which the transplant heals with surrounding tectum without obvious scarring, the projection onto the transplant is appropriate for the transplant and discordant with the surrounding projection (Yoon, 1975; Levine and Jacobson, 1975).

Although there may be room for discussion on the magnitude of the contribution made by each of these processes during normal development, there now seems little doubt that each can make a contribution. Thus, the end product, the properly aligned retinotopic projection, represents an integration of at least three kinds of processes. The future may bring more complexity from the discovery of new kinds of processes or the further analyses of those already known. At present, there are enough interacting factors to raise some serious questions about the past approach to the problem. It has been usual to formulate a broad explanatory model that then prompts experiments designed to test it. To be most useful, such tests should be formulated on the basis of strong unambiguous predictions from the model. Even with simpler past models, making such predictions has been a problem. For example, Sperry's original chemoaffinity model does not provide an unambiguous prediction for size disparity experiments (e.g., half retina innervating an intact tectum or an intact retina innervating a half tectum) because the parameters of the fiber–tectum affinity are left unspecified. Depending on the selectivity and strength of the affinity, one might obtain specificity, wherein fibers terminate at normal tectal sites, or plasticity, wherein fibers terminate at new tectal sites. If specificity is experimentally obtained, chemoaffinity is supported; however, if plasticity is obtained, the result cannot be considered a counterexample. The latter conclusion requires some arbitrary assumptions about the affinities (see Hunt and Jacobson, 1974a; Meyer and Sperry, 1976).

Failure to appreciate this problem has been a major source of confusion and has generated a great deal of semantic argument.

The prediction problem is dramatically compounded with the more complex models now required by the data. With the three aforementioned processes, and perhaps with an additional factor like competition or interfiber interactions, it seems intuitively obvious that nearly every experimental result can, in principle, be accounted for. By the same token, it is difficult to make a prediction that will clearly prove or disprove such a complex model. One reason is that there are a large number of parameters that are not empirically specified. It is a widely accepted truism of engineering and information science that with enough freely chosen parameters a model can mimic practically any behavior. It follows that it is possible to precisely formulate a complex model with arbitrary parameters, which in numerical (computer) simulations will give many of the results in the literature. This only tells us that the model could be right. It would be no surprise if a different model also described experimental observation, and it could not be concluded that a model is wrong if it didn't describe all the data. The latter might be largely correct, having only an incorrect or missing variable. One can already point to several computer models that explain a large segment of the data and yet are dramatically different in their assumptions about how the system works (Prestige and Willshaw, 1975; Hope *et al.*, 1976; Willshaw and von der Malsburg, 1976; Fraser and Hunt, 1980). Of course, models are not equally meritorious and have definite uses (see Fraser and Hunt, 1980a). However, there are fundamental problems associated with this general approach.

There is also a related problem that may be exacerbated with complexity. Past models, by their very nature, have overstepped the available data and so, strictly speaking, must be considered unproved, though a great deal of evidence might support them. The lack of proof has led to a general feeling of ignorance about how the system works. It is distressingly common to read or hear that "we know nothing about how fibers form ordered connections." In fact, we know many significant facts about the system, but these have usually been presented and interpreted in the context of a particular model. When the literature is given a summary evaluation, these facts are often thrown out with the unproved theory.

Two things would be useful. One is a way to describe and to discuss data without, as far as possible, resorting to theoretical constructs. There is a need to talk in more empirical, more certain terms about the way in which the system works. In particular, there should be a way to talk about certain kinds of processes for which there is good evidence but for which the underlying mechanisms remain unknown. The other and related need is to deal in concrete terms with the interactions between these processes. If the system is complex, with many interactive components, a major advance in our understanding is likely to require characterization of the information flow between these components.

II. Operational Analysis

A. THE APPROACH

Analyzing complex systems is an old problem in engineering and has been successfully approached with various systems analysis techniques. Systems analysis generally requires that the components be identified (boxes drawn), the input and output lines determined (arrows between boxes), and the inputs and outputs quantified. From this, transfer functions characterizing each component can be derived and subsequently the system's behavior described. In the retinotectal system, it is difficult to meet precisely any of these requirements, making a literal application of system analysis premature. At the present time, for instance, we do not know to what extent the aforementioned processes operate sequentially or in parallel. Nevertheless, a first step can be taken in this direction with a logical, qualitative analog of this approach.

What will be the equivalent of boxes will be any clear, identifiable process or behavior of the system that plays a role in the patterning of axonal connections. These boxes will be defined operationally. In other words, they will be strictly inferred from the available data and will be stated in empirical terms. To borrow a term from signal detection theory, these features will be called the operating characteristics (Op Chs) of the system. In essence, each operating characteristic will represent a kind of transfer function that relates a particular set of initial conditions, normal or experimental, to the subsequent growth pattern of optic fibers. It should be emphasized that operating characteristics are not mechanisms in the sense of representing the underlying cellular biology or biochemistry of the phenomena, but are on the same level of analysis as the observations from which they are inferred. In a sense, operating characteristics are empirically derived rules but are ones that include the system's dynamic behavior in addition to predicting its end state. They are more properly thought of as processes. Their theoretical content is more than that of experimental facts because they are inferences, but less than that of hypotheses because they are tied to available data. Although this language is perhaps more formal than customary in other fields, it is felt to be necessary in order to avoid confusion with previous theoretical proposals, particularly systems matching (Gaze and Keating, 1972).

It is no easier to give hard and fast rules for deriving the operating characteristics than it is for drawing any other inferences, but some general guidelines can be stated. Grouping data according to experimental design is not especially beneficial because in most types of experiments several different processes operate. What will be done is to assemble all the data, regardless of experimental manipulation, that lead to one common inference, the operating characteristic. For clarity, the Op Ch (operating characteristic) will first be outlined and then the supporting data summarized. To be comprehensive, as many Op Chs as can be justified will be described. In other words, two Op Chs will be distinguished even though one composite Op Ch might be imagined that could cover all the data under consideration. As long as it is reasonable to suppose that different mechanisms may be operating in the two cases, it seems preferable to formulate separate operating characteristics. In this way, the possibility that two very different processes may be involved remains obvious, and thus the identity of mechanism is not prematurely decided. There is nothing irrevocable in this decision, so that with further evidence consolidation of the two may be warranted. Greater complexity will result, but this can be kept in hand by limiting the scope of the problem. Only the patterning of fiber connections and, in particular, the formation of retinotopography will be discussed. Early developmental stages involving the differentiation of neurons, acquisition of positional information, and initial outgrowth (see Hunt, 1975) will not be addressed.

B. OPERATING CHARACTERISTIC 1: FIBER–TECTUM PREFERENCE

This operating characteristic can be described as a strong tendency for optic fibers to terminate in tectum or in particular layers of tectum in preference to terminating in other nonoptic regions. In other words, fibers express a choice and favor tectum over nonvisual nuclei (Fig. 1). Although relevant experiments are fewer than those that test for ordering within tectum, the evidence is nevertheless persuasive. From simple anatomy it can be concluded that optic fibers have the capacity to grow to many different parts of the brain. Even in fish, over a dozen different optic target regions can be distinguished (Reperant and Lemire, 1976), the innervation of which requires fibers to grow past thousands of other neurons. Clearly, fibers are not mechanically prevented from growing to nonoptic regions. This is underscored by thymidine labeling and cell count studies in fish (Johns and Easter, 1978; Meyer,

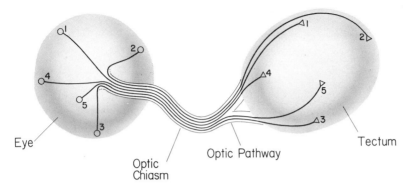

FIG. 1. Diagrammatic illustration of the retinotectal projection. The clockwise numbers 1–5 in retina designate five different ganglion cells, and in tectum these numbers label their corresponding terminals. For simplicity, the actual anterior–posterior axis has been reversed so that the two sets of numbers can be superimposed. Note the orderly arrangement of fibers in their paths to tectum and the retinotopography of the projection. The point here is that there is a significant distance between tectum and chiasm, which fibers grow through in order to terminate preferentially in tectum.

1978b) and amphibians (Straznicky and Gaze, 1971), which show that new optic fibers grow into the central nervous system for up to a year or more. Experimentally delaying ingrowth of fibers in frogs until after metamorphosis (Straznicky, 1978) or transplanting an eye into an adult, genetically blind cave salamander (Hibbard and Ornberg, 1976) produces an apparently normal innervation. Clearly, optic fibers do not terminate on optic nuclei simply because they are the only ones there at the time fibers grow in.

Experimental manipulations fall into two groups: those in which the amount of tectum is reduced and those in which the path of fibers is directly altered. Removing the tectum (colliculus) in newborn hamsters prior to significant optic innervation produces a dramatic increase in the optic innervation of other visual nuclei and causes fibers to innervate regions that they normally never innervate (Schneider and Jhaveri, 1974). This abnormal innervation does not occur when tectum is present, even if these sites are denervated. This again indicates choice and further suggests that the preference is hierarchical rather than all or none. Similar results have been obtained in regeneration experiments. In goldfish, ablation of one tectum leads to a mass invasion of the underlying tegmentum. Eventually many fibers find their way to the tectum on the opposite side (Easter et al., 1978). Removal of the posterior half of tectum produces a transitory invasion of tegmentum by the damaged fibers, which is later completely retracted in favor of a hyperinnervation of the remaining tectum (Meyer, 1976). Analogously, if a piece of cerebellar tissue is transplanted into a large hole made in the middle of tectum, it will not be innervated by the surrounding optic fibers even though a piece of tectal tissue is readily invaded (Yoon, 1979).

In experiments on frog embryos, fibers have been forced to grow into the nervous system at unusual sites by deplanting the eye rudiment posteriorly or dorsally (Sharma, 1972a; Giorgi and Van de Loos, 1978). In spite of their anomalous and often highly circuitous routes, these fibers can still innervate tectum by growing past many regions normally never encountered. In experiments in which the normal eye is left intact and a third eye implanted to a posterior position, a similar finding is obtained, though the tectum is now hyperinnervated (Constantine-Paton and Law, 1978). Thus, this result is not explicable in terms of denervation of tectum. Other misrouting has been done within tectum using goldfish. In normals, optic fibers are principally confined to one major lamina near the tectal surface. It is possible surgically to direct fibers into deep tectum. Invariably, regenerating fibers return to their appropriate layers (Meyer and Sperry, 1976). Similarly, when a piece of tectum is turned upside down, fibers will grow down to its optic layer (Yoon, 1975).

The overall conclusion, defined here as the first operating characteristic (Op Ch 1), is that fibers show an active preference for tectum and other optic nuclei over nonoptic sites. In other words, fibers can and normally do have the opportunity and physical capacity to invade nonvisual nuclei but will "choose" tectum. The operational nature of this statement means that its generality is tied to the data. In particular, Op Ch 1 does *not* say that optic fibers will prefer tectum over all other neurons or tissue. It does *not* say that a preference will be expressed under every possible experimental condition. It does *not* say that fibers will always find tectum no matter where they are placed. What it does say is that fibers can express a preference for tectum over those nonoptic sites that the fibers encounter in normal development (plus a few others tested, like cerebellum), provided tectum and such sites are at comparable developmental stages and conditions (such as degree of innervation). Eye deplantation experiments in which optic fibers enter the spinal cord and fail to find tectum (Constantine-Paton, 1978), although important for other reasons, do not conflict with Op Ch 1. There is no good reason to believe that the necessary guidance mechanism for fibers to reach tectum from spinal cord should have evolved.

A logical corollary of Op Ch 1 is that tectum and nearby nonoptic nuclei must differ in some property or properties that can in some way affect the growth and termination of optic fibers. It also seems safe to infer that optic fibers differ from other fibers in their reactivity to this property. Otherwise, it is hard to explain how other fibers that are in close proximity to optic fibers and to tectum do not innervate tectum. However, direct experimental evidence on this last point is scarce. What is not to be inferred from Op Ch 1 is the mechanism of the preferential growth. Though the general idea of active preferential growth is to be credited to Sperry (1944, 1945, 1965), Op Ch 1 is not chemoaffinity. There is no assumption here that the distinguishing property is cytochemical, that it is a permanent product of embryonic field differentiation, or that the preference is mediated by contact chemotaxis. Other possibilities are left open. The distinguishing property could, for the sake of argument, be electrical, metabolic activity, or a derivative from neighboring cells. The preference could be mediated by chemotaxis to a diffusable substance or by wild growth followed by selective survival. This is not to say that all these mechanisms are equally likely. Guided growth through chemoaffinity is by far the best supported (Meyer and Sperry, 1976). In development, fibers clearly do not grow willy-nilly through all parts of the CNS. Although there may be limited exploratory growth, fibers are by and large restricted to specific paths and targets (Rakic, 1977). This is also true of regeneration (Meyer, 1980). Additionally, there is also good evidence against electrical guidance (Harris, 1980). However, no proof positive for any one mechanism exists and so it must be excluded from the Op Ch. There is also another reason for not trying prematurely to decide on the mechanism. At the present state of our knowledge, knowing which mechanism would probably not greatly add to the power of Op Ch 1 to describe the patterning of optic nerve connectivity.

C. Operating Characteristic 2: Fiber–Tectum Locating Function

This characteristic amounts to a preference by optic fibers from different retinal regions for different tectal regions. (Regions refers to the anteroposterior and mediolateral dimension, not depth.) In essence, Op Ch 2 is the transfer function that describes the following input–output relationship: a fiber can initially be located in different tectal regions or tectal cells can be repositioned. Eventually retinal fibers "tend" to end up in the vicinity of their "normal" or original tectal locations.

There is a large body of data supporting this inference. In lower vertebrates, optic fibers can be cut and scrambled and yet will reform a nearly normal projection (Sperry, 1944, 1945; Gaze, 1959; Udin, 1978; Meyer, 1980). Analysis of the paths of these fibers has further shown that many fibers regenerate through highly anomalous routes through tectum and yet arrive at their retinotopically appropriate position (Udin, 1978; Meyer, 1980). In normal goldfish, few fibers have been detected that are similarly anomalous in their paths (Meyer, 1980). Temporal studies of regeneration show that fibers lack refined retinotopic order during the early phase of regeneration into tectum and gradually become retinotopically organized (Gaze and Keating, 1970; Udin, 1978; Meyer, 1980). In the experiments mentioned earlier (see Section II, B) in which fibers were surgically forced to approach tectum from a variety of unusual routes, the projection was found to be appropriately organized. The latter observation has been more recently extended to mammalian systems (Finlay et al., 1979b). One complication of these kinds of experiments is that they do not adequately distinguish whether fibers grow to particular tectal regions or selectively aggregate with one another. However, a minimum conclusion that can be drawn is that the alignment or polarity of the projection is a fiber–tectum interaction.

Less ambiguous experiments have come from more invasive methods. When part of the retina is removed, the remaining fibers have been shown to bypass inappropriate sites in favor of the more appropriate tectal positions. This has been done for regeneration (Attardi and Sperry, 1963; Meyer, 1975b) and for development (Crossland et al., 1974; Frost and Schneider, 1979). In some regeneration experiments discussed later, there is substantial plasticity. However, if the eye lesion is sufficiently large, some degree of preferential innervation is invariably evident (Meyer and Sperry, 1976; R. L. Meyer, unpublished). The most conclusive evidence is from regeneration experiments in which pieces of tectum are rotated or translocated. In the large majority of cases, fibers successfully track down their original tectal positions on the rearranged fragment (Yoon, 1975; Levine and Jacobson, 1974, 1975). The projection onto the fragment is consequently discordant with that onto the surrounding tissue, even in cases in which the fragment heals into tectum with little evidence of scarring (Fig. 2). A similar result has been obtained in "naive" tecta, that is, those deprived of any prior optic innervation (Straznicky, 1978) or those previously innervated by a disordered projection (Fraser and Hunt, 1980a).

One possible shortcoming of some of the preceding evidence is that the contribution of pathway order has not been rigorously excluded. It has been suggested that the selective innervation following retinal lesions is due entirely to order within the optic pathway and that locus-specific innervation of tectal grafts is due to systematic distortion

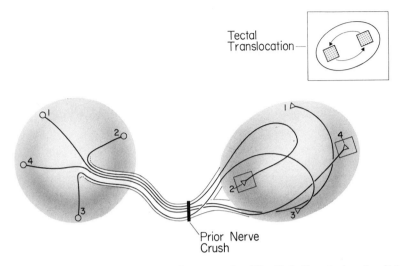

FIG. 2. Diagrammatic illustration of an example of Op Ch 2, fiber–tectum localizing. Here two pieces of tectal tissue were reciprocally translocated as indicated in the inset, and in addition, the paths of fibers were disrupted by a nerve crush. As represented, the fibers grew to their original sites on tectum. Compare with Fig. 1.

of the growth path (Horder and Martin, 1978). While this seems un-likely, it is nevertheless important to clearly rule it out. A recent ap-proach to this has been to use the fiber deflection technique in adult goldfish. This permits a relatively small group of optic fibers to be teased off one tectum and inserted into a specific position of the opposite host tectum. This host tectum can be denervated by enucleation of the host eye. When fibers which normally innervated posterior medial tec-tum were inserted into the anterior medial end of denervated host tectum, these fibers grew into the posterior medial quadrant of host tectum (Meyer, 1978a). This growth was along the normal anterior–posterior route that optic fibers follow through tectum so the pathway argument can still be made. A critical test was to deflect fibers from posterior lateral tectum into anterior medial tectum (R. L. Meyer, un-published). In this case there was no normal route from the insertion site to the appropriate tectal quadrant. In spite of this, a clear preferen-tial innervation of posterior lateral tectum was always found with au-toradiographic and electrophysiological methods. About half of the de-flected fibers traveled a completely novel route (Fig. 3A). This was essentially a short cut through the deep layers of tectum and up into superficial tectum near the lateral edge. There they selectively entered the optic layers and grew posteriorly. The other half of the fibers stayed

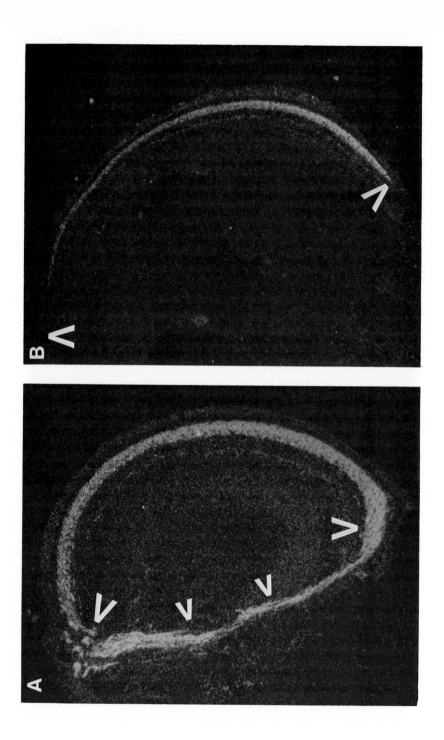

within the optic layers at the insertion point and grew in a posterior lateral direction. The highest density innervation seen autoradiographically was in the posterior lateral quadrant. This selectivity was maintained up to 11 months after the deflection and was observed even when eye enucleation was done as many as 18 months prior to deflection (Fig. 3B). As in previous studies some spillover of grains into adjacent inappropriate regions was seen. This was relatively light and the least labeled region was the posterior medial tectum, which on the basis of fiber path should have been densely labeled.

Another important conclusion from this deflection study is that selective innervation is long-lived and that the tectal property which mediates selectivity is stable in the absence of optic innervation. This contradicts the claim that in goldfish a surgically formed temporal half of retina forms an expanded projection after several months (Schmidt et al., 1978) and that the tectum loses locus specificity after long denervation (Schmidt, 1978). These latter studies have been virtually the only experimental evidence for the notion that during development tectal cells acquire their position-dependent labels (locus specificity) from ordered retinal fibers rather than by independent differentiation. On these terms, selective regeneration becomes a curious artifact. When the fiber deflection findings are considered in conjunction with the studies showing selective innervation during development or during regeneration into tecta deprived of optic innervation in development (Straznicky, 1978; Fraser and Hunt, 1980a), this notion would ap-

---

◀   FIG. 3. Dark field autoradiograms of goldfish tectum. Contralateral eye had been removed and optic fibers from the posterior lateral quadrant of the opposite "donor" tectum were subsequently redirected into the anterior medial region of the denervated "host" tectum. The deflected fibers were subsequently labeled with tritiated proline. Sections are frontal in orientation with dorsal to the top and lateral to the right. (A) Section through the insertion site near the anterior end of host tectum. Upper large arrowhead indicates the insertion site near the medial edge. Lower large arrowhead is positioned at the lateral edge of tectum. The thick band of grains indicated by the small arrowheads are defected fibers that had taken a short-cut from the insertion site to lateral tectum. The arched band of grains above and to the right is the normal optic lamina of tectum. Some fibers also grew through this layer to lateral tectum as evidenced by the grains there, so that much of this medial label was from fasciculated fibers of passage. The overall density of label was heavy to demonstrate fibers of passage. Eye enucleation was 56 days prior to deflection and autoradiography was 51 days after deflection. (B) Section at about the posterior third of tectum. The arrowhead at the top points to the medial edge and the arrowhead at the bottom points to the lateral edge. Note the much greater density of label in the lateral half of tectum and the gradientwise decrement toward the medial half. Enucleation was 533 days prior to deflection and autoradiography was 90 days after deflection. Also note label density was much less than in (A) to permit a density comparison.

pear to be without support, contrary to available data, and requiring implausible mechanisms. The original experiments by Schmidt *et al.* (1978) and Schmidt (1978) remain intriguing, but proper interpretation will require clarification of a number of technical difficulties with these kinds of experiments. The data showing expansion of a half retina were entirely electrophysiological. Three autoradiographic studies (Meyer, 1975; Schmidt *et al.*, 1978; Strumer, 1980) have failed to confirm substantial expansion of a temporal half retina. This suggests the recordings were postsynaptic or did not accurately reflect fiber density. Other discrepancies between anatomy and physiology have been previously reported in attempts at mapping spatially restricted innervation in this system (Levine and Jacobson, 1975). The danger is that when probing a tectal region having little or no optic innervation, the chance of recording from nonretinal visual units or retinal fibers of passage is maximized. Anatomical confirmation is essential. Another difficulty is the likelihood of substantial retinal and tectal cell addition during the course of the study. This occurs normally in juvenile goldfish (Johns, 1978; Meyer, 1978b), but its magnitude and pattern following retinal and tectal surgery is unknown. If a result is found *only* after a long postoperative interval such as the expansion of a half retina after 6 months or more (Schmidt *et al.*, 1978), then one must ask whether cell addition, normal or regenerative, played a role. Another potential complication is the effect of certain surgical procedures on the time and path of fiber ingrowth. In the Schmidt (1978) study one tectum was removed to force fibers into ipsilateral tectum. It has since become apparent that this ipsilaterally directed growth is prolonged and follows a number of different routes (Easter *et al.*, 1978). This growth may be modulated by the kind of innervation existing in ipsilateral tectum. Thus the resulting ipsilateral projection may reflect differential crossing instead of tectal changes. Finally, interpretation must take careful stock of all putative processes, not just the one being tested. In the study in question, both eyes were often made to coinnervate one tectum. Interfiber interactions (see Section II,D) would be extensive and would tend to make fibers from each eye terminate in register. This could explain the absence of expansion from a half retina in one of the two eyes as easily as tectal relabeling. Other processes are the loss of synaptic sites with denervation (Murray, 1976) or possible collateral sprouting from nonretinal fibers. These could produce expansion via decreased density of terminals without altering locus specificity properties themselves. These comments are meant as a cautionary note for this general class of experiments and not just as a rebuttal to the interesting and careful experiments by Schmidt *et al.* (1978) and Schmidt (1978).

This and other evidence (see Meyer and Sperry, 1976) positively identifies a fiber–tectum locating function as opposed to a tectal polarity cue uniformly distributed across tectum (Hope *et al.*, 1976). However, at present it is difficult to characterize this function beyond the preceding general statement. In particular, it cannot be said that a single optic fiber prefers its normal tectal locus. There are two reasons for this reservation. One is that the data do not support such a strong conclusion. Precise tectal targeting has only been demonstrated in experiments with large numbers of fibers and so may be complicated by interfiber effects such as selective interfiber aggregation. Such complexity is implicated in experiments in which the posterior half of the tectum is removed and a piece of anterior tectum rotated. In this case some "abnormal" fibers innervate the fragment, yet do so with a polarity appropriate for that fragment (Yoon, 1977).

The other reason is that optic fibers in frogs and goldfish have no absolute, normal tectal loci. In tadpoles, new cells are continually added to the margin of retina and tectum for most of the larval period (Straznicky and Gaze, 1971, 1972; Gaze *et al.*, 1979); and in goldfish this growth extends to juveniles and possibly adults (Johns, 1978; Meyer, 1978b). However, the pattern of cell addition in retina is annular, whereas that in tectum is horseshoe-shaped, thereby necessitating a shift in the tectal position of optic fibers during this growth (Gaze *et al.*, 1974, 1979; Meyer, 1978b; but, for a complication in frog, see Jacobson, 1976). It is not known whether this shift is produced by a competition by fibers for available tectal space or is a regulated or programmed change in the intrinsic nature of the locating function (for a possible biochemical correlate of the latter, see Gottlieb *et al.*, 1974).

This developmental shift in the projection suggests that our traditional notion of fiber–tectum localization may require modification. In the past, fibers have been viewed as having the strongest preference for their "normal" tectal loci and a preference for neighboring regions, which falls off as a function of distance from the "normal" loci. Thus, a traditional conception of the localizing function would resemble a bull's eye, the rings of which represent isostrength of preference. The developmental data instead suggest a distorted bull's eye in which these rings are stretched posteriorward into ellipses. In effect this means that the localizing function may be more highly determined in the mediolateral axis than in the anterior–posterior axis. Some experimental observations support such a difference. Expansion after eye lesions and compression after tectal ablations is more extensive and homogeneous along the anterior–posterior axis than the mediolateral dimension (Meyer, 1974, 1977). However, alternative explanations such as time–position effects have not yet been eliminated.

What was said about mechanisms for Op Ch 1 goes equally well here. Although it logically follows from Op Ch 2 that retinal fibers and tectal cells have some property or properties that vary as a function of position within retina or tectum and that mediate the locating function, little can be said operationally about the nature of the property and of the interaction. In the tectum, this property is stable enough to survive most translocations, indicating that it is locally autonomous. This and demonstrations of biochemical gradients in retina and tectum lend considerable support to the chemoaffinity notion of cytochemical coding. However, such a mechanism cannot yet be considered proved and so is outside the scope of the operating characteristic. Although the general idea of a regional fiber–tectum preference is derived from chemoaffinity (Sperry 1945, 1965), Op Ch 2 does not specifically postulate a chemoaffinity type of contact chemotaxis but is open to all mechanisms compatible with available data.

D. OPERATING CHARACTERISTIC 3: INTERFIBER LOCATING FUNCTION

The general idea that interfiber interactions mediate retinotopography has been widely discussed for a number of years, particularly by Gaze and associates (Gaze and Keating, 1972; Chung and Cook, 1975; Hope *et al.*, 1976), but only recently has unambiguous evidence for it been obtained (Meyer, 1979a). As an operating characteristic, it can be described as a tendency for fibers to grow or to terminate next to other fibers that originate from neighboring retinal ganglion cells. Operationally, this function is independent of the tectal locating function, that is, in a formal sense, it operates independently of the tectal position of fibers. In other words, fibers possess some correlate of retinal neighborliness that can be read by optic fibers within tectum and translated into a pattern of fiber growth such that fibers tend to terminate next to their retinal neighbors. As for Op Ch 1 and 2, this is an active process and not a passive consequence of an initial temporal and spatial proximity. This operational formulation makes no free assumptions about the nature of the correlate of neighborliness, about the mechanism of the interaction, about the kind of growth response, or whether the interaction occurs directly between fibers or indirectly via tectal cells.

The evidence from which this operating characteristic is deduced is briefly the following. In experiments in which part of retina or tectum are removed there is often a subsequent expansion or compression of the projection into "foreign" tectal regions such that the retinotopography (neighborliness) of fibers is preserved (Gaze and Sharma, 1970; Yoon, 1972a,b; Sharma, 1972b; Meyer, 1977; Schmidt, 1978; Finlay *et al.*, 1979b; Frost and Schneider, 1979).

As an aside, it should be noted that these "plasticities" are fully compatible with Op Ch 2, fiber–tectum locating. Although plasticity has been often depicted as the general case and as being complete and homogeneous in extent, it is becoming increasingly clear that this is not the case. Retinal lesions in chick embryos produce no detectable plasticity (Crossland *et al.*, 1974); and in anurans, compression can be absent, apparently depending on species and experimental conditions (Meyer and Sperry, 1973; Straznicky, 1973; Udin, 1977). The extent and homogeneity of obtained expansion and compression is clearly limited in neonatal hamster (Finlay *et al.*, 1979b; Frost and Schneider, 1979). The most extensive plasticity was thought to occur in goldfish, but this had been measured only with eye-in-air recording methods. More recent reexaminations using autoradiographic tracing (Meyer, 1974, 1975b, also unpublished; Schmidt *et al.*, 1978) and an eye-in-water recording technique (Meyer, 1977) have shown that both expansion and compression are incomplete and are inhomogeneous, that is, appropriate fibers tend to occupy more of their usual tectal space than do the inappropriate fibers. Long postoperative periods ameliorate but do *not* eliminate the incompleteness and inhomogeneity of plasticity in goldfish. Thus, the earlier conclusion that optic fibers entirely disregard tectal position is not supported. These limits to plasticity probably reflect Op Ch 2, the fiber–tectum locating function, though other Op Ch discussed later may also be involved.

Expansion or compression does not unambiguously demonstrate an interfiber locating function. These plasticities can be alternatively explained as a fiber–tectum interaction if tectal positional markers regulate in response to retinal or tectal ablations (Yoon, 1972a; Meyer and Sperry, 1973, 1974). Although regulation appears likely in certain experiments on embryos, it is not the explanation for plasticity in regeneration experiments. In goldfish both expansion and compression can be demonstrated under conditions in which regulation is not possible. When a select group of fibers is surgically deflected into a denervated ipsilateral, tectum expansion ensures (Meyer, 1978a). If both deflected fibers and contralateral fibers regenerate into one tectum, compression occurs (Meyer, 1979b). Furthermore, expansion of a surgically formed half retina can be prevented if fibers from the opposite eye are made to grow into the same tectum (Schmidt, 1978; Sharma and Tung, 1979).

In frogs, when fibers from part of retina are diverted during development into ipsilateral tectum which is denervated by contralateral eye enucleation, they expand across tectum (Fraser and Hunt, 1980b).

Without regulation, Op Ch 2 cannot explain plasticity because Op Ch 2 does not include a formal fiber-to-fiber relationship. Such a relationship is required by the observation that the map is affected by the number of kinds of fibers. However, the preceding data do not distinguish between an interfiber locating function and a fiber–tectum locating function operating in combination with interfiber competition (for such a model, see Prestige and Willshaw, 1975). In other words, fibers could be interacting only by way of competing for available tectal sites for which they have some sort of graded locus specificity preferences. For example, if this preference were an adhesive one, then the various plasticities would simply represent the lowest energy configuration. Evidence that points to a selective interfiber locating function instead of a "best fit" scheme comes from studies in which a piece of tectum is rotated or translocated. In a minority of cases, fibers formed a single retinotopic projection across the entire tectum, including the graft (Levine and Jacobson, 1975; Rho, 1978). This result would be contrary to a best fit scheme provided that the surgery did not alter the number of tectal sites being competed for or the locus specificity of the graft and that pathway variables were not differentially affected. Unfortunately, we do not yet have firm evidence for these latter assumptions. A design more immune to these problems involves deflecting fibers that normally innervate posterior medial tectum in goldfish into the anterior medial end of the opposite *intact* host tectum. The anterior region of host tectum was denervated by a retinal lesion and, consequently, deflected fibers formed an innervation there (Fig. 4). The projection was always highly retinotopic and sometimes normally polarized, but more often polarity was reversed from normal along the anterior–posterior axis (Meyer, 1979a). Thus retinotopography can also be independent of tectal polarity in an intact tectum. Other suggestive evidence comes from studies in which during regeneration (Meyer, 1978a) or development (Fraser and Hunt, 1980b) only a limited number of fibers are allowed to invade an otherwise denervated tectum. Fiber density was apparently reduced (Meyer, 1978a) and retinotopography was markedly degraded. This degradation would be a logical consequence of decreased interfiber interactions which are selective in nature.

What may be the best evidence for a distinct interfiber locating function is pharmacological. By transplanting embryonic axolotl into tetrodotoxin (TTX)-producing newts, it was shown that optic fibers prevented from spiking could grow to tectum, form morphologically

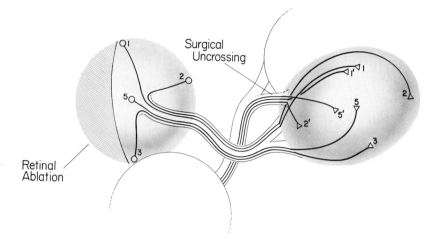

FIG. 4. Diagrammatic representation of an example of Op Ch 3. In this experiment, simplified from Meyer (1979a), the surgically formed half retina on the left innervates its normal half of tectum. Fibers from a comparable region of the opposite eye, designated by primed numbers, were surgically inserted into the denervated part of the tectum. As illustrated, these fibers lined up with homotopic fibers from the normal eye, though this produced a reversal in polarity of the projection.

normal synapses, and produce a roughly retinotopic projection (Harris, 1980). However, the anatomical method and small eye size made it impossible to determine whether refined retinotopography had also developed. For this reason a TTX study was done in goldfish (Fig. 5; R. L. Meyer, unpublished) whose large eye lends itself to a more precise autoradiographic mapping technique used in a previous study of regeneration (Meyer, 1980). The method, referred to here as sector mapping, was to make a small retinal lesion immediately prior to labeling the eye for autoradiography. In an intact projection, the result was normal labeling throughout tectum except for the tectal region corresponding to the retinal lesion at which a small denervated sector was found. When this sector mapping (*simultaneous* point retinal lesion and labeling) was done at various times following optic nerve crush in otherwise normal fish, no denervated sector was seen in tectum at 40 days but one did appear by 60 days (Meyer, 1980, and unpublished; times here are plus or minus a few days). However, gross retinotopography could be demonstrated at 40 days by labeling half of the retina. Thus the projection had an initial rough polarity, but refined retinotopography took time to develop (see Section III,C for other studies on this). When repeated intraocular injections of TTX were given following nerve crush, no difference from untreated fish was seen at 40

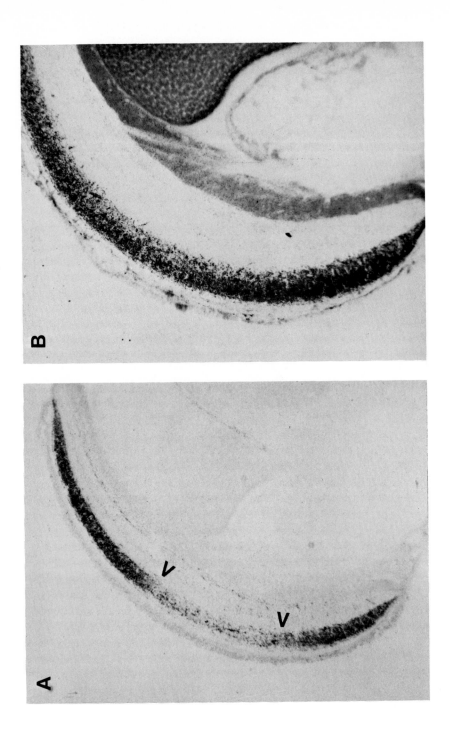

days, but at 60 or 90 days no evidence of refined retinotopography (no denervated sector) could be demonstrated (Fig. 5B) although gross retinotopography was seen. The same inhibition was found when TTX treatment was begun only at 40 days after crush. If TTX blockade was stopped at 80 days and the fish allowed to survive until sector mapping at 120 days, refined retinotopography was found; that is, the TTX effect was not only reversible but could delay the formation of refined retinotopography. Refined retinotopography also formed if continuous but subthreshold TTX injections were given (Fig. 5A) or if TTX blockade was stopped by 45 days. Sector mapping revealed no obvious effect from TTX treatment as long as 90 days on the intact projection. As a further control, lateral posterior fibers were deflected into medial anterior host tectum denervated by eye enucleation. Under TTX deflected fibers grew to lateral posterior host tectum as well and as rapidly as untreated fibers (see Section II,C). It should also be noted that the only known action of TTX is to bind to the voltage-dependent Na^+ channels of the membrane. At the very least then, gross retinotopography can be pharmacologically differentiated from refined retinotopography so that two distinguishable processes are implicated. A further implication is that impulse activity is involved in the genesis of refined retinotopography. Correlated activity between neighboring retinal ganglion cells as found in goldfish (Arnet, 1978) could be translated into neighboring sites of termination, if fibers follow the rule: fibers that fire together, terminate together.

Further evidence along these lines is from studies of ocular dominance columns in lower vertebrates. When fibers from both eyes of goldfish were made to innervate one tectum during regeneration, fibers from each eye formed separate but contiguous patches of innervation (Levine and Jacobson, 1975; Meyer, 1975a). Each patch spanned the depth of the main optic layer and had sharply defined and radially

---

◀ FIG. 5. Frontal autoradiograms near the posterior end of tectum in goldfish. Bright field illumination. Medial is at top and lateral to the left. The dark band of grains near the surface is the main optic lamina. Immediately prior to intraocular injection of proline, a small lesion was made in nasal retina. In fish in which the optic nerve had been crushed and allowed to regenerate for 60 days, this procedure had been reported to result in a small zone of denervation in the middle of this optic lamina. This local denervation has been taken as an index of refined retinotopography. (A) Optic nerve was crushed 80 days prior and biweekly injections of subthreshold tetrodotoxin (TTX) were given intraocularly. A zone of denervation can be seen. Its appearance is identical to that previously reported following nerve crush alone. (B) Same as (A) but a dose of TTX which blocked impulse activity for 4 days was given biweekly. Note the absence of a denervated zone, that is, absence of refined retinotopography.

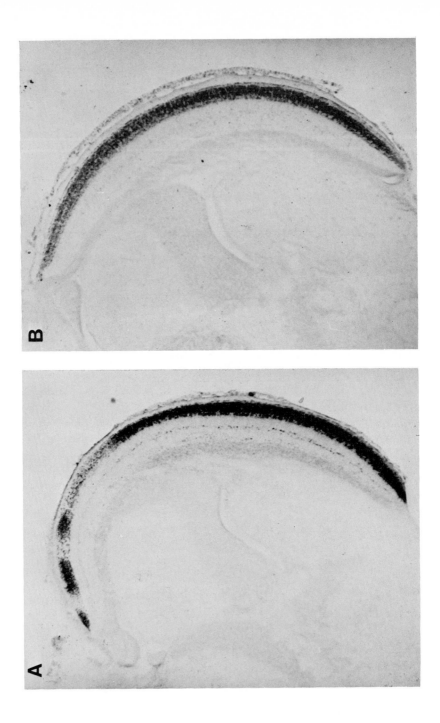

oriented edges; that is, the patches were columnar in structure. It was subsequently shown that when a third eye is transplanted into a frog embryo the resulting projection is a zebra-stripe pattern of alternating columns very reminiscent of cortical columns in mammalian visual cortex (Constantine-Paton and Law, 1978). In one of the goldfish studies (Meyer, 1975a, 1979b), a fiber deflection technique was used to make deflected and host fibers simultaneously regenerate from the same position in host tectum. This was accomplished by inserting posterior medial fibers into a large mediolateral incision across the anterior end of host tectum, thereby simultaneously cutting host fibers. When the progress of regeneration was monitored by autoradiography (Meyer, 1981), it was found that at 30 days both host and deflected fibers overlapped extensively without a hint of columns, but by 60 days well-formed columns appeared. Interestingly, this sequence of overlap then segregation mimics that of mammalian cortex (Rakic, 1977). This definable period for column formation in goldfish enabled a TTX study like that above for nerve crush to be done. Both eyes were repeatedly injected with TTX following the same protocol as above (Fig. 6). The results were remarkably analogous. Continuous TTX treatment prevented the appearance of columns for up to 90 days (Fig. 6B) and was just as effective if begun only at 40 days after deflection. The TTX effect was also reversible, so that cessation of TTX injections at 80 days was followed by columns at 105 days. TTX had no effect on the capacity of deflected fibers to grow from the anterior insertion site into their appropriate posterior quadrant by 30 days and fish with subthreshold TTX injections formed columns at 60 days (Fig. 6A). In addition, TTX blockade of only one eye did *not* inhibit column formation at 60 days. The latter is an important control in that it shows that TTX-treated fibers can form columns. This study again implicates a sorting mecha-

---

◀ Fig. 6. Frontal autoradiograms taken near the posterior end of tectum in goldfish. Conventions same as Fig. 5. but lateral is to the right. Fibers from the medial posterior quadrant of the contralateral tectum were diverted into the anterior medial end of the tectum shown here, the "host" tectum. Host optic fibers were simultaneously severed so that both host and deflected fibers grew in simultaneously. It had been previously shown that after 60 days columns are formed. Consequently, when host fibers were labeled, gaps appeared in medial posterior tectum where columns are located. (A) Subthreshold TTX was injected biweekly into the eye and autoradiography was at 61 days postoperatively. Gaps were found in posterior medial tectum which were identical to those previously reported. (B) A dose of TTX was given that blocked impulse activity for 4 days. After 90 days no columns were found and only an overall lightening of label was noticed in medial posterior tectum.

nism based on impulse timing, and this last observation would further suggest that the absence of activity in one set of fibers can be read as a timing difference between the two groups. A report of TTX blockade of cortical columns in development in cats has also appeared (Stryker, 1981).

However, an alternative to impulse timing can be entertained. There might be a specificity difference between fibers from different eyes which mediates the formation of columns through selective aggregation or adhesivity between fibers from the same eye (Meyer, 1979b; Hunt and Fraser, 1980a). For goldfish, this specificity difference could simply be a left–right one, but for three-eyed frogs a heterogenic difference like histocompatibility antigens might be a factor. Aggregation based on retinal locus specificities could also explain the formation of refined retinotopography (Meyer and Sperry, 1976; Hunt and Fraser, 1980). It might be supposed that TTX somehow blocks the expression of this interfiber aggregation or adhesivity. In other words, impulse activity may be permissive but not instructive. For columns at least, a recent experiment makes this extremely unlikely (C. F. Ide, S. E. Fraser, and R. L. Meyer, unpublished). In *Xenopus* embryos the temporal two-thirds of one eye bud was ablated. The eventual consequence of this is an eye that is morphologically normal but which has a double nasal retinal projection wherein each half of retina projects (as seen electrophysiologically) across the entire tectum (Ide *et al.*, 1979). By labeling only half of retina for autoradiography, it was shown that each half projects across the whole tectum in the form of columns similar to those of three-eyed frogs (Fig. 7). Since these double nasal eyes are isogenic and originated from one side of the embryo, specificity differences between fibers from the two retinal halves are extremely unlikely. The only obvious mechanism is impulse timing. The relevant parameters of activity, however, are not obvious. Maintaining fish in complete darkness, continuous light, or constant strobe illumination within a uniform white visual environment had no apparent effect on column formation at 60 days, even though these conditions dramatically alter the amount and overall pattern of impulse activity (Meyer, 1981). This would suggest that the putative correlation in impulse activity is largely stimulus independent and is instead mediated through neural coupling within retina. This is supported by evidence for correlated dark discharge between neighboring ganglion cells in goldfish (Arnet, 1978). In any event, an interfiber locating function which is distinguishable from a fiber–tectum locating function appears to be on firm ground.

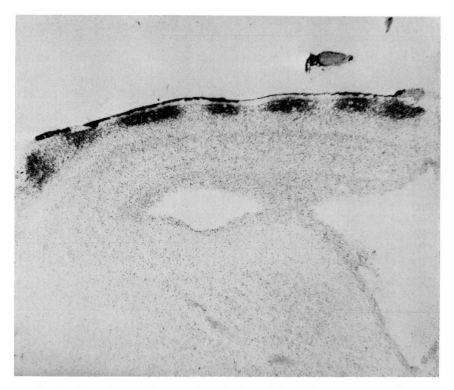

FIG. 7. Frontal autoradiogram through the middle of the optic tectum of a post-metamorphic *Xenopus*. Medial is to the right and lateral to the left. An isogenic double nasal retina was formed by removing the temporal two-thirds of the eye bud in embryonic stage 32. Only the anatomical temporal half of the retina was labeled with tritiated proline. As is evident, label in tectum is grouped into columns. In normal *Xenopus* this same labeling procedure gives continuous labeling.

E. OPERATING CHARACTERISTIC 4: POSITION BIASING

It has long been obvious that fibers often follow precise paths to their target cells, but what role, if any, such positioning has in the patterning of nerve connections has been unclear. For the retinotectal system, it is apparent from the evidence discussed earlier for Op Chs 1–3, that the position of a fiber within the optic pathway is not the prime overriding determinant of its connectivity. On the other hand, there is experimental evidence that termination of a fiber can be significantly influenced by its initial position. Although this evidence is still meager, it is sufficient to warrant a broadly defined operating characteristic. This characteristic can be described as a tendency for a fiber to remain at the tectal position at which it finds itself through developmental circumstance or experimental manipulation. This "inertia" can be expressed as an increased probability that a fiber will terminate in the tectal region into which a fiber initially grows as compared to the probability of ending there when the fiber enters elsewhere. Or it can be expressed as a shift in the site of termination in the direction of initial entry. This Op Ch includes the idea that an optic fiber has the capacity to terminate at many tectal positions but typically does not because of other Op Chs. Formally, the interaction involved in this characteristic is between fibers and tectum, not between fibers themselves. Contrary to the usual treatment of this subject, position biasing is considered to be formally independent of timing effects, which will be dealt with later. It should be emphasized that it is not the fact of normal fiber ordering or positioning that is meant here, but the effect of such positioning on the growth of fibers. Fiber position can be integrated into almost any operating characteristic as the initial condition of the system. Obviously, the effect of position (position biasing) is meaningful only in the context of actual position, so that in practice position biasing must be discussed in terms of developmentally determined and experimentally produced position.

To further clarify this concept, it is worthwhile briefly to contrast position biasing with the previous notions of mechanical pathway ordering. Unlike the latter, position biasing can operate within tectum as well as in the pathways and does not require guidance structures. Also Op Ch 4 does not necessarily imply that relative fiber positions are strictly preserved from retina to tectum but only that there is an inertial effect of position. Position biasing need only be considered to make a limited contribution to the organization of the projection, whereas pathway ordering was usually considered to be the major explanation of the final projection. There is good experimental evidence that the distribution of fibers within the pathways is an active and selective process (Attardi and Sperry, 1963; Straznicky et al., 1979). In this light, pathway patterning would seem better viewed as something to be explained rather than as an explanation. The available data suggest that pathway ordering is complex and may involve many of the processes discussed here under the various Op Chs.

In spite of its long history, the hypothesis that fiber position affects place of termination is supported by data that are more suggestive than conclusive. Part of the reason is that positional effects are usually thought of in combination with timing factors. Consequently, many observations confound position with timing. The classical argument for positional effects is one of correlation. In most animals, optic fibers form a highly ordered array within the nerve and tract and in their paths through tectum. This order is thought to be conducive to the formation of a retinotopically organized projection (for an extensive review, see Horder and Martin, 1978). Highly detailed studies of growing fibers in *Daphnia* have shown that the dynamic position of fibers is extremely precise and may explain a great deal of the connectivity (Lopresti *et al.*, 1973). In apparent contradiction to this are the retinotectal experiments described earlier in which retinotopography is obtained following the disruption of pathway order. Closer inspection of the data, however, leads away from the simplistic conclusion that position, and hence position biasing, play no role in retinotopography. Although usually not explicitly described in detail, the published and so presumably best projections that result from these experimental procedures are invariably degraded in retinotopography. They are topographically noisy. Autoradiographic analysis of the regenerated projection in goldfish indicates that the precision of retinotopography is roughly half that of normal (Meyer, 1980). As far as one can tell from the available data, it also takes significantly longer to regenerate an ordered projection than it does to grow one in normal development (for frog, compare Gaze *et al.*, 1974, with Gaze and Keating, 1970, and Udin, 1978; for goldfish, see Meyer, 1978b). Thus, these experimental results leave room for two possible roles for positioning. Fiber position coupled with position biasing may aide in or be a prerequisite to refined retinotopic ordering. Or, positioning may be required for the rapid formation of a retinotopic projection. It should be emphasized that this does not necessarily mean that position biasing of retinotopically positioned fibers in the pathways directly produces retinotopic order. On the contrary, preliminary anatomical analysis of pathway order in goldfish suggests that it is not sufficiently ordered to account for the degree of order in the final projection (Meyer, 1980). If the same is true of other species with ordered pathways, then position biasing could at most provide assistance or reinforcement to more dominant ordering factors like Op Chs 2 and 3. In *Rana* and cats, which lack apparent ordering in the optic nerve (Maturana, 1960; Horton *et al.*, 1979), the role of position biasing may be more limited. (For the sake of argument, it is assumed that the disorder observed in the adults of these animals is not a secondary derangement of an initial order found earlier in development.) In this case, position biasing is obviously not preserving the relative position of fibers from retina to tectum. But this is not a prerequisite for it to have an effect. In both animals, there is good evidence for pathway order central to the chiasm (Horton *et al.*, 1978; Udin, 1978), and there position biasing may exert a significant influence. Conceivably, the Op Ch may even be important when relative order between fibers is absent by tending to confine optic fibers as a group to the optic paths or by serving as a damping factor to wild growth.

More direct evidence comes from autoradiographic experiments in goldfish in which half of retina was ablated and the nerve severed (Meyer, 1974, 1975b; Strumer, 1981). As in normal development, these fibers grew into tectum at its anterior end. If the fibers were from temporal retina, which normally projects to anterior tectum, then the fibers were largely confined to the anterior half of tectum, especially at early stages. If the fibers were from nasal retina and so had to bypass anterior tectum to project onto their normal posterior sites, then fibers projected to both anterior and posterior halves. Whereas the posterior projection was noticeably denser, the anterior projection was always substantial. Similarly, when fibers were from an inferior half retina, which normally projected to medial tectum, a substantial fraction entered tectum through the lateral brachium of the tract. Whereas most of these errant fibers grew back into the medial half, a substantial number remained in the lateral half. A mirror image pattern was found when the superior half of retina was removed. These results point to position biasing. The possibility remains, though, that the results suggest a complex fiber–tectum locating function operating in combination with competitive processes.

More conclusive evidence comes from a series of autoradiographic experiments in which fibers are surgically introduced into different regions of tectum using the fiber deflection method mentioned earlier. Though some of these experiments are still in progress, the following seems clear: If tectum is denervated, there is a strong positional effect. For example, the far lateral fibers from one tectum were inserted at the medial edge of the opposite tectum, which was denervated by enucleation. Whereas many fibers grew into lateral tectum, a large number remained in medial tectum (Fig. 8). If other fibers were present, the biasing was still evident but was less marked. In other experiments, fibers that normally terminated in medial posterior tectum were inserted into a more lateral and anterior position in the contralateral tectum. If this anterior region was selectively denervated by a retinal lesion, the inserted fibers formed an apparently permanent anomalous innervation there (Meyer, 1979a). The permanency of this innervation apparently resulted from Op Ch 3 (see earlier). If this anterior region was only temporarily denervated by cutting its fibers supply, the deflected fibers formed a temporary innervation. After several months, this anomolous anterior innervation tended to disappear while a cor-

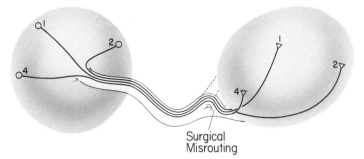

Surgical
Misrouting

FIG. 8. Diagram of an example of Op Ch 4. Fibers 1, 2, and 4 have been surgically forced into the wrong side of tectum. The position of their terminals is consequently displaced toward this side of tectum.

rectly positioned innervation developed (Meyer, 1974, also unpublished). If the entire innervation of contralateral tectum was temporarily disrupted so that both deflected and contralateral fibers regenerated into tectum at the same time, deflected fibers initially dispersed in the region in which they were inserted. By 1 month, deflected fibers became largely confined to their appropriate posterior region, but some fibers remained in incorrect regions. Thus, under conditions that would be expected to widely vary the effect of other factors, a consistent position effect still emerges. Obviously, position biasing needs further elucidation through experiments specifically aimed at this question; however, the general notion that fibers exhibit a positional biasing seems strongly indicated.

F. OPERATING CHARACTERISTIC 5: TEMPORAL PRECEDENCE

The notion underlying this characteristic is, in essence, first come first served. It can be thought of as a resistance by established fibers to displacement by subsequent ingrowing fibers. This resistance can be expressed as a delay, reduction, or inhibition in the innervation formed by fibers that arrive later. In a formal sense, this Op Ch represents an interfiber interaction in contrast to position biasing (Op Ch 4), which is a fiber–tectum interaction. Whereas position biasing can be expressed in terms of a single fiber on tectum, temporal precedence can only be expressed as a relationship between fibers. This relation is only operational or formal and as such does not preclude mechanisms that are fiber-to-tectum in nature. An example of the latter might be the formation of high energy, perhaps synaptic, bonds between fibers and tectum. If early fibers must break these bonds to allow late fibers to form an innervation, the late fibers would be at a relative disadvantage.

Fiber–fiber mechanisms such as interfiber inhibition of growth can also be easily entertained. Of course, to give meaning to this Op Ch as a patterning process, it is necessary to specify the tectal position at which this interfiber interaction occurs. In other words, one would say that ingrowing fibers encounter established fibers at tectal position X but not at Y and so would tend to innervate Y instead of X. Again, it is not temporal difference that is the Op Ch but the effect of this temporal difference, that is, it is the intrinsic process that is meant, not extrinsic circumstance (see similar discussion for Op Ch 4). In practice, it is obviously necessary to talk about both the process and extrinsic temporal differences.

As for Op Ch 4, we need first ask whether the many experimental results against timing leave any role for temporal precedence. Moderately good retinotopography can be generated in regeneration where temporal differences are largely, if not entirely, absent (see Section II,C). In experiments on frog embryos in which two half eyes are fused into one compound eye, fibers grow in sequentially, as in normals (Feldman and Gaze, 1974), but terminate independently of timing. In normal eyes, the first fibers to grow in come to project to midtectum in the adult. In double nasal eyes, the first fibers innervate the anterior end of tectum in the adult; and in double temporal eyes, they innervate the posterior end (for further discussion, see Meyer and Sperry, 1976). However, with regeneration and compound eyes, the retinotopography is decidedly less well ordered than in normals. In regeneration, an ordered projection generally requires a longer time to form (see Op Ch 4). These results suggest that position biasing may assist or refine retinotopography but is not a dominant factor in producing overall organization.

Studies on normal development support a similar conclusion. On the one hand, it is becoming increasingly clear that growing optic fibers do not form a permanent innervation on the first vacant target site encountered. In chicks, for example, the first fibers grow past much of tectum (Crossland et al., 1974); and in lower vertebrates, the site of innervation changes with time (Gaze et al., 1974; Meyer, 1978b). The situation in mammals is similarly complex (Rakic, 1977). On the other hand, time of ingrowth is apparently precisely controlled in development and strikingly similar between species. In general, there is a radial sequence of fiber outgrowth from retina such that peripheral ganglion cells send out fibers after central ganglion cells. In other words, fibers from central retina would already be occupying tectum when fibers from peripheral retina grow in. These central fibers would, by temporal precedence, offer resistance to tectal invasion by later peripheral fibers. This would tend to keep late-arriving fibers at the periphery of tectum, which is exactly where these fibers need to be for retinotopography. It follows that the first ingrowing fibers would not benefit from this Op Ch, and, interestingly, they seem not to be as well ordered as later fibers (Gaze *et al.*, 1974; Rakic, 1977). A more striking correlation comes from developmental studies in the hippocampus of rats. A spatial asymmetry in the distribution of ipsilateral versus contralateral fibers is strongly associated with a difference in their time of ingrowth (Gottlieb and Cowan, 1972). Obviously the role of temporal precedence in normal development is not yet clear. It may only serve as an adjunct to more selective Op Chs by reducing the directional options of growing fibers or by reinforcing or triggering these other Op Chs. Nevertheless, the evidence for the existence of temporal precedence is highly suggestive.

Experimental evidence comes from incidental observations made in various regeneration studies. In adult goldfish, removing the posterior half of tectum without severing the optic nerve leaves the optic innervation of anterior tectum intact. Eventually fibers that originally terminated on the posterior half can displace the existing fibers to form a compressed retinotopic projection. The displacement has been observed with electrophysiological (Gaze and Sharma, 1970; Yoon, 1971; Meyer, 1977) and autoradiographic methods (Meyer, 1976). However, the compression is significantly slower, less extensive, and less retinotopic than the compression that occurs when the nerve is cut (Meyer, 1977, also unpublished) (Fig. 9). [It might be objected that because the initial projection that forms after cutting the nerve is an uncompressed rep-

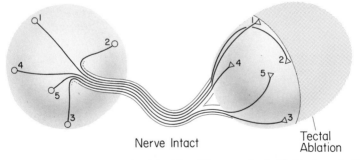

Nerve Intact   Tectal Ablation

FIG. 9. Diagram of an example of Op Ch 5. The posterior half of tectum, indicated by the tone, was removed. The innervation of anterior tectum was initially preserved. As illustrated, the fibers from the part of retina corresponding to the posterior half eventually come to terminate on anterior tectum. However, they project to a smaller area of tectum than fibers from other parts of retina. If the nerve is crushed so that all fibers regenerate at the same time, this asymmetry is greatly reduced.

resentation of the appropriate half of retina (Cook and Horder, 1974), there should be no difference between cutting and not cutting the nerve. This claim, however, has not been replicated (Yoon, 1976). More recent reexamination with eye-in-water electrophysiological mapping and autoradiographic analysis indicates that the initial projection is partially compressed and not well established (Meyer, 1976, also unpublished).] In *Rana*, removing posterior tectum while leaving the optic nerve intact permanently prevents posterior fibers from forming an orderly compression onto the anterior remnant. Instead these fibers form a weak disorganized projection superimposed on that of the anterior half. These same fibers form a compressed projection if the nerve is severed (Udin, 1977).

A similar effect can be detected in electrophysiological experiments in goldfish, where fibers are made to grow into ipsilateral tectum, which they normally never innervate. Although the particular design of these experiments is complex, typically involving various retinal or tectal surgeries, a temporal effect seems clear. If the normal innervation is temporarily disrupted or permanently eliminated, ipsilateral fibers rapidly form a projection in which the entire existing retina is represented. If contralateral innervation is present, then the formation of the ipsilateral projection is delayed and under certain circumstances some parts of ipsilateral retina are inhibited from projecting for many months (Sharma, 1973; Schmidt, 1978). [One caveat for interpreting these uncrossing experiments has come to light in more recent anatomical studies (Easter *et al.*, 1978; Lo and Levine, 1981). The surgical procedures do not simply force fibers to grow ipsilaterally but instead produce a great deal of damage on one side, which often includes removal of tectum. This damage results in fibers finding their way to ipsilateral tectum by numerous different routes. Thus, some of the preceding results may be explained as a differential ability of fibers to find ipsilateral tectum.]

A cleaner method involves teasing a select group of fibers from one tectum and inserting them into an inappropriate part of the opposite tectum. If the nerve to the opposite tectum is simultaneously cut, regenerating deflected fibers form an innervation in the appropriate tectal region in 1–2 months. If the nerve is not cut, deflected fibers require several months to form a comparable innervation (Meyer, 1974, also unpublished).

Although experiments specifically aimed at this question are obviously needed, the existence of a time precedence effect seems clear, if one accepts evidence from other systems. It is well known for neuromuscular systems that fibers that can readily reinnervate denervated muscles are delayed or inhibited from doing so if the muscles are already innervated. This can occur even if the regenerating fibers are the appropriate ones and the established fibers are foreign ones (Slack, 1978; Dennis and Yip, 1978). What is now needed are studies that will elucidate the actual contribution of timing precedence in the normal development of the retinotectal system.

G. OPERATING CHARACTERISTIC 6: FIBER–TECTUM SPREADING

The preceding four Op Chs principally address the question of topography. What they cannot formally account for is the spatial extent of the projection. In the past, the concept of competition has often been invoked to explain the allocation of target space. However, competition is a rather broad notion and has, on occasion, included timing and position factors. To avoid any confusion with Op Chs 4 and 5, the term will not be used here. Instead two processes affecting occupation of space will be distinguished and will be discussed as Op Chs 6 and 7.

The first of these two operating characteristics is a tendency by optic fibers to invade as much tectal areas as possible, that is, to spread out or expand. Operationally or formally, the interaction is between fibers and tectum. In other words, the Op Ch relates fibers to tectal space, not fibers to each other. The mechanism must, again, be left open. It may be a fiber–tectum mechanism, such as stimulation of fiber growth by tectal cells. Alternatively, a fiber–fiber mechanism, such as interfiber repulsion, could be readily entertained.

There is substantial evidence for this Op Ch. When part of retina is removed, fibers from the remaining retina will grow over a larger tectal area than normal, both in regeneration (Yoon, 1972b, Meyer, 1975a,b, 1976) and normal development (Frost and Schneider, 1979). Or, if a minature eye is formed by arresting growth, it will nevertheless project across all of tectum (Hunt, 1977). These kinds of experimental results, however, could be alternatively interpreted in terms of the previous operating characteristics. In particular, the spreading could be considered to be a consequence of the spatiotemporal disruption caused by surgery if position biasing is strong and fiber–tectum localization weak.

This particular possibility is mitigated in experiments where fiber position is controlled. When a small number of fibers from one tectum are inserted into a small incision in an otherwise denervated opposite tectum, fibers have been found to grow in all directions from the point of insertion (Meyer, 1978a, 1979b). Normally fibers grow only in an anterior–posterior direction. Other experiments demonstrate that established normal fibers will invade neighboring areas simply as a result of denervation of this area. When part of the retina is lesioned, the fibers from the remaining retina can spread into the denervated area even though the optic nerve is left intact (Yoon, 1972b; Schmidt *et al.*, 1978) (Fig. 10). Numerous examples are available from other systems. The "reactive synaptogenesis" of the hippocampus and collateral sprouting of neuromuscular fibers are particularly well known (Cotman and Nadler, 1978).

It is tempting to speculate that the same mechanisms that mediate other Op Chs, especially 1 and 3, are also the basis for spreading. In the case of Op Ch 1, a high fiber–tectum affinity that is independent of tectal position and that would stimulate extensive fiber growth within tectum could be postulated. Or, if Op Ch 3 is mediated by a selective repulsion between fibers, spreading would be a consequence. Alternatively, the mechanism could be entirely different from those of Op Chs 1 and 3. The secretion of a growth stimulant by denervated or uninnervated cells is an obvious example. The diversity of plausible explanations again points to the necessity of listing fiber–tectum spreading as a separate operating characteristic.

H. Operating Characteristic 7: Fiber–Fiber Rivalry

The second operating characteristic to speak to the allocation of tectal space is "formally" an interfiber interaction. It is an "aggressive" behavior on the part of growing optic fibers whereby they tend to occupy tectal space in an exclusive manner. It is as if two fibers or two sets of fibers will not readily occupy the same terminal space. The underlying biology may involve a fiber–tectum interaction, but, operationally, rivalry is an interfiber interaction.

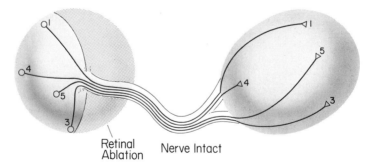

Fig. 10. Diagram of an example of Op Ch 6. As indicated by the tone, half of the retina was removed without severing fibers from the remaining half. These remaining fibers eventually expanded their projection across a greater extent of tectum than they normally project to. Compare with Fig. 1.

The evidence for rivalry also seems convincing. In virtually every experiment where it has been closely examined, rivalry within terminal space has been demonstrated. (For a discussion of the general phenomena, see Schneider and Jhaveri, 1974.) In goldfish, for instance, removal of the posterior half of tectum leads to an invasion of anterior tectum by posterior fibers which exclude anterior fibers from regions of anterior tectum (Gaze and Sharma, 1970; Meyer, 1977). Because this can occur when anterior innervation is left intact and can occur at considerable distances from the cut edge of tectum (Fig. 11), it cannot be explained as position biasing or timing precedence. Also, when fibers from one eye are made to grow into ipsilateral tectum, the innervation of which is intact (presumably), they eventually displace the previously existing optic fibers from regions of tectum (Levine and Jacobson, 1975). But could not such apparent rivalry merely be the product of Op Ch 2 or 3? One set of fibers might exclude another because its fiber–tectum locating function predominates at that tectal position or because its fiber–fiber locating function eliminates the anterior fibers as incompatible neighbors. Synapse elimination (Op Ch 9) might also produce this effect, although its standard interpretation would not readily explain the displacement of an existing innervation. At the moment there is no conclusive evidence to rule out the possible congruence between the mechanisms underlying Op Ch 2, 3, or 9 and the present one. However, it is easy to imagine that separate mechanisms are involved. Some quite plausible processes that could underly Op Ch 7 and not be directly involved in the other ones are: Optic fibers could be secreting a growth inhibitor that acts on each other in a local fash-

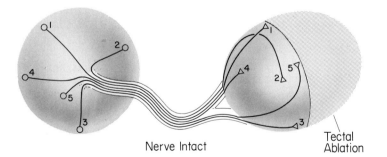

Nerve Intact   Tectal Ablation

FIG. 11. Diagram of an example of Op Ch 7. The posterior half of tectum was removed without severing the nerve. Fibers corresponding to posterior tectum eventually displaced normal fibers from regions of anterior tectum. As shown by fiber 2, this displacement can take place in violation of retinotopography. In other cases (not illustrated), retinotopography is preserved.

ion. Fibers could exhibit contact inhibition of extension like the transient retraction that can take place upon contact between growing fibers in tissue culture (Dunn, 1971). Another possibility is that postsynaptic sites in tectum are fixed in number or at least limited. It might be supposed fibers have the capacity to make many more synapses than they normally do but compete for limited terminal space. While the other Op Ch might specify which fibers will be more competitive at a given site, these other Op Chs as presently formulated do not necessarily require competition per se. None of the other Op Chs would tell us that an infinite number of identical silent fibers would not take up infinite space; that is, there is a logical requirement for Op Ch 7. It may eventually be possible to merge this Op Ch with another, but this will require data not yet available.

I. Operating Characteristic 8: Stereotypic Guidance

This characteristic is essentially mechanical guidance but without any connotation that it is the predominant patterning force in development. Because mechanical guidance is a well-known effect, its application to the retinotectal system will be only briefly discussed. The Op Ch is simply the tendency of optic fibers to grow along or through guidance structures. The best described structures are the long glial tubes or channels found in the optic nerve and tract (Turner and Singer, 1974; Murray, 1976). Other candidates are blood vessels and the shafts of existing fibers, including pioneering and other optic fibers. The direction of growth is strictly dependent on the shape of the guidance structure and is nonselective with respect to the retinal origin of fibers. In other words, the guidance an optic fiber receives from a particular structure is the same as that of any other optic fiber, though, as a group, optic fibers may exhibit a preference for their normal guidance structures. Position-dependent fiber growth would be better viewed in terms of Op Ch 2 or similar processes. It should be noted that this does not mean that every optic fiber will actually show the same pattern of growth, but only that they will tend to do so. Fibers are not rigidly confined to these guidance structures and so can be influenced by other processes.

Although much of the critical evidence for this Op Ch comes from observations on growing neurites in culture, its existence in the retinotectal system is strongly indicated. In the nerve, tract, and stratum opticum of tectum, fibers are ensheathed by large glial tubes and channels. During development, these channels appear in retina and in the optic stalk just prior to fiber outgrowth (Silver and Sidman, 1980; see also Singer et al., 1978) and are entered and followed by growing fibers.

When the optic nerve is severed in adult amphibians and fish, these channels survive and are followed by regenerating fibers through the nerve and stratum opticum (Turner and Singer, 1974; Murray, 1976). Autoradiographic studies (Meyer, 1980) indicate that some regenerating fibers that normally grow into medial tectum enter channels in lateral tectum. Nevertheless, they appear to follow the anterior–posterior path of the channels and only grow back to medial tectum after exiting from them. In this case, the guidance structures may be inhibiting the formation of retinotopography. One might wonder whether the normal role of these channels is not simply to instruct fibers about their proper pathway as is generally supposed but also to isolate fibers from some surrounding tissue. This could reduce time-consuming or error-producing interactions, which in normal development may be unnecessary. There would also seem to be some limits on the instructional role these channels can have. They may be essential for getting fibers from retina to the central nervous system and possibly even into tectum. (It should be noted that no evidence of the latter possibility has yet been found.) The organization of the channels, however, appears insufficiently precise to account for the high order of retinotopography in the projection (Silver and Sidman, 1980), though a rough initial order may be imparted.

J. Operating Characteristic 9: Fiber–Synapse Elimination

In the development and regeneration of neuromuscular connections, it is well known that each fiber initially innervates an unusually large number of muscles. Subsequently, many of these connections are withdrawn to leave the normal number of contacts (Brown et al., 1976). This elimination process is apparently not simply competition for available sites. When a small number of fibers innervate a muscle, synapse elimination still takes place even though some muscles are left without innervation (Brown et al., 1976; O'Brien et al., 1978). A similar elimination process also appears to occur in parts of the central nervous system (Conradi and Ronnevi, 1977; Crepel et al., 1976). There is some suggestive evidence that a similar elimination process occurs in the retinotectal system. Because these data are meager, fiber–synapse elimination is proposed only as a candidate Op Ch.

The most direct evidence comes from EM studies on the optic nerve during regeneration in lower vertebrates (Turner and Singer, 1974; Murray, 1976). In the early phase of regeneration, the number of fibers within the nerve is increased severalfold over normal, apparently as a result of fiber bifurcation near the site of injury. This number later declines during the period in which the topography of the retinotectal projection is being formed. Paralleling this is autoradiographic evidence for extensive overlapping of fibers in the early stage of tectal reinnervation followed by a restriction of fibers to a limited area of tectum (Meyer, 1980, 1981). This restriction may, in part, represent elimination of extra branches. A similar sequence of overlap and restriction that has been observed in the normal development of the retinogeniculate and retinocollicular projections of mammals may have the same explanation (Rakic, 1977; Calvalcante and Rocha-Miranda, 1978). Certain aberrant retinal projections that form after various ablations in neonatal hamsters and rats appear to follow a similar sequence and may reflect the same processes (Schneider and Jhaveri, 1975; Lund, 1978; So, 1979).

Other suggestive evidence comes from TTX studies. TTX is known to prevent synapse elimination in the neuromuscular system (Lømo and Jansen, 1980), so in effect it freezes development or regeneration at the early stage of widespread innervation. The TTX studies on columns and refined retinotopography (see Section III,D) also show a similar cessation of regeneration at the early stage of dispersed innervation and so by analogy implicate synapse elimination. In fact, Op Ch 3 may be *selective* synapse elimination, that is, Op Ch 3 and Op Ch 9 may represent one underlying process. If confirmed, it may not be necessary to distinguish these two Op Chs, but, on the other hand, it may still be useful to deal with them as two Op Chs.

Obviously, direct evidence is needed to establish and characterize fiber–synapse elimination. One question to be addressed is whether "elimination" really represents a net loss of fibers and synapses. There is some evidence from neuromuscular studies that no net loss of synapses take place (Lømo and Jansen, 1980), so that what may be happening here is a restructuring of axonal branching and a recording or translocation of synaptic connections.

K. OPERATING CHARACTERISTIC 10: CELL DEATH

In some systems, such as the ventral horn of spinal cord and the ciliary ganglion of the autonomic system, massive death of neurons takes place during normal development. This cell death temporally precedes fiber–synapse elimination and is generally considered to be a separate process. Although the function of cell death is not entirely clear, there is evidence that it regulates the number of neurons and may possibly help mediate selectivity of connections (see Jacobson, 1978). Because the evidence for cell death in the retinotectal system is rather limited and is not clearly tied to the patterning of the projection, cell death is proposed only as a possibility.

Quantitative electron microscopy on the developing optic nerve of chick has shown that between day 10 and hatching some 40% of the fibers disappear (Rager and Rager, 1978). This appears to be roughly paralleled by cell degeneration in the ganglion cell layer of retina. Thus, it seems likely that there is a massive death of ganglion cells during the formation of the retinotectal projection. Conceivably this may help determine or enhance retinotopography by eliminating cells that send fibers to the wrong part of tectum. Other processes such as trophic effects by fibers on the development and survival of tectal cells are other potential candidates for Op Chs, but their relation to the patterning of the projection is, at present, more obscure than that of cell death.

III. Concluding Remarks

It would be nice to end the chapter by casting each of the preceding Op Chs into a mathematical expression and writing a series of equations to describe the system. After all, this is the ultimate goal of the present approach. To attempt this seems premature for two reasons. One is that the Op Chs have not yet been characterized with sufficient precision. The other is that the mathematics for relating the Op Chs seem neither obvious nor trivial. We might then ask what we have presently gained by this analysis? At the very least, Op Chs are a way of summarizing what we know about the system. To the extent that this is true, Op Chs are, in a sense, permanent. This does not mean they will always exist in their present form. Future work may create new Op Chs, split some Op Chs into multiple ones, or consolidate other Op Chs into single ones. In each case, the essence of an Op Ch will have to be conserved. As long as an Op Ch reflects valid empirical observations, it cannot be discarded. In addition, Op Chs can also be thought of as criteria to be met by models. To be valid and useful, a model must satisfy and explain the Op Chs. To be complete it must, in some way or another, incorporate all the Op Chs. But if the Op Chs are merely a distillation of the behavior of the system, then they might be considered as just a formal device with which to organize a review. In the case of the retinotectal system, this formality was felt to be useful, but it hardly seems essential for a synopsis. The main point of the Op Chs has to do with the following problem. In a multiprocess and multielement system, at what level of analysis do you base an explanation? Do we really want to try to explain developmental neurobiology in terms of gauge theories or quantum chromodynamics? It would be considered ridiculous. An explanation based on quantum mechanics is only an amusing possibility. But when we come to chemistry, we find serious advocates. Some would argue that knowing the underlying chemistry is essential for any real understanding. But how useful will this be? Is it a case of looking for the message in the chemistry of the ink? What does chemistry have to say about the position of cells, the time of differentiation, and axonal outgrowth? We find ourselves introducing new concepts or emergent properties in order to arrive at a complete explanation. These new properties are symptomatic of a higher level of complexity and are warning signs that the preceding level of analysis is becoming less useful. To be practical, the explanation may need to be based on more complex phenomena, here referred to as Op Chs. These Op Chs are really shorthand for a lot of underlying chemistry in the same way that standard chemistry is shorthand for underlying quantum mechanics. Such a shorthand is in practice frequently used by developmental neurobiologists when they speak of contact guidance, chemoaffinity, and cell death, for example, but there is often the intimation that such language is a temporary necessity, a stop-gap measure before chemical interactions explain it all.

The purpose of invoking Op Chs is to show that we can talk in terms of suprachemical or higher order processes as a valid and permanent basis on which to build an explanation. In truth, the motivation for the chapter is a frequently asked and increasingly disturbing question: What molecules are responsible for target recognition by growing nerve fibers? Certainly this is important to know. What is worrisome is the assumption that the chemical processes will explain the development of neural connections. If the goal is a general understanding of neural development, we must also ask how this molecule controls cell movement, how it affects target cells, how it interacts with other developmental processes, what contribution it makes to the patterning of the system, whether it competes or cooperates with other molecular processes, and, most importantly, what other relevant processes are also occurring. In a word, just how useful will it be to explain the spatiotemporal dynamics of the system in terms of this molecule? Would we not be better off using an Op Ch that describes the selective recognition as a molar process and try to see the big picture? It seems worth a try.

REFERENCES

Arnet, D. W. (1978). *Exp. Brain Res.* **32**, 49–53.

Attardi, D. G., and Sperry, R. W. (1963). *Exp. Neurol.* **7**, 46–64.

Brown, M. C., Jansen, J. K. S., and Van Essen, D. (1976). *J. Physiol. (London)* **261**, 387–422.

Calvalcante, L. A., and Rocha-Miranda, C. E. (1978). *Brain Res.* **146**, 231–248.

Chung, S., and Cook, J. (1975). *Nature (London)* **258**, 126–132.

Chung, S. H., Gaze, R. M., and Stirling, R. V. (1973). *Nature (London), New Biol.* **246**, 186–189.
Conradi, S., and Ronnevi, L. O. (1977). *J. Neurocytol.* **6**, 195–210.
Constantine-Paton, M. (1978). *J. Comp. Neurol.* **158**, 31–43.
Constantine-Paton, M., and Law, M. I. (1978). *Science* **202**, 639–641.
Cook, J. E., and Horder, T. J. (1974). *J. Physiol. (London)* **241**, 89P–90P.
Cotman, C. W., and Nadler, J. V. (1978). "Neuronal Plasticity." Raven, New York.
Crepel, F., Mariani, J., and Delhaye-Bouchaud, N. (1976). *J. Neurobiol.* **7**, 567–578.
Crossland, W. J., Cowan, W. M., Rogers, L. A., and Kelly, J. P. (1974). *J. Comp. Neurol.* **155**, 127–164.
Dennis, M. J., and Yip, J. W. (1978). *J. Physiol. (London)* **276**, 299–310.
Dunn, G. A. (1971). *J. Comp. Neurol.* **143**, 491–508.
Easter, S. S., Schmidt, J. T., and Leber, S. M. (1978). *J. Embryol. Exp. Morphol.* **45**, 145–159.
Feldman, J. D., and Gaze, M. (1964). *J. Embryol. Exp. Morphol.* **27**, 381–387.
Finlay, B. L., Wilson, K. G., and Schneider, G. E. (1979a). *J. Comp. Neurol.* **183**, 721–740.
Finlay, B. L., Schneps, S. E., and Schneider, G. E. (1979b). *Nature (London)* **280**, 153–155.
Fraser, S. E. (1980). *Dev. Biol.* **79**, 453–464.
Fraser, S. E., and Hunt, R. K. (1980a). *Annu. Rev. Neurosci.* **3**, 319–352.
Fraser, S. E., and Hunt, R. K. (1980b). *Dev. Biol.* **79**, 444–452.
Frost, D. O., and Schneider, G. E. (1979). *J. Comp. Neurol.* **185**, 517–568.
Gaze, R. M. (1959). *Q. J. Exp. Physiol. Cogn. Med. Sci.* **44**, 290–308.
Gaze, R. M., and Keating, M. J. (1970). *Brain Res.* **21**, 183–195.
Gaze, R. M., and Keating, M. J. (1972). *Nature (London)* **237**, 375–378.
Gaze, R. M., and Sharma, S. C. (1970). *Exp. Brain Res.* **10**, 171–181.
Gaze, R. M., Keating, M. J., and Chung, S. H. (1974). *Proc. R. Soc. London, Ser. B* **185**, 301–330.
Gaze, R. M., Keating, M. J., Ostberg, A., and Chung, S.-H. (1979). *J. Embryol. Exp. Morphol.* **53**, 103–143.
Giorgi, P. P., and Van de Loos, H. (1978). *Nature (London)* **275**, 746–748.
Gottlieb, D. I., and Cowan, W. M. (1972). *Brain Res.* **41**, 452–456.
Gottlieb, D. I., Merrell, R., and Glaser, L. (1974). *Proc. Natl. Acad. Sci. U.S.A.* **71**, 1800–1802.
Harris, W. A. (1980). *J. Comp. Neurol.* **194**, 303–317.
Hibbard, E., and Ornberg, R. L. (1976). *Exp. Neurol.* **50**, 113–123.
Hope, R. A., Hammond, B. J., and Gaze, R. M. (1976). *Proc. R. Soc. London, Ser. B* **194**, 447–466.
Horder, T. J., and Martin, K. A. C. (1978). *Symp. Soc. Exp. Biol.* **32**, 275–358.
Horton, J. C., Greenwood, M. M., and Hubel, D. H. (1979). *Nature (London)* **282**, 720–722.
Hunt, R. K. (1975). *Ciba Found. Symp.* [N.S.] **29**, 129–157.
Hunt, R. K. (1977). *Biophys. J.* **17**, 128a.
Hunt, R. K., and Jacobson, M. (1974a). *Curr. Top. Dev. Biol.* **8**, 203–259.
Hunt, R. K., and Jacobson, M. (1974b). *Dev. Biol.* **40**, 1–15.
Ide, C. F., Kosofsky, B. E., and Hunt, R. K. (1979). *Dev. Biol.* **69**, 337–360.
Jacobson, M. (1976). *Brain Res.* **103**, 541–545.
Jacobson, M. (1978). "Developmental Neurobiology." Plenum, New York.
Jacobson, M., and Levine, R. L. (1975). *Brain Res.* **88**, 339–345.

Johns, P. R. (1978). *J. Comp. Neurol.* **176**, 343–358.

Johns, P. R., and Easter, S. S. (1978). *J. Comp. Neurol.* **176**, 331–342.

Levine, R. L., and Jacobson, M. (1974). *Exp. Neurol.* **43**, 527–538.

Levine, R. L., and Jacobson, M. (1975). *Brain Res.* **98**, 172–176.

Lo, R. S., and Levine, R. L. (1981). *Brain Res.* **210**, 61–68.

Lomo, T., and Jansen, J. K. S. (1980). *Curr. Top. Dev. Biol.* **16**, 253–281.

Lopresti, V., Macagno, E. R., and Levinthal, C. (1973). *Proc. Natl. Acad. Sci. U.S.A.* **70**, 433–437.

Lund, R. (1978). "Development and Plasticity of the Brain, An Introduction." Oxford Univ. Press, London and New York.

Maturana, H. (1960). *J. Biophys. Biochem. Cytol.* **7**, 107–135.

Meyer, R. L. (1974). *Diss. Abstr. Int. B* **35**, 1510B (University Microfilms No. 74-21603).

Meyer, R. L. (1975a). *In* "Developmental Biology, Pattern Formation, Gene Regulation" (D. McMahon and C. F. Fox, eds.), pp. 257–275. Benjamin, Menlo Park, California.

Meyer, R. L. (1975b). *Anat. Rec.* **181**, 427.

Meyer, R. L. (1976). *Caltech Biol. Annu. Rep.* pp. 106–107.

Meyer, R. L. (1977). *Exp. Neurol.* **56**, 23–41.

Meyer, R. L. (1978a). *Brain Res.* **155**, 213–227.

Meyer, R. L. (1978b). *Exp. Neurol.* **59**, 99–111.

Meyer, R. L. (1979a). *Science* **205**, 819–821.

Meyer, R. L. (1979b). *J. Comp. Neurol.* **183**, 883–902.

Meyer, R. L. (1979c). *Caltech Biol. Annu. Rep.* p. 145.

Meyer, R. L. (1980). *J. Comp. Neurol.* **189**, 273–289.

Meyer, R. L. (1981). *Neurosci. Abstr.* **7**, 405.

Meyer, R. L., and Sperry, R. W. (1973). *Exp. Neurol.* **40**, 525–539.

Meyer, R. L., and Sperry, R. W. (1974). *In* "Plasticity and Recovery of Function in the Central Nervous System" (D. G. Stein, J. J. Rosen, and N. Butters, eds.), pp. 45–63. Academic Press, New York.

Meyer, R. L., and Sperry, R. W. (1976). *In* "Studies on the Development of Behavior and the Nervous System" (G. Gottlieb, ed.), Vol. 3, pp. 111–149. Academic Press, New York.

Murray, M. (1976). *J. Comp. Neurol.* **68**, 175–196.

O'Brien, R. A. D., Ostberg, A. J. C., and Urbova, G. (1978). *J. Physiol. (London)* **282**, 571–582.

Prestige, M. C., and Willshaw, D. J. (1975). *Proc. R. Soc. London, Ser. B* **190**, 77–98.

Rager, G., and Rager, U. (1978). *Exp. Brain Res.* **33**, 65–78.

Rakic, P. (1977). *Philos. Trans. R. Soc. London, Ser. B* **278**, 245–260.

Reperant, J., and Lemire, M. (1976). *Brain, Behav. Evol.* **13**, 34–57.

Rho, J. H. (1978). *Biophys. J.* **21**, 137a.

Schmidt, J. T. (1978). *J. Comp. Neurol.* **177**, 279–300.

Schmidt, J. T., Cicerone, C. M., and Easter, S. S. (1978). *J. Comp. Neurol.* **177**, 257–278.

Schneider, G. E., and Jhaveri, S. R. (1974). *In* "Plasticity and Recovery of Function in the Central Nervous System" (D. G. Stein, J. J. Rosen, and N. Butters, eds.), pp. 65–109. Academic Press, New York.

Scott, M. Y., and Sperry, R. W. (1975). *Brain, Behav. Evol.* **11**, 60–75.

Sharma, S. C. (1972a). *Nature (London), New Biol.* **238**, 286–287.

Sharma, S. C. (1972b). *Exp. Neurol.* **34**, 171–182.

Sharma, S. C. (1973). *Exp. Neurol.* **41**, 661–669.

Sharma, S. C. (1975). *Brain Res.* **93**, 497–501.

Sharma, S. C., and Tung, Y. L. (1979). *Neuroscience* **4**, 113–119.

Silver, J., and Sidman, R. L. (1980). *J. Comp. Neurol.* **189,** 101–111.

Singer, M., Nordlander, R. H., and Egar, M. (1978). *J. Comp. Neurol.* **185,** 1–22.

Slack, J. R. (1978). *Brain Res.* **146,** 172–176.

So, K. (1979). *J. Comp. Neurol.* **186,** 241–257.

Sperry, R. W. (1944). *J. Neurophysiol.* **7,** 57–69.

Sperry, R. W. (1945). *J. Neurophysiol.* **8,** 15–28.

Sperry, R. W. (1965). *In* "Organogenesis" (R. L. Dehaan and H. Ursprung, eds.), pp. 161–186. Holt, New York.

Straznicky, K. (1973). *J. Embryol. Exp. Morphol.* **29,** 397–409.

Straznicky, K. (1978). *Neurosci. Lett.* **9,** 177–184.

Straznicky, K., and Gaze, R. M. (1971). *J. Embryol. Exp. Morphol.* **26,** 67–79.

Straznicky, K., and Gaze, R. M. (1972). *J. Embryol. Exp. Morphol.* **28,** 87–115.

Straznicky, C., Gaze, R. M., and Horder, T. J. (1979). *J. Embryol. Exp. Morphol.* **50,** 253–267.

Strumer, C. (1981). *In* "Lesion-Induced Neuronal Plasticity in Sensorimotor Systems" (H. Flohr and W. Precht, eds.), pp. 369–376. Springer-Verlag, Berlin and New York.

Stryker, M. P. (1981). *Neurosci. Abstr.* **7,** 842.

Turner, J. P., and Singer, M. (1974). *J. Exp. Zool.* **190,** 249–268.

Udin, S. B. (1977). *J. Comp. Neurol.* **173,** 561–582.

Udin, S. B. (1978). *Exp. Neurol.* **58,** 455–470.

Willshaw, D. J., and von der Malsburg, C. (1976). *Proc. R. Soc. London, Ser. B* **194,** 431–445.

Yoon, M. (1971). *Exp. Neurol.* **33,** 395–411.

Yoon, M. (1972a). *Exp. Neurol.* **37,** 451–462.

Yoon, M. (1972b). *Am. Zool.* **12,** 106.

Yoon, M. (1975). *J. Physiol. (London)* **252,** 137–158.

Yoon, M. (1976). *J. Physiol. (London)* **257,** 621–643.

Yoon, M. (1977). *J. Physiol. (London)* **264,** 379–410.

Yoon, M. (1979). *J. Physiol. (London)* **288,** 211–225.

CHAPTER 5

MODELING AND COMPETITION IN THE NERVOUS SYSTEM: CLUES FROM THE SENSORY INNERVATION OF SKIN

Jack Diamond

DEPARTMENT OF NEUROSCIENCES

MCMASTER UNIVERSITY

HAMILTON, ONTARIO, CANADA

I. Introduction

A particularly fascinating feature of the development of the nervous system is that a major portion of it begins only after the axons of postmitotic neurons have grown to appropriate target tissues. It is at this time that the modeling of neuronal connections occurs. In the central nervous system (CNS) this modeling is based upon the devel-

CURRENT TOPICS IN
DEVELOPMENTAL BIOLOGY, VOL. 17

opment of the two most characteristic morphological features of neuronal phenotype: the *dendritic fields* and the *terminal fields* (the latter result from the often exuberant sprouting of axons that occurs in the immediate vicinity of the target tissue). In the periphery of vertebrates only the terminal fields are present, and it is the accessibility of these that permits the modeling of neuronal connections to be conveniently studied in skin and in muscle. In addition to the processes involved in axonal and dendritic sprouting, there seem to be specific mechanisms that ensure an appropriate distribution of nerve endings at the growing target tissues, often leading to the development of characteristic associations between these endings and specialized sites ("synaptogenesis"). If to all these activities is added the regression of nerve endings, we have all the important ingredients needed at least to begin to understand how the fine modeling of connections in the nervous system can be brought about. There is another aspect to such an understanding. Because animal behavior continues to evolve and develop into and throughout maturity, we must accept the possibility that an underlying neuronal circuitry also continues to develop; in the mature animal this could probably only occur at the morphological level by remodeling within terminal and dendritic fields.

A fundamental feature of the modeling of neuronal (or synaptic) connections is the phenomenon of competition, the inevitable outcome of too many nerves (or their terminals) chasing too few targets or target sites. The capacity of nerves to produce sprouts may, however, not always outstrip the ability of a target tissue to accommodate them (Raisman, 1977), and, as will be seen, competition could also be based upon other determinants, including the possibility that target tissues themselves may regulate the number or availability of target sites. Our investigations have been directed toward discovering the mechanisms involved in modeling and in competition, utilizing as target tissue the skin of salamanders, rabbits, and rats; the nerves we studied contribute to the low- and high-threshold mechanosensory innervation of skin ("light touch" and "nociceptive" modalities, respectively). Some of the results help in the understanding of how nerve territories develop; the most well characterized of these in the peripheral nervous system of vertebrates are the sensory dermatomes.

II. The Modeling of the Sensory Dermatomes

A. The Sensory Dermatomes

The innervation of vertebrate skin is a clear example of *patterning* in the nervous system; it is organized territorially in the form of sensory dermatomes, each of which represents the cutaneous receptive field of an entire dorsal root ganglion (drg) (e.g., Haymaker and Woodhall, 1953). The dermatomes are generally bilaterally symmetrical. The innervation pattern is seen most clearly in the "serial banding" of the trunk dermatomes (Fig. 1A). In contrast, the shapes and dispositions of the dermatomes in the fully developed limbs are rela-

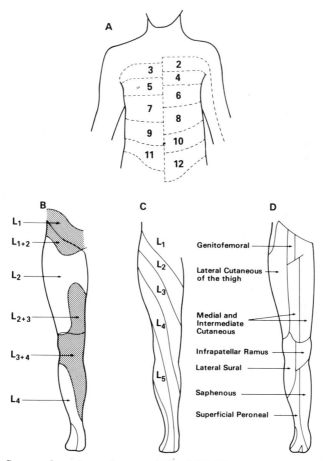

Fig. 1. Sensory dermatomes in man. (A) and (B) Diagrams adapted from Foerster (1933), showing the thoracic and limb "touch" dermatomes of intact dorsal roots whose adjacent roots were cut. The results of another method of "dermatome mapping" are illustrated in (C), which shows the areas of reduced sensitivity after cutting single roots to the limb, but leaving the adjacent roots intact. (Adapted from Keegan and Garrett, 1948.) (D) Peripheral nerve fields. (Adapted from Ranson and Clark, 1959.) Note that in both (B) and (C) the dermatome borders are continuous across two or more peripheral nerve fields.

tively complex (Fig. 1B and C). Nevertheless, a somewhat imaginative extrapolation to the early embryo (Hamilton and Mossman, 1972) (Fig. 2) from the known dermatomal organization in older animals gives a picture of the sensory innervation that suggests that a single overall process may be responsible for its development. The exact size of the dermatome (i.e., the area of skin that contains the functional endings derived from the axons of one drg) often varies according to the sensory modality examined (e.g., light touch, temperature sense, pain), and the dermatomes of different modalities overlap to different extents (Head and Sherren, 1905; Foerster, 1933).

The experimental determination of the size and shape of a dermatome can present problems except when the afferent discharge evoked by the appropriate (or "adequate"; Sherrington, 1947) physiological stimulus can be recorded directly from the dorsal root or from the entire segmental spinal nerve (e.g., Fig. 4). When sensory dermatomes are determined less directly (e.g., by behavioral techniques), their overlapping in particular may be hard to deal with (Cole et al., 1968). More of a problem, however, is that any measure of the sensory innervation that involves the CNS (as do all behavioral methods) is vulnerable to changes in ongoing neural activity; thus, the usual experimental maneuvers involved in the examination of dermatomes, such as the cutting of dorsal roots, in some species can cause immediate

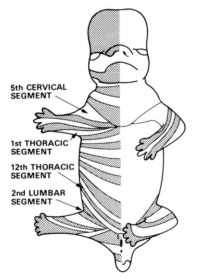

FIG. 2. Serial banding of dermatomes during development. Diagrammatic representation of the dermatomes in the human embryo. (From Hamilton and Mossman, 1972.); the overlap between adjacent dermatomes is not indicated (compare Fig. 1A).

and sometimes dramatic changes in apparent sizes, and even locations, of the cutaneous receptive fields of neurons within the CNS (Kirk and Denny-Brown, 1970; Kirk, 1974; Wall, 1977), presumably because of the consequent changes in the relative inhibition and excitation impinging upon those neurons. This point becomes especially important in the experimental study of how nerve fields can be changed either by sprouting or by regression of nerve endings. Problems can arise, however, even when the direct method of afferent nerve recording is used. For example, a receptive field that has unambiguously enlarged, suggesting the development of new physiologically functional peripheral nerve endings, could, in the absence of morphological evidence, be explained by the unattractive, but admissible, possibility that the field was previously surrounded by a physiologically "silent" fringe of endings that became "competent" as a consequence of the experimental procedures (Wall, 1977).

B. Can the Development of the Dorsal Root Ganglia Account for the Pattern of the Dermatomes?

The usual view of dermatomal development (e.g., Cole *et al.*, 1968; Hamilton and Mossman, 1972) takes account of the initial condensation of neural crest cells (that may or may not be already committed to a sensory role; Le Douarin, 1980) into discrete aggregates—the presumptive drgs—located along either side of the neural tube. The drgs develop in response to the segmentally developing somites; when the number of these on one side is experimentally increased or decreased, there is an addition to or reduction in the number of drgs on that side (though not necessarily by a corresponding amount; Detwiler, 1936). The central axonal projections of the drgs toward the adjacent developing spinal cord—the dorsal roots—probably proceed along nonneuronal cellular pathways (Tennyson, 1970), although the presumed guiding role of the latter is not proved. The peripherally growing axons, most of which are destined to reach the integument and contribute to the sensory dermatomes, penetrate the somites; they could initially follow the segmentally projecting outgrowth of motor fibers from the spinal cord (Taylor, 1943) or the blood vessels (Bennett *et al.*, 1980). However, the possibility exists that there may be nonneuronal guidelines along which axons grow during primary development (Weiss, 1955). Held (1909), in a little-quoted but most interesting discussion, refers to his own observations in favor of cellular pathways and, seemingly successfully (and certainly courteously), refutes Ramón y Cajál's somewhat meretricious dismissal of these observations. Their disagreement may have had more to do with Held's suggestion as to how nerves actually proceeded along such cellular pathways than with their existence.

Ironically, the potential implications of such pathways might well have been disregarded simply because their description, during the first two decades of this century, could have been regarded as yet another attempt to support an alternative to the "neuron doctrine" (see Hughes, 1969, for a relevant discussion). Ramón y Cajál's reputation notwithstanding, Held's observations survive and are a pertinent reminder that the mechanisms responsible for axon guidance are still not understood, that they must exist, and that nonneuronal cells therefore are likely to be involved (cf. Katz and Lasek, 1979; Singer *et al.*, 1979).

With the exception, presumably, of those that will become muscle and joint afferents, most of the drg fibers must reach the presumptive dermis, which has the embryological designation of "dermatome" (not to be confused with the sensory dermatomes, which refer to the geographic locations on the skin of the receptive fields of the drgs, and thus describe a feature of drgs and not of skin as such). In considering the segmental nature of sensory dermatomes, it should be noted that somites give rise to dermis only in the dorsal region of the organism; the presumptive dermis of the lateral and ventral body wall and of the limbs arises from the lateral plate mesoderm (Rawles, 1955; Chevallier *et al.*, 1977). This structure was thought not to be segmented during development, but scanning electron microscopic (EM) evidence favors the possibility that some segmentation of the lateral plate does occur (Meier, 1980). However, the cutaneous epidermis, which is the specific target tissue for axons subserving modalities such as light touch, originates in the embryonic ectoderm, which nowhere undergoes obvious segmentation during development. It is still debatable, then, whether the sensory dermatomes are segmental entirely because the "parent" drgs develop segmentally or whether there is some prior compartmentalization of the skin. In any event, the commonest assumption has been that the ganglia's peripheral outgrowths reach the nearest available skin, ramifying there to create territories that neighboring segmental sensory outgrowths are somehow inhibited from invading (Weiss, 1955). (Our results that are described later explain how an apparent "inhibition" can come about.) First, it is pertinent to review further the possibility that the skin itself may help to impose a segmental pattern on its innervation and in particular to comment on the recently discovered "domains."

C. Problems of Overlap and of Nerve Plexuses

In the trunk the relatively simple sensory innervation pattern persists essentially little changed throughout life; the more complex dermatomal pattern observed in the limbs of the mature animal is presumed to result from regional variations in their growth and orien-

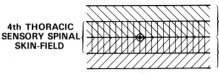

FIG. 3. Overlap of dermatomes. The position of the nipple is illustrated within skin supplied by the third, fourth, and fifth thoracic spinal roots of the monkey. (From Sherrington, 1893.) A single root was "isolated" in separate animals by cutting roots adjacent to it; the remaining dermatome was defined as the area from which reflex responses were evoked by touching or pinching the skin. Maps taken from individual monkeys were superimposed to yield the map shown. The dermatomes of T3 and T5 overlap with that of T4, and the nipple is supplied by fibers from all three roots.

tation (Cole *et al.*, 1968; Hamilton and Mossman, 1972). However, there are observations that the classical explanation of the origin of dermatomes does not readily explain. For example, some adjacent dermatomes overlap only slightly, whereas others do so to a much greater extent; indeed, there can be a sharing of skin among even three sequential drg projections, as in the nipple region of the primate (Sherrington, 1893) (Fig. 3). Also, it is not immediately obvious why the degree of dermatomal overlap can be larger for one sensory modality than for another. As will be seen later, some understanding of these findings may come from a consideration of the influences that regulate the distribution of the individual nerve endings in skin.

The most intriguing problem concerning dermatomes arises from the relative smoothness and continuity of their borders in the limbs, which are supplied from a nerve plexus. For.example, the cutaneous innervation of the lower limb of the human is derived from the lumbar and sacral drgs. Each anatomically distinct nerve (e.g., the lateral sural or the saphenous) has an equally distinctive receptive field, which, interestingly, does not greatly overlap with those of neighboring nerves (Fig. 1D). Most of these nerve trunks, however, are "polysegmental," i.e., they contain fibers from at least two drgs. The dermatomal borders, by whichever techniques they are determined (see Fig. 1B and C), invariably cut across two or more of these nerve fields. This means that fields supplied by, for example, two drgs are divisible into two variably sized subfields, each supplied exclusively by the fibers of one drg (except in the marginal overlap regions). This suggests an ordering of fibers within each nerve (Horder, 1974), possibly based on a mutual adhesivity of axons (Weiss, 1941) deriving from the same drg (irrespective of the modalities subserved[1]). However, such a mech-

[1] This is interesting because the central projections of the same drg neurons eventually segregate on the basis of *modality* also; the pain and temperature axons, for example, taking a quite different trajectory within the spinal cord from the touch and pressure fibers (Brodal, 1969).

anism does not explain the finding that the dermatomal border between one pair of such subfields is smoothly continuous with that between the corresponding subfields of a neighboring nerve. The problem is to understand how, for example, the L2 and L3 subfields in the medial, intermediate, and lateral cutaneous nerve fields shown in Fig. 1 become appropriately oriented, vis-à-vis each other, so that the "internal" L2 and L3 borders become aligned to form relatively smooth and continuous dermatomal frontiers.

D. Solutions to the Plexus Problem

Three possible kinds of solutions can be considered:

1. The axons of each drg connect to the skin before the lumbar nerve plexus forms. Thus, it is only after the sensory dermatome has developed that the progressive subdividing and relocating of axon bundles and a condensing of bundles from one drg with those from another begins and eventually results in the appearance of the characteristic plexus and peripheral nerves (Bennett *et al.*, 1980).

2. The plexus and the defined polysegmental nerves form first, followed by a possibly random ingrowth of fibers into the skin. Thereafter *remodeling* occurs, i.e., sprouting and regression of endings (see following). For example, adjacent or nearby axonal projections of the same drg could "recognize" each other and by some mutually supportive influence persist and continue to ramify, whereas axonal projections that become largely surrounded by "foreign" projections might tend to regress. (Compare the sorting out that occurs in mixed, dissociated cells *in vitro*; Townes and Holtfreter, 1955.) A process like this could end with all the terminal fields of axons from the same drg becoming closely related territorially, to the extent of occupying discrete areas. Even so, to produce the segmental dermatomes, there would need at the very least to be a drastic "smoothing" of the borders between such areas as they approached the frontiers between nerve fields, and this is difficult to envisage.

3. The skin is itself segmentally specified in some way that can be recognized by the invading nerves. The axons of any one drg, despite a possible mixing with "foreign" axons in the plexus, then, in effect, seek out (or are guided to) that skin with which they are "matched"; other axons find the skin uncongenial and do not invade it. Were an anteroposterior gradient of "positional information" (Wolpert, 1971) to exist throughout the entire developing organism, nerves originating at any given anteroposterior level of the neuraxis could recognize posi-

tional labels on tissues that developed at the corresponding level, and innervation would occur appropriately.

There have been conflicting views on the critical issue of whether major plexuses form before nerves reach their target tissues or afterward (e.g., Piatt, 1957; Hughes, 1969). However, recent experiments involving horseradish peroxidase (HRP) uptake techniques suggest that plexus formation in the chick does indeed begin before any axons could have reached the integument (L. Landmesser and M. Honig, personal communication). If so, then solutions (2) and (3) are the most likely candidates to explain the development of sensory dermatomes, and the possibility that the "domains" now to be described may have a significance in the embryo becomes an interesting one.

E. "Domains": A Segmental Regulation of Nerve Territories

The situations we are considering have in common the requirements for mechanisms that regulate (or at least influence) the sprouting and distribution of the individual nerve terminals within the target tissue and probably for mechanisms that can regulate the territory "belonging" to individual terminals, perhaps even to parent axons (i.e., terminal fields). The classical view is that during development nerves inhibit other nerves from invading their territories (Weiss, 1955), presumably by secreting substances that would therefore probably need to act at the level of the growing nerve terminals. Evidence will be presented later that is consistent with a view that is quite different, both conceptually and "operationally." This view assumes that target tissues produce sprouting stimuli and that nerves provide a peripheral influence that reduces the local availability of these sprouting agents (Aguilar et al., 1973). Nevertheless, it is unclear whether mechanisms such as any of these, superimposed on the segmental origin of drg outgrowths, can adequately explain all the characteristics of sensory dermatomes. During our studies of low-threshold mechanosensory ("light touch") nerve sprouting, we observed that the behavior of axons when presented with denervated skin differed according to whether the skin was within the region of the body normally supplied by the parent drg of those axons or was located outside it; the results suggest the existence of some kind of spatial regulation of sprouting. In the context of dermatome development, this is clearly of interest; it must be noted, however, that our evidence comes from studies on postembryonic animals and thus could relate only to the likelihood that such constraints could influence possible remodeling of nerve fields.

1. Mechanosensory Field Mapping and Collateral Sprouting

When target tissues both within the CNS and outside it are partially or even totally denervated, remaining intact nerves will often begin to sprout branches into the deprived regions (Diamond *et al.*, 1976). Most of the investigations referred to in this chapter take account of the possibility that the mechanisms involved in both the initiation and in the subsequent regulation of this collateral sprouting might be related to those that operate when axonal sprouting occurs normally (during development) and perhaps even in the mature organism.

In these experiments we used electrophysiological techniques to determine the presence of functional low-threshold nerve endings. A fine bristle drawn across the skin while the activity in a selected afferent nerve is recorded evokes showers of impulses in the nerve as the bristle enters and traverses its receptive field, and the discharge ceases abruptly as the bristle leaves the field (Fig. 4). This technique permits unambiguous mapping of functional low-threshold mechanosensory fields in skin. A progressive expansion of such fields after denervation of adjacent skin is taken to indicate that collateral sprouting is occurring from the intact axons supplying the adjacent fields (Stirling, 1973). An alternative explanation of such expansion could be that in skin adjacent to the functional nerve fields there is a progressive unmasking of a fringe of normally "silent" endings. However, this latter explanation is unlikely; we have found that punctate electrical excitation of salamander skin, which readily excited mechanosensory terminals, always failed to evoke impulses from mechanically insensitive areas (Cooper and Diamond, 1977). Furthermore, in salamander skin, the specific targets for mechanosensory axons, the epidermal Merkel

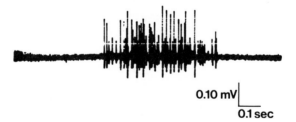

0.10 mV

0.1 sec

FIG. 4. Mechanosensory field mapping. During this single oscilloscope sweep, which displays the recording from a peripheral nerve in the salamander, a fine bristle was drawn across the skin of the hindlimb. The beginning of the burst of afferent impulses indicates the crossing of the border between the skin supplied by neighboring nerves and that supplied by the nerve recorded from; the impulses ceased abruptly as the bristle left the mechanosensory field of that nerve. (From Macintyre and Diamond, 1981.)

cells (see Section III,A,2) survive denervation (Scott *et al.*, 1981a); when expansion of a mechanosensory field occurs into surrounding denervated skin, these nerve-free Merkel cells are seen in the electron microscope to become reinnervated (Macintyre and Diamond, 1981; Scott *et al.*, 1981a). We can therefore dismiss, certainly in the salamander, the existence of "silent" endings, which would need to be not only mechanically insensitive, but electrically inexcitable, and even apparently electron microscopically invisible. In the rat, too (Section IV,E), we have a clear morphological correlate of (in this instance, behaviorally detected) collateral sprouting of remaining nerves.

2. *Preferred Direction of Sprouting*

Figure 5B shows a map of the dorsal surface of a salamander hind limb from which all the innervation was eliminated except for the contribution from the seventeenth spinal nerve (N17) to the cutaneous surae lateralis (CSL) nerves; thus, only the CSL field remained (stippled in the figure). Even this region was partially denervated in such

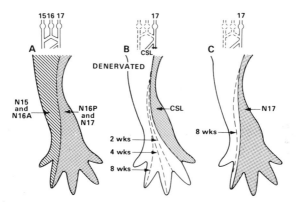

Fɪɢ. 5. Mechanosensory nerve fields in the dorsal skin of the salamander hind limb. The sketch above the limbs shows the pattern of the segmental nerve plexus that supplies the limb and indicates when nerves had been lesioned. (A) The normal innervation divides the skin into two major mechanosensory fields: the anterior, supplied by the fifteenth spinal nerve (N15) and the anterior division of the sixteenth (N16A); and the posterior, supplied by the seventeenth spinal nerve (N17) and the posterior division of the sixteenth (N16P). (B) The field of the cutaneous surae lateralis (CSL) nerves: after elimination of N15 and N16 (including, therefore, the contribution to the CSL nerves of axons from N16P), there was a progressive expansion of this remaining CSL field, now supplied only by axons from N17; the three dashed lines refer to the averaged maps from groups of animals examined, respectively, at 2, 4, and 8 weeks after the initial denervations. (C) Animals in which the hind limb was denervated except for N17. The dotted line shows the extent to which this remaining N17 field enlarged in animals examined up to 8 weeks after the initial operation. (Modified from Macintyre and Diamond, 1981.)

limbs because the CSL nerves also include a contribution from axons of the sixteenth spinal nerve (the posterior division, N16P) that had been cut (see Fig. 5A). As will be described later, within 2–3 weeks of such denervation the density of mechanosensory nerve endings is restored within the normal area of the CSL field by sprouting of remaining N17 axons (Cooper *et al.*, 1977). In addition, sprouting of these fibers occurs into the adjacent, totally denervated skin (Fig. 5B). It is evident, however, that this expansion of the CSL field was not uniform. Over the postoperative period of 8 weeks illustrated, the invaded territory was almost entirely that that was formerly supplied by the other axons of N17 and those of N16P; the immediately adjacent denervated skin that was formerly supplied by axons of the fifteenth spinal nerve (N15) and of the anterior division of N16 (N16A) seemed to have been largely ignored. This apparent reluctance of N17 axons to sprout outside N17 territory was particularly evident in experiments like that shown in Fig. 5C; in this instance, the entire N17 was left intact, whereas N15 and N16 were both eliminated. Except for a small amount at the frontier, there was little expansion of the N17 field during the 8 weeks when the isolated CSL field so clearly would extend into "parent" N17 territory (Macintyre and Diamond, 1981).

Similar findings were obtained with the low-threshold mechanosensory axons of the other spinal nerves that we examined in the salamander (N15, N16, N18, and a trunk nerve, N14). We have also investigated low-threshold mechanosensory nerve sprouting in rat skin. As will be described later (Section IV,D), this sprouting was essentially confined to an early "critical period" that ends at about 20 days of age; during this period we found a spatial constraint similar to that observed in the salamander. A branch of a segmental cutaneous nerve supplying the back skin would sprout into adjacent denervated skin formerly innervated by other axons of the same segmental nerve but only slightly into the apparently equally available skin of the sensory dermatomes rostral and caudal to its own parent dermatome (Fig. 6). Again, when an entire dermatome was left with its innervation intact, we detected no sprouting of these low-threshold nerves into the denervated skin of the neighboring dermatomes (Jackson and Diamond, 1981).

3. "Domains" and Their Borders

We have called these preferred territories within which intact axons will sprout freely *domains* (Diamond *et al.*, 1976); the borders of the domains, at which the hindrance to sprouting is expressed, usually correspond fairly closely to those of the segmental dermatomes, al-

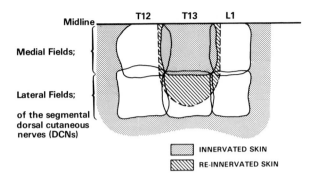

FIG. 6. Low-threshold mechanosensory nerve sprouting in the rat. The outlines indicate the borders of the low-threshold mechanosensory fields of the medial and lateral branches of dorsal cutaneous nerves (DCNs), which supply the back skin of rats; this sketch refers to typical findings in 20-day-old rat pups. In one group of pups, both branches of the T12 and L1 DCNs and the lateral branch of DCN-T13 had been cut 10 days earlier, leaving the medial field of DCN-T13 "isolated" (surrounded entirely by denervated skin). In these 20-day animals, the T12 and L1 DCN fields were "empty," as shown; the remaining medial DCN-T13 field, however, had enlarged (hatched area) well beyond its normal size, as measured from the group of control, unoperated pups at 20 days (shown stippled). Clearly this expansion was directional, occurring largely into the denervated skin within the territory of the "parent" dermatome (that of the T13 dorsal root ganglion). (From Diamond, 1981.)

though there are interesting exceptions that are discussed later. There seems to be no obvious physical barrier to sprouting at the borders of domains. A most revealing indication of this comes from an experiment involving the region of the salamander hind limb where two dermatomes (e.g., those of N15 and N16 in Fig. 5A) partially overlap, one extending further across the limb than the other (see Section IV,B). By totally denervating the skin to one side of the frontier of the smaller (N15) dermatome, the two overlapping segmental nerve fields are caused to "end" at the same border line—in this example, the normal border of N15. The N15 axons subsequently fail to sprout beyond this border, but the remaining axons of N16, without any apparent constraint, cross to invade the adjacent denervated skin (Macintyre and Diamond, 1981) (see also Fig. 15).

There are other situations in which nerves will grow across domain borders. First, nerves regenerating after a cut or crush, both in the salamander (Macintyre and Diamond, 1981) and in the rat (Diamond and Jackson, 1980), freely transgress these borders when they invade formerly denervated skin (see Fig. 14). Second, in salamander experiments in which N17 was the only nerve remaining in the hind limb, the "domain constraint" (apparently fully effective for about 8 weeks) relatively suddenly seemed to disappear, and the axons now sprouted across the frontier into foreign territory (other spinal nerves, however, were constrained for much longer periods than N17). Finally, we have not observed any suggestion of domains for high-threshold nerves in the rat, the sprouting of which will be described later.

Domains appear to relate to the limb as a whole rather than specifically to the skin. We tested for skin preferences by excising skin flaps in salamander hind limbs and reimplanting them after rotation by 0–180°, often in the limbs that were partially denervated (Macintyre and Diamond, 1981). In about half of these limbs, a remaining segmental nerve sprouted into the flap to reestablish its former territory *with respect to the coordinates of the limb*, regardless of whether within the flap it might innervate its former or "foreign" skin (see Fig. 7A and B). In the remaining cases, the invasion of the flaps was indiscriminate (e.g., Fig. 7C), and we interpreted this as being due to nerves that were regenerating after having been cut during excision of the flap. Our conclusion is that the borders of domains are defined with respect to the

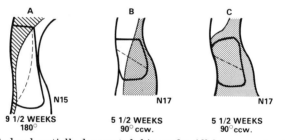

FIG. 7. Rotated and partially denervated skin grafts. All figures are tracings of results of field mapping done in limbs that were partially denervated at the time of skin manipulations. In (A), (B), and (C), the remaining intact nerves are identified, as are the postoperative periods when the mapping was done, and the amount of rotation (and direction for the 90° experiments; ccw. = counterclockwise). In (A) and (B), the line that marked the original boundary of the remaining intact segmental nerve field with respect to the limb coordinates was reestablished in the graft, despite the new orientation of the skin. The frontier between the two major mechanosensory fields within the flap was readily recognized as "normal" by its orientation and smooth continuity with the corresponding frontier in the surrounding skin; a further check of this conclusion was given by a comparison between the fields in the experimental and the control limbs. (C) is typical of the minority group of animals in which apparently "indiscriminate invasion" occurred into the skin of the grafts, and sometimes, as here, into the surrounding skin. The dashed line in all of the figures indicates the orientation of the frontier that would be expected if the nerves had simply reestablished in the skin flap the identical territories they had before the manipulations were made; in all instances, the proximal portion of the graft would have been expected to have become innervated, the more distal part remaining denervated. Results similar to those in (C) were also obtained with 0° rotation. (After Macintyre and Diamond, 1981.)

coordinates of the limb (or the trunk) as a whole. Segmental nerves seem to "view" skin as appropriate or foreign only in regard to its location in "body space" (Diamond *et al.,* 1976). If skin is within the domain of a nerve, it can be invaded by sprouts of intact axons even though the skin was formerly supplied by another segmental nerve; and, conversely, intact axons will not freely sprout into even their former skin if it is relocated to an adjacent region on the "wrong" side of their domain borders. We have discussed elsewhere the extent to which skin may or may not share a domain (i.e., a "positional") specificity with the rest of the tissues within the limb (Macintyre and Diamond, 1981). The results, however, are quite consistent with the skin itself acting as a source of regionally nonspecific sprouting agents (see later).

It seems very clear that the existence of domains (so far identified only in the postembryonic animal) offers a means of regulating the area of target tissue within which sprouting of certain nerves can occur, both during modeling and possible remodeling of nerve connections (see Diamond *et al.,* 1976). If this is true, then the limitations described earlier in the effectiveness of domain constraints must presumably have proved to be of survival value for the organism. The usefulness to the animal of confining the sprouting of segmental nerves (or at least the axons subserving certain modalities) to domains and the possibility that domains represent a general phenomenon that might regulate nerve territories elsewhere in the organism have been discussed by Diamond (1979) and Macintyre and Diamond (1981). Here it is more pertinent to inquire into the possible relationships between domains and dermatomes and between domains and the now well identified compartments that seem to be such an important feature of development in insects and possibly other kinds of organisms.

F. DOMAINS, DERMATOMES, AND COMPARTMENTS

1. Are "Domain Constraints" Responsible for Dermatomes?

The borders of the segmental domains in the rat and in the salamander are sufficiently similar to those of the dermatomes to prompt the question: Do the spatial constraints responsible for domains arise secondarily to the development of the dermatomes or are they actually involved in this development? To support the latter possibility, a key experiment would seem to be the removal of a drg before the skin innervation became established; one would expect that subsequently there might well be a sensory-deprived area corresponding to the missing dermatome. Experiments of this kind have been done in lower vertebrates (Detwiler, 1936; Miner, 1956), and though not addressed explicitly, the results were consistent with there being no obvious signs of insensitive skin in the more mature animal. Recently, direct evidence has been obtained that drgs adjacent to one that was removed in the bullfrog tadpole subsequently supply an extra area of skin of the frog leg, skin that would otherwise have been deprived (E. Frank and M. Westerfield, personal communication). However, in early life, drgs contain many neurons destined to die during the normal course of development (Hamburger and Levi-Montalcini, 1949; Prestige, 1965), and there is evidence suggesting that such neurons can be "rescued" by the provision of extra target tissues at an early enough stage (Detwiler, 1920; Hollyday and Hamburger, 1976). Therefore, we cannot conclude that, in the drg-removal experiments referred to earlier, the axons that contribute normally to form their appropriate dermatomes are also responsible (by sprouting) for the innervation of the neighboring "foreign" skin. It could be that, within drgs specified by their relative position in the anteroposterior axis and "matched" thereby to similarly specified regions of the organism, including the skin (see Section II,D), there are neurons less rigorously committed to respond to specific positional cues and perhaps normally at a disadvantage in competition with better-matched neurons; these disadvantaged neurons could be those rescued by the provision of neighboring skin deprived of its own segmental innervation. Perhaps it is a similar kind of neuron that can more readily innervate a territory supplied by another drg, thus accounting for overlap between adjacent dermatomes.

2. Do Domain and Dermatome Borders Coincide?

A relevant consideration is the extent to which domain and dermatome borders coincide. Because segmental nerve sprouting almost always would transgress the dermatome boundary to a small extent, domains are always somewhat larger than dermatomes; however, we observed one kind of situation in which the discrepancy was a highly significant one, in a direction that supports the possibility that domains do not arise secondarily to dermatomes. As has been already indicated, in the hind limb of the salamander (*Ambystoma tigrinum*), the commonest pattern of innervation is that (Fig. 8A) in which the dorsal skin essentially is divided into two major mechanosensory fields: the anterior one supplied by N15 plus the anterior division of N16 (N16A), and the posterior by N17 plus N16P. The frontier between these major fields also defines the domain border of N15 and of N17 on the dorsal surface (Macintyre and Diamond, 1981). There is, however, another, less common pattern of innervation in which on the dorsal surface the N15 and N17 fields are "small" and fail to abut (Fig. 8B). We observed this pattern particularly in the animals we procured in

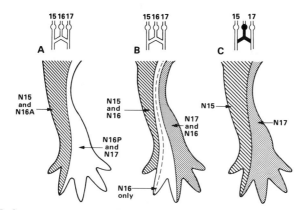

FIG. 8. Mechanosensory nerve fields in salamander hind limbs. (A) The usually observed innervation pattern, as described earlier in Fig. 5. (B) The pattern in the minority group of animals in which the fields of N15 and N17 did not abut on the dorsal surface; however, in these instances, the fields of N16A and N16P did abut, along the dotted line. (C) The situation observed 2–3 weeks following either section of N16 or its treatment by colchicine as described in the text; the expansion of the N15 and N17 fields caused them to abut along a line very close to that defining the normal border between N16A and N16P. (After Diamond *et al.*, 1976. Copyright 1976 by the American Association for the Advancement of Science.)

our earlier studies (Aguilar *et al.*, 1973). Had we known of domains at that time, we would have investigated the behavior of a remaining N15 or N17 in such animals. However, indirectly, we do have a good indication of what the result would have been, and it is an intriguing one.

We showed in those and in later studies (Cooper *et al.*, 1977) that an appropriate application of colchicine to a spinal nerve trunk would, even in the absence of axonal degeneration, result in sprouting of untreated axons into the skin supplied by the colchicine-treated ones (see Section III,B). Thus, colchicine treatment of N16 was as effective as section of N16 in evoking sprouting of N15 and N17 (Figs. 8C, 9A, and 9B). Moreover, the colchicine-treated axons did not themselves sprout, even in response to adjacent denervation. Thus, when N16 was sectioned and N15 treated with colchicine in the "small-field" animals, N15 did not sprout (Fig. 9C); however, the amount by which the N17 field enlarged was no different from that when N15 was present and had itself sprouted. Because colchicine would cause the N15 territory to become "available" for invasion by adjacent intact nerves (see Section III,B,1), the result indicated that the expansion of the "small" N17 field must have ceased when the N17 field border was approximately coincident with the usual border between N16A and N16P; at this point there would still be denervated skin beyond this border line. This line in most animals delineates both the dermatome and the do-

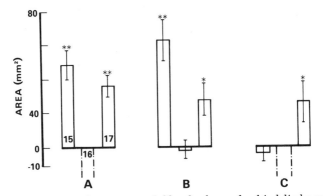

Fig. 9. Changes in mechanosensory fields of salamander hind limbs produced by operative procedures on N16. Each column represents the mean (± SEM) of the difference between field sizes in the right and left hind limbs for the spinal nerve indicated by the number (15, 16, or 17). Significance of difference between right and left sides is shown by $*p < 0.05$, $**p < 0.01$. Increases in field size on the right (treated) side are indicated by a positive value. The results were obtained from animals treated 6–28 days previously. (A) Section of N16 in the right hind limb (30 animals). (B) Colchicine (0.1 M) applied to the right N16 for 30 minutes (21 animals). (C) Section of right N16 (as in A) combined with application of 0.1 M colchicine to right N15 (13 animals). The dashed columns indicate the total loss of mechanosensitivity from the sixteenth nerve fields after section of these nerves. (From Aguilar et al., 1973.)

main border of N17. One interpretation of this result is that segmental nerves are allotted territories (domains) during primary development and normally occupy them more or less fully to create the dermatomes. However, on occasions, as in this minority group of animals, the developing segmental nerves fail for some reason to do so, thus producing significantly smaller dermatomes than domains (in contrast to the normal small discrepancy mentioned earlier). Another view would be that regression of fields occurred after an initial occupancy of the domain. Both views point to the domains we have identified as evidence of a compartmentalization of "body space" (Diamond et al., 1976) during development; normally the projection of the segmental nerve onto the skin bordering that space creates the dermatome.

3. Compartments and Domains

If dermatomes were indeed to result from the innervation of predefined skin territories by segmental outgrowths from the neuraxis, it would seem to indicate that domains are a feature of primary development, relating to regions with positionally specified identities. Other, more precisely defined, regionally determined parts of organisms have been discovered to be a feature of insect development known as *compartments*. Could the existence of domains indicate that an analogous compartmentalization occurs during vertebrate development? Compartments have a number of defining characteristics (García-Bellido *et al.*, 1979), but of great relevance to the findings that led us to propose the existence of domains are the following properties: (1) Compartments can be territories that are morphologically often smaller than identifiable segments of the organism and (2) although segmental boundaries that are also compartment boundaries may be respected by nerves (Lawrence, 1975), nerves can grow across compartment boundaries that occur within segments (Palka, 1981). Thus, a spatially limited behavior of nerves may reveal nothing about compartments as such but may relate only to those compartment boundaries that are also segmental ones. Our findings, therefore, may say no more about vertebrate development than that some sort of segmentation might occur in which different segments have different positional values that are used by the organism as part of a mechanism for the regulation of nerve territories. Even if segmentation occurred generally (cf. Meier, 1980), and if innervation of the periphery were to be based upon preferential selection by neural outgrowths of positionally appropriate, segmental regions of body space, there is no evidence whatsoever supporting a genetic basis for the developmental sequence of events leading to domains. Results of investigations of cell lineage in *Xenopus* embryos (Jacobson and Hirose, 1981) have been interpreted as indicating that compartments ("clonal domains") are a feature of the development of the vertebrate CNS, because relatively small numbers of HRP-injected cells appear to give rise to defined regions of the CNS that were determined by the location of their marked founder cells. It is still too soon to judge whether domains relate to (insect) compartments or to these clonal domains in *Xenopus*, and even to be absolutely sure of their relationship to dermatomes, as pointed out earlier. However, such relationships are not positively excluded by any of the available data, and at present the possible functional implications of domains as spatial constraints on nerve sprouting seem to be a more rewarding line of inquiry (Macintyre and Diamond, 1981).

III. Axonal Sprouting and the Distribution of Nerve Endings

A. The Roles of Target Tissues and Cells

1. The Provision of Sprouting Agents

Axons *in vivo* quite commonly remain unbranched for most of their length, although for some types of neurons there are usually characteristic locations where one or more daughter axons will arise (Dogiel, 1908; Williams and Warwick, 1975). In general, however, there is only one place where profuse collateral sprouting will have occurred in the history of any particular neuron and that is in the immediate vicinity of its target tissue. Ramón y Cajál (1919) was struck by this phenomenon during the development of epidermal innervation and suggested that the obvious explanation was the production by the target tissue of nerve sprouting factors. It is significant that the only such factor that has been well characterized—nerve growth factor (NGF)—has now—more than 60 years after Ramón y Cajál's suggestion—been shown to occur in many of the target tissues of the neurons that are especially sensitive to its action—the postganglionic sympathetic nerve cells (Ebendal et al., 1980). However, although a significant proportion of drg neurons are NGF-sensitive in the embryo (Johnson et al., 1980), they lose this sensitivity well before the period when sensory nerve sprouting in skin still occurs (Fitzgerald, 1961). Presumably other nerve growth factors must exist for these and other NGF-insensitive neurons, to evoke their sprouting into growing target tissues throughout development.

It makes sense that axonal sprouting should occur at the target tissues rather than along the axons themselves, for reasons of space. If within nerve trunks (and central tracts) all the axons were to produce even as few as five branches (with similar conduction velocity, and thus diameter, to that of the parent fiber), the resulting cross-sectional sizes of peripheral nerves and of central pathways would become incompatible with the normal gross anatomy of organisms as we know them. The solution that has evolved, of course, has been the restriction of the necessary branch production to the terminal portions of the nerves (at the target tissues), where the nerve sprouts are freed from the need of relatively high conduction velocities, and so can remain very fine for the short distances that remain to be traversed. The most obvious means whereby sprouting could be made appropriate for the needs of the target tissue would be for sprouting agents to be made and liberated by the target tissue itself, as mentioned earlier; the provision of such agents would then be approximately in proportion to the amount of target tissue present. As will be seen later, Ramón y Cajál also astutely foresaw the need for such a process to have an additional mechanism to regulate the amount of sprouting agent that could persist locally; otherwise, there would be an inexorable tendency for it to accumulate and cause an excess of sprouting.

Despite the lack of identified nerve growth factors other than NGF, it is now clear from a variety of observations and experiments (e.g., those described in Section II,E,1) that at the phenomenological level the influence of uninnervated target tissues on nerves in vivo is clearly compatible with these tissues acting as a source of sprouting factors (Diamond *et al.*, 1976). Tissue explants can influence sprouting of neurons *in vitro* (Ebendal *et al.*, 1980) in a manner consistent with the known ability of added tissues or tissue fragments to increase the sprouting of intact nerves *in vivo* (e.g., Bueker, 1948; Levi-Montalcini and Hamburger, 1951; Olson and Malmfors, 1970; Fitzgerald *et al.*, 1975).

That a variety of nerve sprouting factors originate from target tissues now seems a highly acceptable proposition and one that is supported by other evidence (mentioned later) on the probable role of the nerves themselves in regulating the local availability of sprouting agents. However, the sprouts require mechanisms to regulate how they become distributed within, or at, the target tissue. Our studies on the sensory innervation of skin have thrown some light on this problem.

2. *Specific Target Cells That Determine the Distribution and Function of Endings: The Merkel Cells*

In general, sensory nerves are of two kinds: those whose peripheral terminals are specialized to respond to (usually specific) sensory stimuli in the absence of associated sensory cells and those that terminate at recognizable end organs. It is now known, particularly because of the investigations of Zelena and others, that a common feature of many of the latter structures is that they are induced to develop by the arriving nerves (Stone, 1940; Zelena, 1957, 1976; Farbman, 1965; Werner, 1974). How the endings of the latter nerves, and the former "free endings," might become distributed is considered later. There is one mechanism, however, that can most efficiently regulate the distribution of endings; this is the provision by the target tissue of specific target cells (or even target sites) with which the nerve endings are required to "synapse" if they are to become functional. (Of course, this general "solution" leaves unsolved the problem of how preexisting targets themselves come to be appropriately distributed.) We have found by correlative physiological and electron microscopic techniques (Parducz *et al.*, 1977) that the endings of the low-threshold mechanosensory axons (responding to light touch) in salamander skin are associated with specialized sensory cells in the epidermis, known as Merkel cells (Merkel, 1875) (Fig. 10) that are targets for these nerves (Scott *et al.*, 1981a; see Fig. 12). Thus, we have a fortunate exception to the more usual rule of end-organ induction that operates generally in the integument.

A number of observations now indicate that in a variety of vertebrate species, which includes the salamander and the human, epider-

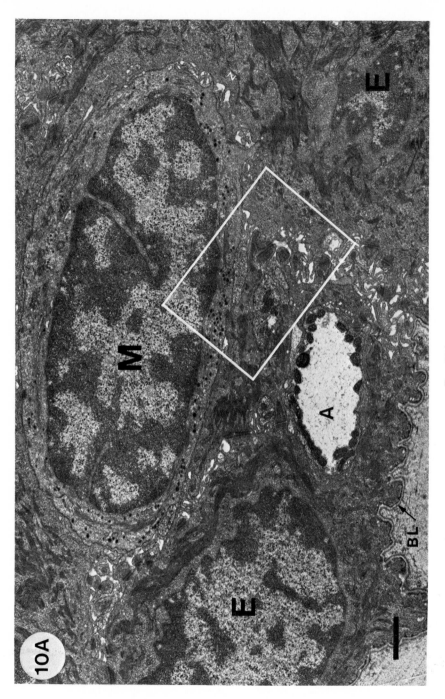

Fig. 10A.

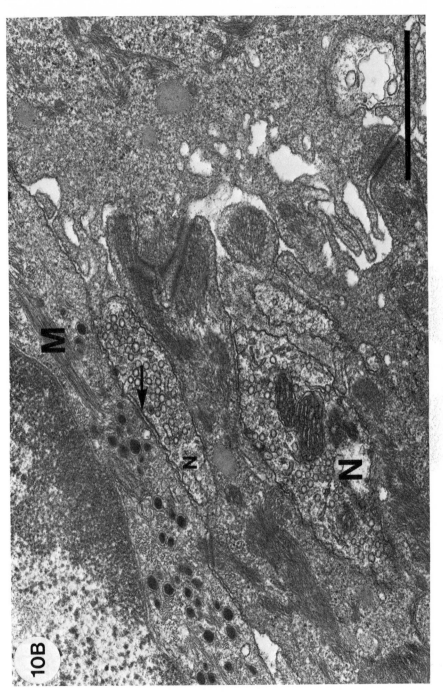

Fɪɢ. 10B. See legend on page 170.

mal Merkel cells can differentiate before nerve endings arrive in their immediate vicinity (Lyne and Hollis, 1971; Hashimoto, 1972; Tweedle, 1978; Call and Bell, 1979). In our salamander studies, we showed that in the mature animal Merkel cells appeared in skin that had regenerated (in place of an excised portion of skin)—even in a totally denervated hind limb—and that they could be detected in such skin as early as 1–2 weeks, when only a thin transparent epidermis was present (Scott *et al.*, 1981a). When nerves were allowed to grow into such regenerating skin (this would occur only after 3 or more weeks, when *dermis* had formed), normal mechanosensitivity gradually developed in the regenerated area, and at that time the Merkel cells within it were found to have become innervated.

Merkel cells survive denervation, and in this situation, too, they act as targets for arriving nerves (Scott *et al.*, 1981a). One especially intriguing result that we had obtained earlier by purely physiological techniques (Cooper *et al.*, 1977) was that in partially denervated salamander skin the remaining intact nerves sprouted new functional endings by an amount that always restored the density of "touch spots" to the level that had existed in the skin originally (Fig. 11) (i.e., equal to their density in the corresponding region in the skin in the opposite hind limb, which we showed was a valid control for such side-to-side measurements). The survival of denervated Merkel cells explains this quantitative recovery of touch spots. Significantly, in normal skin, in regenerated skin that had become innervated, and in reinnervated skin that had been previously denervated, we obtained no evidence of the existence of mechanosensory endings that were not associated with Merkel cells. It seems that sprouting within skin ceases when all the available Merkel cells become innervated and/or that endings that are unsuccessful in finding their specific targets remain "silent," or disappear. The latter might involve "regression" of endings rather than their degeneration; such regression has been observed directly (Speidel, 1942) and inferred from an absence of signs of degenerating endings in

---

FIG. 10. Neurite–Merkel cell complex in salamander skin. (A) The basal region of epidermis showing a Merkel cell (M), portions of epidermal cells (E), an unmyelinated axonal profile (A), and the basal lamina (BL and arrow) between the epidermis and the dermis below. The rectangle includes the region of nerve endings at the Merkel cell, shown at higher magnification in (B). Bar: 1 μm. (B) The region within the rectangle in (A), showing the Merkel cell cytoplasm (M) with its characteristic dense-cored granules, nerve endings (N), one of which abuts on the Merkel cell, and at one place makes a simple "synaptic" contact (arrow). Note the clear vesicles within the nerve endings (compare Fig. 28). Bar: 1 μm.

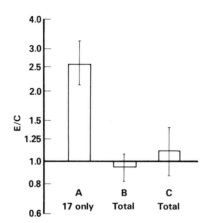

FIG. 11. Quantitative sprouting after partial denervation of the hind limb of the salamander. The percentage of low-threshold receptors feeding into the sixteenth and seventeenth nerves from a shared region of skin was measured, and the values compared between right (E) and left (C) limbs. Column A refers to a group of animals in which the right sixteenth nerve had been sectioned 3 weeks previously and shows the right : left ratio for the number of seventeenth nerve touch receptor population only. An increase in seventeenth nerve receptors is clearly seen. Column B shows, for the same group of animals, the right : left ratio for the total receptor population (that is, the seventeenth on the treated side, the sixteenth plus the seventeenth on the control). Column C shows right : left ratios for the total population of touch receptors in a control group of animals, with sixteenth plus seventeenth nerves intact on both sides. There is no significant difference ($p > 0.2$) between column B and column C, indicating that the increase in seventeenth nerve receptors on the right side of the experimental group had quantitatively made up the loss due to sixteenth nerve section (vertical bars equal SEM). (From Cooper et $al.$, 1977.)

situations where disappearance of functional terminals was known to have occurred (Korneliussen and Jansen, 1976; Kuffler et $al.$, 1980).

Could specific target cells be a source of sprouting agents? Within the dense-cored granules of Merkel cells, for example, is a material whose identity and function is unknown (Smith, 1967). One report suggests that Met-enkephalin is present in certain Merkel cells (Hartschuh et $al.$, 1979); we ourselves have confirmed (Nurse et $al.$, 1981), for a variety of tissues that contain Merkel cells, a report suggesting that in $Xenopus$ these cells selectively take up the fluorescent dye quinacrine, possibly signifying the presence of purines (Crowe and Whitear, 1978). However, there is no evidence that either of these classes of substances has a sprouting effect. We do not know whether growing mechanosensory nerves are attracted to Merkel cells by diffusible agents released from the cells or simply recognize the cells upon contact. In cocultures of drg neurons and $Xenopus$ skin, we obtained indications of preferential growth of axons to epidermal explants (Mearow *et al.,* 1981), but we have not yet ascertained whether this is related to the presence in the epidermis of Merkel cells. In the salamander it is clear that some important information is exchanged when an ending finds and synapses with an available Merkel cell, because the individual sprout then ceases its growth (although other endings of the same parent axon presumably continue to grow until they, too, find a Merkel cell), and mechanosensory function develops. Moreover, as mentioned earlier, endings that fail to contact Merkel cells may not survive. [An additional and important outcome of the development of a successful contact (see later) is that the Merkel cell loses its target character.] The mechanosensory nerve terminals in the salamander contain "packaged" information, as do the Merkel cells; in the endings, this is in the form of clear vesicles. There must be other strategies available, however, to achieve an appropriate distribution of nerve endings, because sensory terminals that do not have specific target cells seem to "know" when to stop growing, when to develop their characteristic sensitivities, and/or when to induce the differentiation of characteristic end organs.

3. *Generalization from the Skin*

All these phenomena have their counterparts in the CNS. For example, certain brain nuclei or cell layers may constitute target tissues (cf. Jones *et al.,* 1981); individual neurons are obviously target cells for the axons of other neurons; dendritic spines may form on some neurons as specific target sites before the afferent nerves arrive (Herndon *et al.,* 1971; Rakic and Sidman, 1973); and on other neurons, spines seem to be induced to develop by the arrival of the nerves (Globus and Scheibel, 1967; Valverde, 1967). Moreover, it is clear that there must often be a mosaic of surface membrane specialization, invisible in the EM, because on any one neuron there is frequently a selective association of certain afferent inputs with particular locations on the cell (e.g., Schwartz and Kane, 1977). Even on Merkel cells there is almost always a preferred region where nerve endings make their contacts. Another possible explanation of the distinct aggregation of different inputs observed on neurons such as the Mauthner cell (Bodian, 1942; Diamond, 1968) is that a remodeling of synaptic connections occurs after an initial, and possibly random, innervation of the cell develops. It is difficult to attribute these selective regional associations of particular nerve endings on one cell to the operation of specific growth factors. The picture that emerges from our data on the innervation of skin is one of sprouting agents that are effective locally and that may or may not help in guiding the growing endings to specific target sites; this may be followed by competition among endings that are able to recognize specific targets or target sites (these may differ for different types of nerve fibers, or nerve modalities), and finally synaptogenesis occurs. On this basis, sprouting agents would be expected to be diffusible substances, whereas recognition sites would involve molecules "fixed" either to target cell surfaces or to nearby matrix or substrate surfaces (Moscona, 1974; Burger, 1979; Weber, 1979; Gottlieb and Glaser, 1980). Unsuccessful endings are those that fail to recognize, or to form a stable association with, a target site, or that are displaced by other endings (see later).

B. REGULATORY ACTIONS OF NERVES AND THE ROLE OF FAST AXOPLASMIC TRANSPORT

A simple hypothesis that can explain a number of phenomena is that nerves themselves are involved in regulating the peripheral mechanisms that, as indicated earlier, probably direct the sprouting and distribution of nerve endings at target tissues. Ramón y Cajál (1919) was the first to propose such a function when he recognized the need to limit nerve sprouting as well as to evoke it. The results of our studies support and extend this concept.

1. Regulation of Sprouting Factors and the Creation of Individual Terminal Territories

The sprouting of intact nerves in salamander skin can be evoked by colchicine treatment of neighboring nerves just as effectively as if the latter nerves were sectioned (see Section II,F,2). We did those experiments to test our hypothesis that partial denervation leads to the sprouting of remaining axons because of the reduction at the target tissue of an influence exerted normally by intact nerves; specifically, we proposed that the effectiveness (or production) of a sprouting factor made by skin was continuously counteracted by the release from the nerve endings of "neutralizing" agents conveyed to the periphery by fast axoplasmic transport (Aguilar et al., 1973). Another possibility we have considered is that nerve endings continuously remove the assumed sprouting agent by an uptake mechanism whose functioning is similarly dependent upon fast axoplasmic transport (Diamond et al., 1976). Our ongoing in vitro studies may eventually distinguish between these two proposed neural mechanisms. The sprouting of untreated nerves that we observed in the colchicine experiments was postulated to result from the buildup of the tissue-derived sprouting agents to higher than normal levels; this sprouting would be expected to continue until the influence (perhaps the actual number) of the nerve terminals within the target tissue was again adequate to balance the effects of the sprouting stimulus. The results of these experiments were consistent with our hypothesis for nerve sprouting and its regulation. We were later able to demonstrate that the colchicine treatment (which we showed reduced fast transport without obvious interference with slow flow; Aguilar *et al.*, 1973; Holmes *et al.*, 1977) could evoke the sprouting of untreated nerves at a time when the mechanosensory function of the treated ones was quantitatively unchanged (Cooper *et al.*, 1977). The mechanosensory threshold is a very sensitive indicator of the state of these nerves, and our physiological analysis was much more satisfactory in giving information about the possibility of degeneration of mechanosensory axons than a histological examination could be. Nor did we have any indication that other, nonmechanosensory axons were degenerating in these experiments (see Aguilar *et al.*, 1973), although it was not possible to exclude this entirely.

Our results were thus in accord with Ramón y Cajál's original suggestion of the influence of nerves during primary development; they also explained "denervation sprouting" without the necessity of having to invoke sprouting agents created either as a product of nerve degeneration or of nonneuronal cellular changes associated with degenerating nerves (Hoffman, 1950). The results also explained why the sprouting occurred at the target tissue rather than in the nerve trunks where intact and degenerating axons might lie side by side. There are other implications of the hypothesis (Diamond, 1979). It is of particular interest that the mechanisms that we suggested would provide a means of regulating the sprouting of axons not associated with preexisting targets. In effect, each nerve terminal would acquire its own local "territory" within which the level of sprouting agent would be kept continuously low by the influence of that ending, and other endings would thus not be induced to sprout into it; there would be a tendency for "free" endings, therefore, to become fairly evenly distributed.

Essentially similar effects of colchicine treatment have now been shown to occur in mammalian muscle (Guth *et al.*, 1980) and in mammalian brain (Goldowitz and Cotman, 1980). A most interesting finding is that of Ebendal *et al.* (1980), who showed that NGF levels in smooth muscle that is normally a target tissue for postganglionic sympathetic axons is high when the muscle is denervated and is much reduced when it is innervated. This is probably the only direct evidence supporting our hypothesis that nerves regulate the level, or availability, of a sprouting agent deriving from a target tissue.

2. Regulation of Target Characteristics

We have consistently observed in our quantitative studies of the distribution of mechanosensitive spots in salamander skin that, within the limits of resolution afforded by our techniques, a single "touch spot" appeared always to be innervated by the endings of only one axon. This was true of normal skin (Cooper and Diamond, 1977), of skin that had been denervated and subsequently reinnervated by the regenerating nerves or by sprouts from intact nerves, and of newly regenerated skin after it became innervated (Scott *et al.*, 1981a). We have correlated touch spots with Merkel cell locations in essentially all these situations (Fig. 12) (Parducz *et al.*, 1977; Scott *et al.*, 1981a). Therefore, these findings indicate that when the endings of one axon successfully innervate a nerve-free Merkel cell, its target character is suppressed; other growing endings seem to ignore that cell. In experiments in which nerves were encouraged to compete for territory within denervated skin (Scott *et al.*, 1981b), the results were always consistent with the explanation that this suppressive effect was achieved by the first endings to arrive at an available Merkel cell. However, in the colchicine experiments, a different result was suggested; the untreated axons

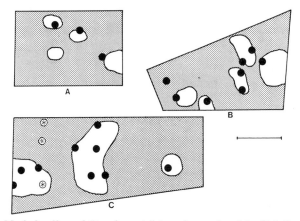

FIG. 12. Merkel cells and "touch spots" in salamander skin. Relationship between Merkel cells (black spots) whose distribution was obtained from a morphological study, and the regions (white areas) that were found to be highly sensitive to mechanical stimulation of the same region of skin. The maps of the Merkel cell locations were superimposed (after appropriate correction for shrinkage) on the maps of the mechanosensory areas; the correlation between the two in all examples is highly significant ($p < 0.001$). (A) and (B) Normal skin. (From Parducz *et al.*, 1977.) (C) Skin that had regenerated in place of a flap that was excised some months earlier and had become innervated from nerves in the surrounding skin. (From Scott *et al.*, 1981a.) Scale: (A) and (B) 200 μm; (C) 250 μm.

sprouted and hyperinnervated the skin supplied by the treated nerves (Aguilar *et al.*, 1973). In preliminary quantitative experiments, it seemed that many touch spots in this skin were now supplied by at least two axons, the original untreated one, and a treated axon that had sprouted (Fig. 13) (Cooper *et al.*, 1977). We have not yet, however, confirmed this result at the level of identified Merkel cells.

The simplest explanation of these findings is that the target character of the Merkel cells supplied by axons in which fast transport was reduced became similar to that of denervated Merkel cells; the mechanosensory function of these presumably "doubly innervated" neurite–Merkel cell complexes was, however, normal in both nerves. We conclude that some component of fast transport is normally continually involved in masking the target character of the Merkel cell. We are now looking in the EM to see whether the unmasking produced by colchicine is related to some structural change in the mechanosensory complex, e.g., a reduction in area of contact. However, as Fig. 10 indicates, normally only a very small amount of the Merkel cell membrane is apposed by nerve endings in salamander epidermis. This is not

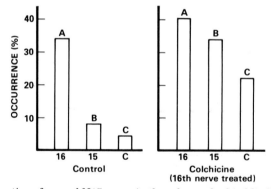

FIG. 13. Sprouting of normal N17 axons in the salamander hind limb after colchicine treatment of N16. In a single animal, the touch receptor density (i.e., the percentage occurrence in the "low-threshold" range; see text and Cooper and Diamond, 1977) in a region of skin shared by N15 and N16 was investigated after the sixteenth nerve on one side was treated with colchicine. There was no loss in the population of receptors feeding into the treated sixteenth nerve, compared to the sixteenth nerve in the control hind limb (columns A). In this animal, the number of touch receptors associated with N15 was only a small proportion of the total. However, on the treated side, N15 now supplied an extra population of receptors almost equal to the number associated with N16 (columns B). The third columns, C, indicate the amount of overlap between the fields of the low-threshold receptors (i.e., the percentage occurrence of touch spots that appeared to be supplied by axons of both N15 and N16). (From Cooper *et al.*, 1977.)

true of the Merkel cells in the (adult) mammalian touch dome, however (Smith, 1968; English *et al.*, 1980). Moreover, it seems that in the rat the target masking of previously denervated Merkel cells that have become innervated by sprouts cannot be total (see following).

3. *Fast Transport and Axonal Sprouting*

The results of our colchicine studies also indicate that some component of fast axoplasmic transport is required for a nerve to respond to the presence of a sprouting agent. When, in addition to denervation of skin (into which neighboring nerves would normally sprout, see Section II,F,2) one neighboring nerve was also treated with colchicine, that nerve failed to sprout (Fig. 9C). The mechanosensory function of this treated, "nonsprouting" nerve, however, remained apparently unchanged. It seems then that in addition to regulating the local availability (or effectiveness) of tissue-derived sprouting agents in the vicinity of their endings, axons also possess a means of controlling their own sensitivity to such sprouting factors.

All the mechanisms of regulation mentioned in the preceding sections (1, 2, and 3) are dependent upon fast axoplasmic transport and all involve a component of it that apparently is not crucial, at least in the short term, for the sensory functioning of the axon terminals.

IV. Determinants of the Outcome of Competition among Nerves for Target Sites

We can now propose how some of the processes described earlier affect competition among nerve endings for target sites. Additionally, there are two unexpected phenomena that we have come across in the mammal and that will be briefly described (Section IV, D and E), because both seem basic to the problem of how modeling and remodeling of innervation patterns could be influenced. Not included, but also clearly of importance, is the likelihood that there will often be a matching of some specific qualities (in addition to possible "positional" ones) between certain nerves and targets, e.g., motor axons and muscle, and sensory axons and skin, which could totally bias a potential competition in favor of one class of competing nerves (cf. Gutmann, 1945; Weiss and Edds, 1945).

A. Timing of Arrival of Endings at Specific Targets

By producing a variety of situations in the salamander hind limb in which either regenerating axons or intact sprouting ones were in competition with other regenerating axons for denervated skin, we showed that it was the sensory endings that arrived first that "captured" the denervated Merkel cells, apparently permanently (Scott *et al.*, 1981b). In Fig. 14, the touch spots (i.e., the neurite–Merkel cell complexes) supplied entirely by regenerated N15 and N17 axons were shared equally between the two segmental nerves in the two regions of skin, although each region was previously supplied by only one of these nerves (in association with axons of N16). There was no discrimination or matching apparent nor was there any subsequent displacement of one class of endings, even of a "foreign" nerve by endings of the appropriate one. In one situation only, in which there must have been a nearly simultaneous arrival of regenerating and sprouting endings, the former seemed to have some advantage over the latter, irrespective of whether or not the Merkel cells were "appropriate" or "foreign" for the regenerating fibers. If the sprouted ones clearly arrived first, however, they were the successful ones in the competition. This slight suggestion of a competitive advantage of regenerating over sprouting axons is of particular interest in view of the quite different results found in analogous experiments in the mammal (see following).

If such a phenomenon were to occur during development, the timing of axonal outgrowth would clearly be crucial in determining the contribution of individual neurons to the overall pattern of endings at a

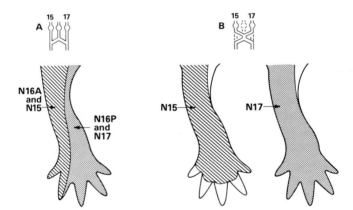

Fig. 14. Simultaneous regeneration of segmental nerves 15 and 17. (A) Normal mechanosensory fields of N15, N16A, N16P, and N17 on the dorsal surface of the control limb. The normal nerve plexus is drawn above. (B) The experimental limb in which the proximal stumps of N15 and N17 were redirected down the distal stump of N16. The mechanosensory fields established by the regenerated nerves 14 weeks later are shown separately for clarity. Both nerve fields overlap extensively, and the touch spots, too, were shared equally between N15 and N17; each individual touch spot was supplied by only one axon, however (see text). (From Scott *et al.*, 1981b.)

shared target tissue. This simple strategy of resolving a potential competition among nerve endings seems likely to be heavily utilized in the developing organism (e.g., Romanes, 1941).

B. DISCRIMINATION AT DOMAIN BORDERS

The special relevance of domains (Section II,E and F) to competition is that the spatial constraint at a domain border can operate selectively (Macintyre and Diamond, 1981). Figure 15 shows the results of an experiment in the salamander, in which by cutting appropriate nerves to the hind limb one half of the dorsal skin was totally denervated; the border between innervated and denervated regions was coincident with the dermatomal border of N15 but effectively ran across the "middle" of the N16 dermatome (on one side of the line the skin was supplied by N16A axons and on the other side by N16P axons; Fig. 15A and B). The eventual outcome of this situation is seen in Fig. 15C; the axons of N15 were essentially confined to their domain and hardly invaded the denervated skin at all, whereas the N16 axons showed no signs of constraint, freely sprouting across the boundary to innervate the available territory on the other side, which, of course, was within the N16 domain. This means of influencing a potential competition between two segmental nerves would clearly be of importance if any

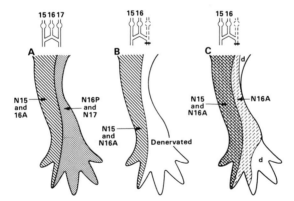

FIG. 15. Discrimination at a domain border. (A) This limb shows the typical pattern of innervation of the dorsal skin of the salamander hind limb. (B) The remaining innervation after removal of the seventeenth drg and section of the nerve indicated in the diagram above; this operation denervates all the posterior part of the dorsal skin. (C) Six weeks later the fifteenth spinal nerve essentially remained confined to its original territory, whereas the anterior division of N16 had sprouted across to fill almost the entire surface of the dorsal skin, that which had previously been innervated by N17 and by the other division of the sixteenth spinal nerve. [(C) taken from Macintyre and Diamond, 1981.]

remodeling (by sprouting and regression of endings) were to occur; if, however, domains are a true feature of primary development, the same mechanisms would presumably contribute to the production of the eventual overall innervation pattern within a target tissue such as skin, but perhaps within other regions, too (Diamond, 1979; Macintyre and Diamond, 1981).

C. VARIATIONS IN FAST AXOPLASMIC TRANSPORT

From the description given in Section III,B, it can be inferred that appropriate alterations of all, or specific components, of fast axoplasmic transport in nerves might markedly influence the outcome of competition among them. For example, a reduction of such transport in one of a pair of axons that arrive more or less simultaneously at a target tissue would disadvantage that axon on three counts, relative to the other: (1) It would respond less vigorously to the tissue-derived sprouting stimulus. (2) It would be less efficient in reducing the local availability of the sprouting stimulus, so permitting the endings of the other axon to grow into its territory. (3) It would be less able to mask the target character of any specific target cells or sites it had contacted, so becoming vulnerable to displacement by sprouts of the other axon. It could be speculated, too, that the ability of the first axon to induce the differentiation of end organs might also be impaired, though this has not been shown. Another importance of the fast transport mechanism is that it could be the means whereby some other processes (e.g., see Section IV,E) might be effective in influencing competition. Alterations in substances conveyed by fast axoplasmic transport would certainly be a very rapid way for the organism to respond to appropriate information and to alter the bias in favor of one or the other of the axons that were potentially in competition at a target tissue. At present, these results on fast axoplasmic transport are the only clues we have as to possible mechanisms involved in the regulation of axonal sprouting and in the ability to gain or lose an advantage in competitive situations among axons.

D. "CRITICAL PERIODS" FOR SPROUTING

In Section II,E,2, the "domain character" of the sprouting of low-threshold mechanosensory nerves in rat skin was described (Fig. 6). These results were obtained from rats younger than about 20 days of age at the time of denervation. Figure 16 shows the results of these (Fig. 16B) and similar experiments done on rats of 20 days of age (Fig. 16C); there was no detectable sprouting in the latter group or in older animals (Diamond and Jackson, 1980). It seems that there is a "critical period" for the sprouting of low-threshold mechanosensory axons in the

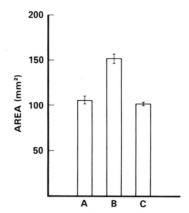

FIG. 16. Lack of nerve sprouting in adult rats. The histograms (± SEM) show the sizes of the fields of the medial branch of T13 dorsal cutaneous nerves in groups of rats, all aged 40 days. In the B and C groups, the back skin of the animals had been denervated over a large area by section of the DCNs of T10, T11, T12, L1, L2, and L3, and including section of the *lateral* branch of DCN-T13. The control animals of group A were unoperated prior to being mapped. The DCNs of the group B rats were sectioned at 10 days of age and those of the group C rats at 20 days of age. The remaining mDCN-T13 fields of the latter were no different from the controls at 40 days, i.e., no functional sprouting of the mDCN-T13 axons had occurred. In contrast, in the animals denervated at 10 days of age (Group B), the mDCN-T13 fields were significantly increased, showing that there had been sprouting of the spared axons.

rat that ends at about 20 days of age. Not only can sprouting of intact low-threshold mechanosensory nerves no longer be evoked by adjacent denervation after this age, but when sprouting is initiated in the rat pup, it ceases at about 20 days. The evidence for this result came from a study of the distribution of the cutaneous "touch domes." These sensory structures in the hairy skin of most mammals, readily made visible by depilation of the skin (Fig. 17), are low-threshold mechanosensory structures that produce an irregular, slowly adapting train of impulses in the one to three axons that supply them when a maintained stimulus is applied (Fig. 18) (Iggo and Muir, 1969; Smith, 1977). We now know from a study of domes made visible by prior administration to the animal of quinacrine, which fluoresces when illuminated by UV light, that probably all the touch domes are present within a very few days of birth (Diamond *et al.*, 1981); as the animal grows they get farther apart. Each cutaneous nerve thus has a characteristic number of touch domes in its field after the first few days of life, a number that does not change as the area of the field increases thereafter (Fig. 19). When sprouting of a selected nerve was evoked by appropriate denervations in a 10-day-old rat pup, the remaining nerve innervated extra touch

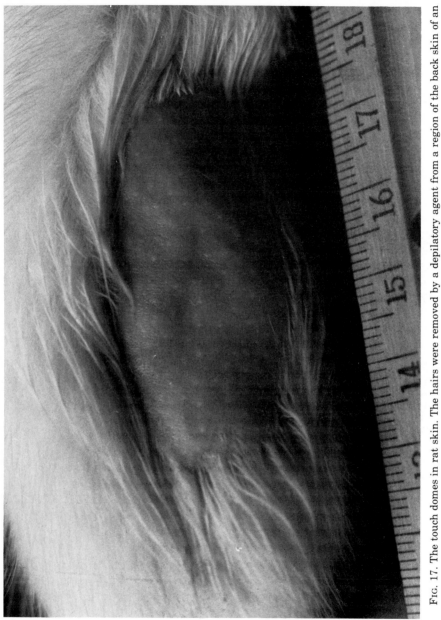

Fig. 17. The touch domes in rat skin. The hairs were removed by a depilatory agent from a region of the back skin of an adult rat. With oblique illumination, the touch domes are then clearly revealed as pale-pink raised spots (note that there are many fewer domes than hair follicles).

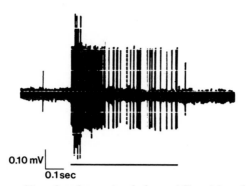

0.10 mV

0.1 sec

FIG. 18. Discharge of impulses from a touch dome. Afferent impulses were recorded from a DCN; during a single sweep of the oscilloscope, a single touch dome was stimulated by pressure from a blunt probe during the period indicated by the horizontal bar. When the probe was similarly applied to skin just to one side of the dome, no impulses were evoked.

domes in the adjacent skin (the touch domes survive denervation; English, 1977). However, this acquisition of touch domes ceased at about 20 days of age (Table I), after which the now-extended field of the "isolated" nerve, although still surrounded by denervated skin, simply enlarged *pari-passu* with the growth of the animal (Fig. 20; Table I) (Jackson and Diamond, 1981).

Our preliminary results indicate that this sprouting into skin that was denervated at 10 days of age probably does not begin until about 15 days of age; thus it appears that there is a brief interval during

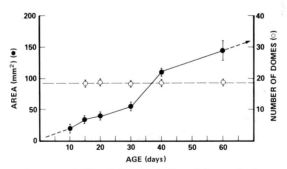

FIG. 19. Area of a nerve field and its population of domes during growth. The low-threshold mechanosensory field areas (± SEM) of the medial branch of DCN-T13 was mapped in groups of animals whose ages were those indicated on the abscissa. The number of touch domes (± SEM) within each field was also determined. Although over the period illustrated (15 to 60 days) the area of the field (●) increased about 8-fold, the number of its domes (○) remained constant.

TABLE I

Numbers of Domes in the Medial Field of the T13 Dorsal Cutaneous
Nerve in the Rat

	Days of age				
	15	20	30	40	60
Control	18.6 ± 1.5	19.0 ± 2.4	18.1 ± 1.8	18.6 ± 4.7	18.8 ± 2.0
	$n = 8$	$n = 27$	$n = 18$	$n = 7$	$n = 11$
The "isolated" field	19.0 ± 3.4	28.5 ± 2.0	28.0 ± 0.8	27.7 ± 1.5	28.4 ± 3.2
	$n = 4$	$n = 8$	$n = 4$	$n = 3$	$n = 11$
After regeneration of the cut nerves					19.6 ± 2.2
					$n = 8$

[a] From Jackson and Diamond, 1981.

which sprouting of these nerves can occur. The exact timing of this
critical period will probably vary from one species to another, but its
occurrence (for low-threshold mechanosensory nerves) seems likely to
be general in mammals, and it is certainly true, for example, in the
rabbit (Jackson, 1980). Because nerves that are regenerating after a
cut or crush will recognize and innervate Merkel cells within denervated touch domes, even in the adult rat, it is clear that the target
character of those Merkel cells that do survive (we are unsure how
many this is) persists. It is a disappearance either of sprouting stimu-

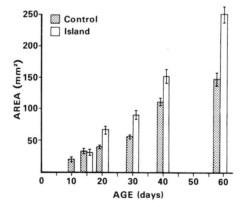

Fig. 20. Sprouting and its cessation at 20 days of age in the rat. The histograms show
the areas (± SEM) of the low-threshold mechanosensory fields of the medial branch of
DCN-T13 in control rats (stippled) and in animals in which the surrounding skin was
denervated at 10 days of age (isolated "island" fields, open columns). (From Jackson and
Diamond, 1981.)

lus per se or of the ability of the nerves to respond to it that must mark the end of the critical period.

It seems possible, therefore, that during development one class of nerves might well "run out" of time for sprouting into available target tissue before another, i.e., that critical periods will vary among modalities and that some will not have a critical period at all (see following). In competitive situations, then, the nerves with the critical periods that end earlier will be at a disadvantage relative to others.

E. IMPULSE ACTIVITY AND SPROUTING

It has long been supposed that impulse activity (i.e., "use") would be of great importance to the "stabilization" of synapses and to the establishment of permanent connections (Hebb, 1949; Young, 1951; Jacobson, 1969; Changeaux and Danchin, 1976). Synaptic mechanisms have indeed been shown to be capable of being influenced in a long-lasting manner by activity (Kandel, 1979). A vivid example, albeit an indirect one, of the possibility that activity can alter the terminal fields of fibers competing for the same target tissue is provided by the organization of "binocular columns" in the mammalian cortex, which clearly depends on the character of the visual inputs during a critical period of development (Hubel et al., 1977). In our investigations of the high-threshold innervation of skin, we have come across a very clear-cut example of how a competitive advantage in target tissue innervation can be conferred by impulse activity in the nerves.

Because the axons subserving high-threshold sensibility are too fine for impulses in them to be reliably resolved in electrical records of whole nerve activity like those we use to examine low-threshold sensory fields, we used behavioral techniques to test for the presence of high-threshold nerve endings in skin. In the anesthetized rat, a moderate pinch delivered by fine forceps to the skin of the back elicits a localized (bilateral) puckering of the skin due to the reflex excitation of the cutaneous trunci muscle (CTM), a skeletal muscle located just under and inserted into the dermis of the back and flank skin (Theriault and Diamond, 1981). We have shown that this reflex excitation results from activation of two groups of sensory axons in the dorsal cutaneous nerves (DCNs): the Group III or Aδ fibers and the C (unmyelinated) fibers. Excitation of Group II low-threshold mechanosensory axons supplying the same back skin does not evoke this reflex response (Nixon et al., 1981).

After appropriate denervations, even in the adult rat, we found a progressive expansion of the skin supplied by a remaining DCN branch, indicated both by a spread of the area within which localized

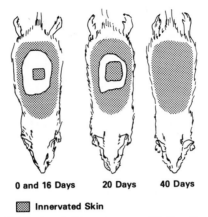

0 and 16 Days   20 Days   40 Days

▓ Innervated Skin

Fig. 21. Recovery of high-threshold mechanosensitivity after partial denervation of adult rat skin. (A) Within the area deprived by section of DCNs at day 0 (shown "empty") was an "island" of skin innervation supplied by the spared medial branch of DCN-T13 (see also Fig. 6). By postoperative day 16, there was no detectable recovery of sensibility in the denervated skin. (B) About 20 days after the denervations, high-threshold sensitivity had begun to return, as indicated by enlargement of the sensitive island and a shrinking of the outer border of surrounding sensitive skin. (C) By the fortieth postoperative day, there was no detectable region of anesthetic skin. (From Diamond, 1981.)

pinching would evoke the reflex response (Fig. 21) and by the associated reappearance of histologically detectable axons in the previously empty Schwann tubes (Fig. 22) (Nixon *et al.*, 1980). The spread of behavioral sensitivity was clearly due to collateral sprouting into the deprived areas, because the sensitivity disappeared from most of the recovered region when the remaining nerve was cut. In another series of experiments, we did not leave the animals unexamined until the final testing period as in those just described, but we tested them periodically, i.e., we pinched and noted the changes in behavioral sensitivity every 4 days, for a period of some weeks. Remarkably, in this group of rats, there was a striking reduction in the time taken for the sprouting to occur (Fig. 23) (Jackson and Diamond, 1979). We showed

Fig. 22. Reappearance of nerves in initially denervated skin that had recovered high-threshold sensibility. Frozen and silver-stained sections (30 μm thick) of adult rat dermis. (A) Normal skin. (B) Skin whose DCNs were cut 16 days earlier; "empty" Schwann tubes are visible, but no axons remain. (C) Skin in which high-threshold (but not low-) sensitivity had recovered after DCNs were sectioned >20 days earlier; the skin up to, but not beyond, the newly established border of the behaviorally recovered region was now supplied by neighboring intact DCNs, which had enlarged their fields in the intervening postoperative period. Most of the Schwann tubes in the dermis of this sensitive region now contained axons. Bar: 50 μm. (From Diamond, 1981.)

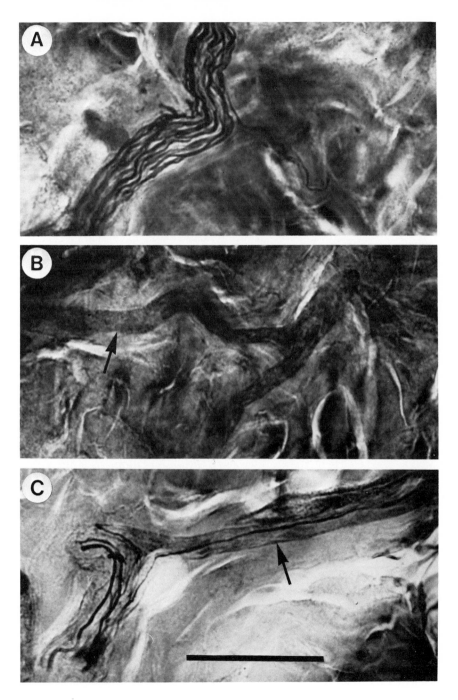

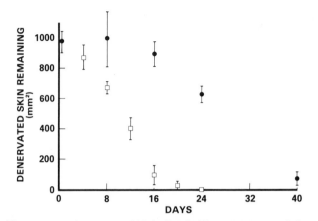

F IG. 23. Time course of recovery of high-threshold sensitivity in adult rat skin. The denervations were those described in Fig. 21. The ordinate shows the area of remaining insensitive skin, the abscissa the days after the denervations were done. The circles (± SEM) show the results for individual groups of animals examined once only, at the postoperative time indicated. The squares (± SEM) show the findings on four similarly operated rats that were each examined repeatedly, at the times indicated. (Based on Diamond, 1981.)

as before that there was an excellent correlation between the advance of the behaviorally sensitive border into denervated skin and that of the area of dermis that could be observed to contain silver-stainable nerve axons; thus, we were dealing with collateral sprouting of remaining intact nerve fibers (Nixon *et al.*, 1980). That the effects of the repeated behavioral examination were indeed attributable to the initiation of impulses in the nerves and not to some ancillary phenomenon such as the release of growth factors in the skin or damage to surviving nerve endings leading to their regeneration was shown by the use of tetrodotoxin (TTX); this drug abolished the apparent acceleration of sprouting caused by repeated pinching (Fig. 24).

We confirmed that impulses were effective in causing this acceleration by exposing a remaining nerve at the time when the denervations were done and electrically exciting it directly; a mere 10 minutes of excitation (at 20 Hz) was adequate to cause the isolated field entirely to fill the available denervated area at a time when normally, without electrical or physiological stimulation, sprouting would usually not even be detectable (Fig. 25). When we compared the time course of normal sprouting with that of the sprouting of a nerve that had received this one brief period of stimulation, it was clear that the most striking effect of the impulses was greatly to reduce the latency of the

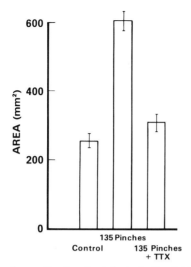

Fig. 24. Acceleration of sprouting by physiological stimulation, and its prevention by TTX. The back skin of three groups of adult rats were denervated except for the medial branch of DCN-T13 on the left side (see Fig. 21). All groups were examined 16 days later; the presence of high-threshold endings supplied by the remaining nerve branch was indicated by the reflex response of the underlying cutaneous trunci muscle to forceps pinching of skin, and the source of the endings was confirmed by the disappearance of the response after section of mDCN-T13. The ordinate shows the area of high-threshold sensitivity attributable to this nerve. In two groups of animals, this remaining sensitive field was pinched lightly 135 times at the time of the denervations and while the animals were still anesthetized; in one of these groups, however, (the results of which are on the right) TTX (31 μM) was applied to the exposed mDCN-T13 prior to the pinching, but after the field border had been approximately defined (by pinching a few spots in the region expected to be sensitive). Clearly the increased sprouting caused by the 135 pinches was largely abolished by the TTX; the slight increase in sprouting in the TTX group as compared to the control group is probably attributable to the few pinches given before TTX was applied and to the manipulations of the nerve.

onset of the sprouting into the denervated skin. Although the actual rate of the collateral reinnervation was not much changed, it began at about 4–5 days after the denervation (and the stimulation period) instead of at the usual 15 or more days.

The relevance of these findings to competition between nerves is clear. In unstimulated animals, the deprived skin was always eventually shared between the surrounding intact nerves sprouting inward, and the intact nerve (that supplied an innervated "island") sprouting outward (as in Fig. 21). When the latter nerve had been stimulated as described earlier at the time of the initial denervations, it was now able to reinnervate virtually the entire area that had been denervated ini-

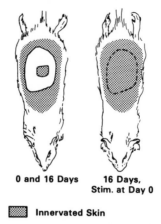

0 and 16 Days   **16 Days,**
Stim. at Day 0

▨ **Innervated Skin**

FIG. 25. Acceleration of sprouting by electrical excitation of nerves. Rats were prepared as in Fig. 21, with the back skin denervated except for the mDCN-T13 on the left
side. In one group of animals, at the time of denervation (while the animals was still
anesthetized) this nerve was lifted on to stimulating electrodes and excited by a stimulus
adequate to fire off the Aδ (Group III) and C fibers, at 20 Hz for 10 minutes. The intact
nerve was then replaced *in situ*, and the animal was sutured and allowed to recover. A
typical result from such an animal is shown on the right, and one in which an identical
procedure was carried out, except that no stimulation was given, on the left. In the
control animals, the high-threshold mapping at 0 days had not changed 16 days later in
controls, whereas in the stimulated group there was full recovery of skin sensitivity to
forceps pinching by 16 days.

tially at a time when the surrounding unstimulated nerves had barely
begun their invasion (as in Fig. 25). Presumably, in the manner described earlier, the sprouted nerves were able subsequently to preclude
invasion (by the surrounding nerves) of their newly acquired territory
by reducing the availability of the sprouting stimulus in the skin they
now innervated. We have also shown that when an essentially similar
electrical stimulation experiment was done in the absence of available
target tissue (i.e., without performing any denervations at all), there
were no subsequent changes in the existing nerve fields, again suggesting that it is the sensitivity of the nerve to available sprouting stimulus that is responsible for the phenomenon. From our electrophysiological studies, we know that certainly the small myelinated Aδ fibers are
involved in this acceleration phenomenon, although the C fibers may be
similarly affected. Because the sprouting of low-threshold afferents is
restricted to the critical period mentioned earlier, we have not yet been
able to examine whether it, too, would be affected by electrical excitation, but we have shown that after the critical period is over, stimula-

tion of low-threshold afferents does not evoke their sprouting into de-nervated skin.

It is interesting to note that there were no detectable "domain constraints" operating on the sprouting of intact high-threshold nerves comparable to those that influence the low-threshold ones.

These experiments indicate that at least in one situation impulse activity in axons could be an important determinant on the outcome of competition among them, particularly when the nerves are approaching a relatively sparsely innervated or noninnervated target tissue. Possibly, too, the ability of nerves to resist displacement by later-arriving ones might also be enhanced by increased impulse activity. Our preliminary findings, in studies of TTX blocking of activity in DCN axons, show that the impulses are required to proceed centrally (i.e., toward the cell body) for the acceleration of sprouting to be evoked and that the response is absent if the electrically excited impulses are only allowed to conduct to the nerve endings, i.e., toward the target tissue. It seems likely that the effect, therefore, depends upon the elaboration in the cell body of some appropriate substance that is then conveyed to the terminals by fast transport, apparently then increasing the responsiveness of the endings to the local sprouting agent. The time scale of the phenomenon definitely excludes the possibility that the transmission of the "trophic" information from the cell body to the terminals depends upon slow flow.

V. Modeling and Remodeling of Nerve Connections

A. REGENERATING NERVES VERSUS SPROUTED ENDINGS

In our investigations of the reinnervation of rat skin, we obtained results that were strikingly different from those that we had observed in the salamander. When, in the rat, regenerating nerves arrived at skin that earlier had been successfully invaded by sprouts (during the critical period for low-threshold axonal sprouting that was described earlier), the regenerating endings functionally replaced the sprouted ones (Jackson and Diamond, 1981). The mechanosensory function of the axons within the enlarged fields were perfectly normal for at least 60 days (see Fig. 20). However, soon after regenerating nerves arrived in that region of the skin (this occurred at 40–45 days, i.e., approximately 30 days after the initial denervation), the enlarged field reverted to the control size; this is seen in the 60-day "regeneration" group in Fig. 26. Moreover, the touch domes that had been taken over during the early sprouting were now captured by regenerated axons

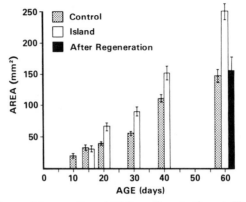

FIG. 26. Sprouting and its cessation at 20 days of age in the rat. The histograms show the areas (± SEM) of the low-threshold mechanosensory fields of the medial branch of DCN-T13 in control rats (stippled) and in animals in which the surrounding skin was denervated at 10 days of age (isolated "island" fields, open columns); the third column at 60 days of age shows the results for a similarly denervated group of animals, in which the originally lesioned adjacent nerves regenerated and successfully reinnervated skin after sprouting of the intact nerve was completed. In this group, reinnervation by lesioned nerves occurred at about 45 days of age. (From Jackson and Diamond, 1981.)

(Table I). A similar functional replacement has been observed for regenerating high-threshold nerves that reach skin into which sprouting had occurred earlier (Devor *et al.*, 1979).

We have now begun examining touch domes excised at various stages during this competitive situation. The axons that supply the domes end upon typical Merkel cells in the basal epidermis (Figs. 27 and 28) (Smith, 1967; Winkelmann and Breathnach, 1973; Munger, 1977); but in contrast to the scattered distribution of these sensory cells in salamander skin (Parducz *et al.*, 1977), the Merkel cells in a touch dome occur in large numbers (50–170) in the form of a flat elliptical annulus (Nurse *et al.*, 1981). At least a proportion of Merkel cells survive denervation (Smith, 1967). We have shown that during the competitive takeover by regenerating nerves there is one period when impulses can be evoked in both the sprouted nerve and the regenerated nerve; after a further day or two, impulses can no longer be evoked in the sprouted nerve and the regenerated nerve has then successfully taken over the dome. We are interested in whether or not the competition is occurring at the level of the individual Merkel cell. Our preliminary observations suggest that there are occasions during the competition when two sets of nerve endings are visible on individual Merkel cells; we showed this by cutting one of the competing nerves (at a time when impulses from the identified dome were recorded from both

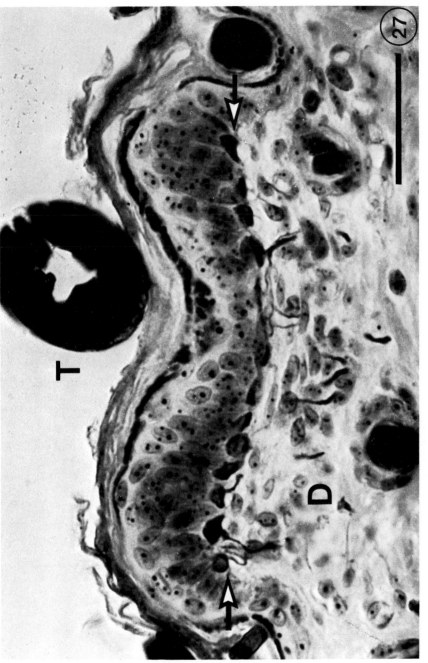

Fig. 27. Histology of a touch dome in rat skin. This silver-stained section shows numerous neurites in the dermis (D); many of these terminate as "end plates" on distinguishable cells at the base of the thickened epidermis of the dome (between the two arrows). A portion of the associated tylotrich hair (T) is seen above the dome. The animal was 14 days old. Bar: 50 μm.

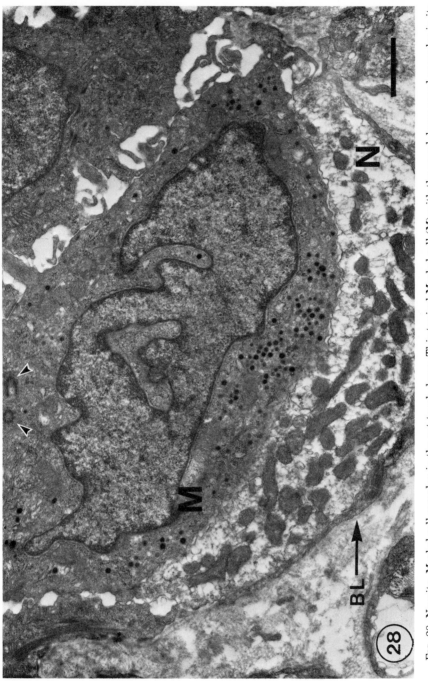

FIG. 28. Neurite–Merkel cell complex in the rat touch dome. This typical Merkel cell (M) with the usual dense-cored granules in its cytoplasm is contacted by a large nerve ending (N) containing numerous mitochondria but few clear vesicles (compare the salamander appearances in Fig. 10). BL and arrow indicates the pale-staining basal lamina, with dermis beneath. The two white arrows above the Merkel cell nucleus indicate the centrioles in this cell. Bar: 1 μm.

nerves) and allowing 24 hours for degeneration; in the EM we often observed that on some Merkel cells only one of the endings was degenerating. However, because of the "modeling" that appears to be going on naturally at this time (described later), we have not yet confirmed that this finding is truly indicative of the one ending taking over from another. Normally, in an adult touch dome the coverage of Merkel cell membrane by the apposed nerve endings is quite extensive (Fig. 28); in the Merkel cells that we have examined in touch domes excised during this competitive situation this is not so, and a "bare" portion of the Merkel cell may therefore have been available to be recognized by the competing (regenerating) nerve endings.

Because the regenerating endings of a nerve that did not originally supply a particular dome were also able to successfully displace sprouted ones, it is not simply a matter of repossession of a target but of displacement of sprouted endings by apparently any regenerated endings. Such a competitive advantage of regenerating over sprouted endings has been seen in other situations (Brown and Ironton, 1978; Devor et al., 1979; Fangboner, 1979; Wigston, 1980). It has been suggested that it is the "over-extended" condition of the sprouted endings that makes them vulnerable to regenerating ones, which have not yet developed an equivalently large terminal field (Purves and Lichtman, 1978; Kuffler et al., 1980). We do have some preliminary findings that suggest that regenerating nerves are still capable of displacing sprouted ones even when they themselves have already innervated a larger than normal area of skin and number of touch domes. At present, therefore, we cannot conclude that it is a matter simply of size of terminal fields that determines the outcome of such competition. One possibility may be that the sprouted endings are in somewhat "foreign" territory (even though they are still within the segmental domain) and thus are somewhat ill-matched to the skin they occupy. As described earlier, regenerating nerves are not susceptible to the positional (domain) constraints that operate on sprouting nerves, and perhaps, therefore, they are also able to form more extensive and stable associations on "foreign" (or their original) target cells than are endings that are sprouting outside their original territory.

B. The Innervation of Maturing Target Structures

It is now known that during development a transient hyperinnervation of targets or target tissues can occur that gradually regresses to the normal adult innervation pattern, usually in the first few weeks of life (Fitzgerald, 1966; Redfern, 1970; Crepel et al., 1976; Lichtman, 1977). In the context of the hypotheses discussed earlier, we may suppose that the hyperinnervation represents either an initial "overshoot" in the response of nerves to sprouting agents or an initial inadequacy of the perhaps immature axons to "neutralize" the effectiveness of these agents. The conversion of the initial pattern of innervation to the mature one, however, is a good example of remodeling in the nervous system (Purves and Lichtman, 1980) and may give us insight into modeling mechanisms generally.

We also have been studying the development of touch domes and their Merkel cells in the rat. Our studies have been greatly facilitated by the use of the fluorescent dye quinacrine (Crowe and Whitear, 1978; Nurse *et al.*, 1981). Our preliminary results indicate that postnatally there is an increase in the number of fluorescent cells, which we expect to be labeled Merkel cells or their precursors. However, although we find that there are frequently in excess of 100 such cells in the adult dome, the estimates that are available based on nerve terminal counts has suggested that there are only about 25 neurite–Merkel cell complexes per adult dome (Smith, 1967). In our EM examination of domes taken from animals as young as 1 day to as old as many weeks, we find not only normal Merkel cells, but abnormal Merkel cells, "transitional" cells (English *et al.*, 1980), and an occasional "dead" Merkel cell; however, any of these might have normal, abnormal, or no nerve endings associated with them. It seems that there may be some continual process of "modeling" going on in the touch dome. We are currently investigating what seems to be an elimination of nerve endings from touch domes (Diamond *et al.*, 1981); it would be surprising if a loss of dome endings were occurring in animals younger than 20 days, however, when axon(s) supplying a dome (see later) can sprout to innervate denervated domes in adjacent skin.

We can probably discard the possibility that entire axons (with their associated endings) are eliminated from touch domes during this modeling period. We have used electrophysiological techniques (Horch, 1979) to study in animals of various ages how many touch domes are supplied by a single axon, as compared with "polyneuronally" innervated domes; the results to date indicate that the proportion of polyneuronally innervated domes is no greater in the younger animals than in adults (30–40%). It seems then that the remodeling involves principally the nerve endings and their target cells.

The outcome of these studies is still uncertain, but they do suggest the possibility that within a particular target tissue like the touch dome, with specific target cells, modeling (and remodeling) may be going on not only in the young animal but conceivably even in the adult. Furthermore, the target tissue itself must be considered as a possible active participant in the process of the modeling of synaptic connections (see also Mugnaini, 1970), perhaps by providing more or fewer target cells, in addition to the involvement of the nerve in providing and probably withdrawing axonal endings.

C. HYPOTHETICAL MOLECULAR MECHANISMS FOR COMPETITIVE DISPLACEMENT AT TARGET SITES

An important item in considering how the eventual displacement of one axonal ending by another could be brought about is whether the influence of an axon is exerted over a distance of some tens or even hundreds of micrometers, or whether the competition occurs between immediately adjacent nerve elements. It is now generally assumed that the development of intercellular adhesion involves an interaction, leading to binding, between appropriate surface molecules (or clusters of these) located on the two apposed cell membranes (Weiss, 1947; Tyler, 1946; Roseman, 1970). However, not all nerves form typical synaptic relationships—the axons subserving most cutaneous modalities, for example, have no specific targets, and autonomic fibers may end near to, but not directly abutting, the cells of their target tissues (Gabella, 1976). Such nerves might best be regarded as interacting with an extracellular matrix containing localized populations of appropriate binding sites (see Culp, 1978). Even the postsynaptic binding receptors of muscle fibers may be located on their closely associated basement membrane (Sanes et al., 1978; Sanes and Hall, 1979). In our studies of competition between regenerating and sprouted fibers in rat touch domes, we have preliminary indications that at a certain stage in the process two axons might each make a functional contact with the same Merkel cell, and that these contacts are immediately adjacent. We have not, however, excluded an additional possibility, namely, that later-arriving endings may synapse with unoccupied Merkel cells (Diamond et al., 1981), and as a consequence resident endings that have already formed synapses on neighboring Merkel cells in the same touch dome are somehow induced to regress.

One mechanism that might be involved, especially in the type of replacement occurring between immediately neighboring endings, simply requires that the binding between the appropriate surface molecules of the nerve endings and the complementary receptors of the target site be reversible (see, e.g., Pricer and Ashwell, 1971; also Roseman, 1970; Roth, 1973). [Synaptic binding must be reversible if, for example, the regression of endings at synapses is a real phenomenon (see Section III,A).] Given this condition there are at least two ways in which an arriving ending could replace a resident one.

1. By having surface binding sites with a higher overall affinity for the target receptors than do the binding sites of the resident ending. This is a kind of "differential affinity" hypothesis akin to Steinberg's differential adhesion one (Steinberg, 1978). The binding molecules of the arriving ending would more successfully compete for the target receptors that would periodically become exposed at the *border* of the synapse than would those of the resident ending, and would also be able to occupy any "extrajunctional" target sites (cf. Miledi, 1960). Consequently, the invading membrane would gradually form more stable contacts with postsynaptic surface, from the periphery inward, at the expense of the membrane of the resident ending, thus making a successful replacement by the invading one inevitable.

2. By having a higher surface *density* of binding sites than the resident nerve ending, although the molecular nature of the sites could be identical (see Steinberg, 1963; Barondes, 1980). In such a case the competition by the more concentrated population of binding molecules of the invading ending would eventually lead to replacement of the sparsely distributed ligands of the resident axonal surface at the target sites, beginning, as in (1), at the periphery of the synapse and extending progressively inward.

Competitive displacement at a distance could be achieved by a somewhat more complicated but essentially similar mechanism to the above. An influence exerted via the target cell [e.g., a muscle fiber (Lomo and Jansen, 1980)] has been suggested; if, however, resident endings do vacate Merkel cells that are themselves *not* directly contacted by the invading axons, then the release of diffusible agents by these axons seems required. Such an agent could also be postulated as binding reversibly with postsynaptic target sites. Interestingly, surface membrane components, including glycoproteins and proteoglycans, that are often proposed as likely candidates to mediate cell adhesion (Hughes, 1976) can be spontaneously shed into the extracellular medium (e.g., Moscona, 1962; Hausman and Moscona, 1975; Pessac and Defendi, 1972; Kapeller *et al.*, 1973; Hughes *et al.*, 1975; Rutishauser *et al.*, 1976).

Consider the situation of two neighboring Merkel cells, one in the process of being successfully taken over by an invading terminal, the other already occupied by a resident ending. If the invaders were releasing a diffusible agent that could bind to Merkel cell target sites more effectively than the binding molecules of the resident ending, this ending would gradually be replaced by the diffusing molecules in the manner described above. If the effective binding affinity of these diffusing molecules were less than that of the surface ligands of the invading endings that released them, then were any such invader to arrive, it could displace these occupiers of the receptor sites on the target membrane and so capture the Merkel cell. It also follows from this model that even if two competing endings were to arrive more or less simultaneously at the same target cell, whose specific receptor sites might already be saturated with one or other of their released binding agents, the ending with the surface binding molecules of higher affinity would always eventually be the one successfully maintained upon the target.

One attractive feature of mechanisms like these is that the density, the rate of turnover, and the character (e.g., the affinity) of the binding molecules of an axonal membrane could presumably be continually regulated by its neuron by way of the fast axoplasmic transport system. The likelihood that such transport is involved in neuron–target interactions has already been discussed. Incidentally, similar mechanisms could explain, for example, how glial cells might actively detach degenerating nerve endings at synapses (Raisman and Field, 1973). Even a relatively low affinity of glial membrane binding molecules for postsynaptic receptors would eventually suffice to replace a degenerating nerve membrane, in which the production and maintenance of surface ligands would be gradually running down.

VI. Summary and Conclusions

Based on our studies of the sensory innervation of skin, one can itemize a number of the likely processes and conditions that could influence the outcome of potential competition among nerves for target tissues or specific target sites. The first seven items of the following list have received either direct or indirect support from our own investigations.

1. Nerves that arrive first at available targets may capture them permanently.

2. Spatial constraints may operate to create "domains" whose boundaries can hinder certain intact nerves from sprouting across them while allowing others to do so freely.

3. Temporal constraints may produce "critical periods" for collateral sprouting that vary among different classes of nerves and target tissues; thus, with increasing age one class of nerves might rapidly find itself at a disadvantage in competition with another for a potentially common target tissue.

4. Agents conveyed to nerve endings by fast axoplasmic transport seem to be involved in the following processes: (a) "Neutralizing" the

local effectiveness of target-derived sprouting factors. A variation of this transport in one axon would thus affect the likelihood of other axonal endings growing into the territory of the first. (b) "Masking" the features that permit targets to be recognized by nerve endings. An alteration in the delivery of the necessary factors to the nerve endings would influence the ability of later-arriving endings to make appropriate connections. (c) Allowing nerves to sprout in response to the stimulus provided by available target tissue. A change in the level of the transported agents involved in this "permissive" function would presumably influence the amount of sprouting an axon will undergo or its sensitivity to sprouting agents.

5. The amount of impulse activity in certain nerves can influence the rapidity with which they respond to the sprouting stimulus provided by available target tissue and, therefore, their ability to compete for it with other nerves. In such instances, the more active axons will have a competitive advantage over less active ones.

6. "Metabolic" or "growth" states may influence the capacity of a nerve to recognize a target, to synapse with it, or to resist competition from other axons. This may explain the instances when regenerating axons display a clear competitive advantage over sprouted endings. The phenomenon draws attention to the possibility that neurons can vary in their apparent "drive" to grow and capture territory, as in their susceptibility to spatial (domain) constraints.

7. Target tissues may vary the number (or trophic character) of their target cells or sites; this activity of a target tissue may be a response to the presence of nerves and thus could vary for different classes of afferent inputs.

8. There may be an intrinsic "matching" between nerves and any one target, which varies for different classes of nerves; the less well matched pairs would be at a competitive disadvantage.

9. Larger neurons probably can respond to sprouting influences by producing many more branches than smaller ones and thus would have a competitive advantage in establishing territory.

10. Nerves may secrete substances that adversely affect other nerves, possibly inhibiting their sprouting. (There is no evidence for such a process at present that cannot be explained by the process mentioned in 4a.

It seems likely that a number of the preceding influences might operate more or less simultaneously and not necessarily all in the same direction. As has been mentioned earlier, some of the itemized "influences" may actually be components of others, for example, the effects of impulse activity (5) may be achieved by a consequent modification in the levels of some axoplasmically transported factors (4a or c). Similarly a "critical period" (3) may be dependent upon the progressive disappearance from a neuron of some particular component of fast transport (e.g., 4c). It would be particularly interesting if these influences were available in the mature organism (and certainly some of them seem to be) and thus capable of being involved in the process of remodeling in the nervous system.

ACKNOWLEDGMENTS

It is a pleasure to acknowledge the help and useful criticisms I have received in the writing of this chapter from my collaborators, namely, Elizabeth Theriault, Bruce Nixon, Karen Mearow, Lynn Macintyre, Anne Foerster, Mike Holmes, Colin Nurse, and Bert Visheau. I am grateful to Prof. Sir Bernard Katz, F.R.S., for drawing my attention to the discussion by Held (1909) of his findings and of Ramón y Cajál's objections; Sir Bernard kindly translated and clarified the relevant portions of Held's monograph.

Most of our investigations were supported by grants from the Medical Research Council [Canada] and the Multiple Sclerosis Society of Canada.

REFERENCES

Aguilar, C. E., Bisby, M. A., Cooper, E., and Diamond, J. (1973). *J. Physiol. (London)* **234,** 449–464.

Barondes, S. H. (1980). *In* "Cell Adhesion and Motility. The Third Symposium of the British Society for Cell Biology" (A. S. G. Curtis and J. D. Pitts, eds.), pp. 309–328. Cambridge Univ. Press, London and New York.

Bennett, M. R., Davey, D. F., and Nebel, K. E. (1980). *J. Comp. Neurol.* **180,** 335–357.

Bodian, D. (1942). *Physiol. Rev.* **22,** 146–169.

Brodal, A. (1969). "Neurological Anatomy." Oxford Univ. Press, London and New York.

Brown, M. C., and Ironton, R. (1978). *J. Physiol. (London)* **278,** 325–348.

Bueker, E. D. (1948). *Anat. Rec.* **102,** 369–390.

Burger, M. M. (1979). *In* "The Role of Intercellular Signals: Navigation, Encounter, Outcome" (J. G. Nicholls, ed.), pp. 97–118. Dahlem Konferenzen, Berlin.

Call, T. W., and Bell, M. (1979). *Anat. Rec.* **193,** 495.

Changeaux, J. P., and Danchin, A. (1976). *Nature (London)* **264,** 705–712.

Chevallier, A., Kieny, M., and Manger, A. (1977). *J. Embryol. Exp. Morphol.* **41,** 245–258.

Cole, J. P., Lesswing, A. L., and Cole, J. R. (1968). *Clin. Orthop. Relat. Res.* **61,** 241–247.

Cooper, E., and Diamond, J. (1977). *J. Physiol. (London)* **264,** 695–723.

Cooper, E., Diamond, J., and Turner, C. (1977). *J. Physiol. (London)* **264,** 725–749.

Crepel, F., Mariani, J., and Delahaye-Bouchard, N. (1976). *J. Neurobiol.* **7,** 567–578.

Crowe, R., and Whitear, M. (1978). *Cell Tissue Res.* **190,** 273–283.

Culp, L. A. (1978). *Curr. Top. Membr. Transp.* **11,** 327–396.

Detwiler, S. R. (1920). *Proc. Natl. Acad. Sci. U.S.A.* **6,** 96–101.

Detwiler, S. R. (1936). "Neuroembryology. An Experimental Study." Macmillan, New York.

Devor, M., Schonfeld, D., Seltzer, Z., and Wall, P. D. (1979). *J. Comp. Neurol.* **185,** 211–220.

Diamond, J. (1968). *J. Physiol.* (*London*) **194**, 669–723.

Diamond, J. (1979). *In* "The Neurosciences: Fourth Study Program" (F. O. Schmitt and F. G. Worden, eds.), pp. 937–955. The MIT Press, Cambridge, Massachusetts.

Diamond, J. (1981). *In* "Post-Traumatic Peripheral Nerve Regeneration" (A. Gorio, H. Millesi, and S. Mingrino, eds.), pp. 533–548. Raven, New York.

Diamond, J., and Jackson, P. (1980). *In* "Nerve Repair: Its Clinical and Experimental Basis" (D. L. Jewett and H. R. McCarroll, eds.), pp. 115–129. Mosby, St. Louis, Missouri.

Diamond, J., Cooper, E., Turner, C., and Macintyre, L. (1976). *Science* **193**, 371–377.

Diamond, J., Nurse, C. A., and Visheau, B. (1981). *Soc. Neurosci. Abstr.* **7**, 540.

Dogiel, A. S. (1908). "Der Bau der Spinalganglien des Menschen und der Saugethiere." Fischer, Jena.

Ebendal, T., Olson, L., Seiger, A., and Hedlund, D. (1980). *Nature* (*London*) **286**, 25–28.

English, K. E. (1977). *J. Comp. Neurol.* **172**, 137–163.

English, K. E., Burgess, P. R., and Kavka-Van Norman, D. (1980). *J. Comp. Neurol.* **194**, 475–496.

Fangboner, R. F. (1979). *J. Exp. Zool.* **209**, 355–366.

Farbman, A. (1965). *Dev. Biol.* **11**, 110–135.

Fitzgerald, M. J. T. (1961). *J. Anat.* **95**, 495–514.

Fitzgerald, M. J. T. (1966). *J. Comp. Neurol.* **126**, 37–42.

Fitzgerald, M. J. T., Folan, J. C., and O'Brien, T. M. (1975). *J. Invest. Dermatol.* **64**, 169–174.

Foerster, O. (1933). *Brain* **56**, 1–39.

Gabella, G. (1976). "Structure of the Autonomic Nervous System." Chapman and Hall, London.

García-Bellido, A., Lawrence, P. A., and Morata, G. (1979). *Sci. Am.* **241**, 102–110.

Globus, A., and Scheibel, A. B. (1967). *Exp. Neurol.* **19**, 331–345.

Goldowitz, D., and Cotman, C. W. (1980). *Brain Res.* **181**, 325–344.

Gottlieb, D. I., and Glasser, L. (1980). *Annu. Rev. Neurosci.* **3**, 303–318.

Guth, L., Smith, S., Donati, E. J., and Albuquerque, E. X. (1980). *Exp. Neurol.* **67**, 513–523.

Gutmann, E. (1945). *J. Anat.* **74**, 1–8.

Hamburger, V., and Levi-Montalcini, R. (1949). *J. Exp. Zool.* **111**, 457–501.

Hamilton, W. J., and Mossman, H. W. (1972). "Human Embryology." Heffer, Cambridge, England.

Hartschuh, W., Weihe, E., Buchler, M., Helmstaedter, V., Feurle, G. E., and Forssmann, W. G. (1979). *Cell Tissue Res.* **201**, 343–348.

Hashimoto, K. (1972). *J. Anat.* **111**, 99–120.

Hausman, R. E., and Moscona, A. (1975). *Proc. Natl. Acad. Sci. U.S.A.* **72**, 916–920.

Haymaker, W., and Woodhall, B. (1953). "Peripheral Nerve Injuries." Saunders, Philadelphia, Pennsylvania.

Head, H. H., and Sherren, J. (1905). *Brain* **28**, 116–338.

Hebb, D. O. (1949). "The Organization of Behaviour." Wiley, New York.

Held, H. (1909). "Die Entwicklung des Nervengewebes bei den Wirbeltieren." Barth, Leipzig.

Herndon, R. M., Margolis, G., and Kilham, L. (1971). *J. Neuropathol. Exp. Neurol.* **30**, 557–570.

Hoffman, H. (1950). *Aust. J. Exp. Biol. Med. Sci.* **28**, 383–397.

Hollyday, M., and Hamburger, V. (1976). *J. Comp. Neurol.* **170**, 311–320.

Holmes, M. J., Turner, C., Fried, J. A., Cooper, E., and Diamond, J. (1977). *Brain Res.* **136**, 31–43.

Horch, K. (1979). *J. Neurophysiol.* **42**, 1437–1449.

Horder, T. J. (1974). *J. Physiol. (London)* **241**, 84P–85P.

Hubel, D. H., Wiesel, T. N., and LeVay, S. (1977). *Philos. Trans. R. Soc. London, Ser. B* **278**, 377–409.

Hughes, A. F. W. (1969). "Aspects of Neural Ontogeny." Academic Press, New York.

Hughes, R. C. (1976). "Membrane Glycoproteins". Butterworths, London.

Hughes, R. C., Laurent, M., Lonchampt, M.-O., and Courtois, Y. (1975). *Eur. J. Biochem.* **52**, 143–155.

Iggo, A., and Muir, A. R. (1969). *J. Physiol. (London)* **200**, 763–796.

Jackson, P. C. (1980). Ph.D. Thesis, McMaster University, Hamilton, Ontario, Canada.

Jackson, P. C., and Diamond, J. (1979). *Soc. Neurosci. Abstr.* **5**, 2135.

Jackson, P. C., and Diamond, J. (1981). *Science* **214**, 926–928.

Jacobson, M. (1969). *Science* **163**, 543–547.

Jacobson, M., and Hirose, G. (1981). *J. Neurosci.* **1**, 271–284.

Johnson, E. M., Gorin, P. D., Brandeis, L. D., and Pearson, J. (1980). *Science* **210**, 916–918.

Jones, E. G., Valentino, K. L., and Fleshman, J. W., Jr. (1982). *Dev. Brain Res.* **2**, 425–431.

Kandel, E. R. (1979). *Harvey Lect.* **73**, 19–92.

Kapeller, M., Gal-Oz, R., Grover, N. B., and Doljanski, F. (1973). *Exp. Cell. Res.* **79**, 152–158.

Katz, M. J., and Lasek, R. J. (1979). *J. Comp. Neurol.* **183**, 817–832.

Keegan, J. J., and Garrett, F. D. (1948). *Anat. Rec.* **102**, 409–437.

Kirk, E. J. (1974). *J. Comp. Neurol.* **155**, 165–176.

Kirk, E. J., and Denny-Brown, D. (1970). *J. Comp. Neurol.* **137**, 307–320.

Korneliussen, H., and Jansen, J. K. S. (1976). *J. Neurocytol.* **5**, 591–604.

Kuffler, D. P., Thompson, W., and Jansen, J. K. S. (1980). *Proc. R. Soc. London, Ser. B* **208**, 189–222.

Lawrence, P. A. (1975). *Ciba Found. Symp.* [N.S.] **29**, 3–23.

Le Douarin, N. (1980). *Curr. Top. Dev. Biol.* **16**, 32–85.

Levi-Montalcini, R., and Hamburger, V. (1951). *J. Exp. Zool.* **116**, 321–362.

Lichtman, J. W. (1977). *J. Physiol. (London)* **273**, 155–177.

Lomo, T., and Jansen, J. K. S. (1980). *Curr. Top. Dev. Biol.* **16**, 253–281.

Lyne, A. G., and Hollis, D. E. (1971). *J. Ultrastruct. Res.* **34**, 464–472.

Macintyre, L., and Diamond, J. (1981). *Proc. R. Soc. London, Ser. B* **211**, 471–499.

Mearow, K. M., Nurse, C. A., Visheau, B., and Diamond, J. (1981). *Soc. Neurosci. Abstr.* **7**, 540.

Meier, S. (1980). *J. Exp. Embryol. Morphol.* **55**, 291–306.

Merkel, F. (1875). *Arch. Microbiol. Anat.* **11**, 636–652.

Miledi, R. (1960). *J. Physiol. (London)* **151**, 24–30.

Miner, N. (1956). *J. Comp. Neurol.* **105**, 161–170.

Moscona, A. A. (1962). *J. Cell. Comp. Physiol. Suppl. I* **60**, 65–80.

Moscona, A. A. (1974). *In* "The Cell Surface in Development" (A. A. Moscona, ed.), pp. 67–100. Wiley, New York.

Mugnaini, E. (1970). *In* "Excitatory Synaptic Mechanisms" (P. Andersen and J. K. S. Jansen, eds.), pp. 149–169. Universitetsforlaget, Oslo.

Munger, B. L. (1977). *J. Invest. Dermatol.* **69**, 27–40.

Nixon, B. J., Jackson, P. C., Diamond, A., Foerster, A., and Diamond, J. (1980). *Soc. Neurosci. Abstr.* **6**, 59.2.

Nixon, B. J., Jackson, P. C., Theriault, E., and Diamond, J. (1981). *Soc. Neurosci. Abstr.* **7**, 179.

Nurse, C. A., Mearow, K. M., Visheau, B., Holmes, M. J., and Diamond, J. (1981). *Soc. Neurosci. Abstr.* **7**, 417.

Olson, L., and Malmfors, T. (1970). *Acta Physiol. Scand., Suppl.* **348**, 1–112.

Palka, J., Schubiger, M., and Hart, S. (1981). *Nature (London)* **294**, 447–449.

Parducz, A., Leslie, R. A., Cooper, E., Turner, C., and Diamond, J. (1977). *Neuroscience* **2**, 511–521.

Pessac, B., and Defendi, V. (1972). *Science* **175**, 898–900.

Piatt, J. (1957). *J. Exp. Zool.* **134**, 103–125.

Prestige, M. C. (1965). *J. Embryol. Exp. Morphol.* **13**, 63–72.

Pricer, W. E., and Ashwell, G. (1971). *J. Biol. Chem.* **246**, 4825–4833.

Purves, D., and Lichtman, J. W. (1978). *Physiol. Rev.* **58**, 821–862.

Purves, D., and Lichtman, J. W. (1980). *Science* **210**, 153–157.

Raisman, G. (1977). *Philos. Trans. R. Soc. London, Ser. B* **278**, 349–360.

Raisman, G., and Field, P. M. (1973). *Brain Res.* **50**, 241–264.

Rakic, P., and Sidman, P. L. (1973). *J. Comp. Neurol.* **152**, 103–132.

Ramón y Cajál, S. (1919). "Studies on Vertebrate Neurogenesis" (L. Guth, trans.). Thomas, Springfield, Illinois (reissued, 1960).

Ranson, S. W., and Clark, S. L. (1959). "The Anatomy of the Nervous System." Saunders, Philadelphia, Pennsylvania.

Rawles, M. E. (1955). "Analysis of Development." Hafner, New York.

Redfern, P. A. (1970). *J. Physiol. (London)* **209**, 701–709.

Romanes, G. J. (1941). *J. Anat. (London)* **76**, 112–130.

Roseman, S. (1970). *Chem. Phys. Lipids* **5**, 270–297.

Roth, S. (1973). *Q. Rev. Biol.* **48**, 541–563.

Rutishauser, U., Thiery, J., Brackenbury, R., Sela, B., and Edelman, G. M. (1976). *Proc. Natl. Acad. Sci. U.S.A.* **73**, 577–581.

Sanes, J. R., and Hall, Z. W. (1979). *J. Cell Biol.* **83**, 357–370.

Sanes, J. R., Marshall, L. M., and McMahan, U. J. (1978). *J. Cell Biol.* **78**, 176–198.

Schwartz, A. M., and Kane, E. S. (1977). *Am. J. Anat.* **148**, 1–18.

Scott, S. A., Cooper, E., and Diamond, J. (1981a). *Proc. R. Soc. London, Ser. B* **211**, 455–470.

Scott, S. A., Macintyre, L., and Diamond, J. (1981b). *Proc. R. Soc. London, Ser. B* **211**, 501–511.

Sherrington, C. S. (1893). *Philos. Trans. R. Soc. London, Ser. B* **134**, 641–763.

Sherrington, C. S. (1947). "The Integrative Action of the Nervous System." Cambridge Univ. Press, London and New York.

Singer, M., Nordlander, R., and Egard, M. (1979). *J. Comp. Neurol.* **185**, 1–22.

Smith, K. R., Jr. (1967). *J. Comp. Neurol.* **131**, 459–474.

Smith, K. R., Jr. (1968). *J. Comp. Neurol.* **131**, 459–474.

Smith, K. R., Jr. (1977). *J. Invest. Dermatol.* **69**, 68–74.

Speidel, C. C. (1942). *J. Comp. Neurol.* **76**, 57–73.

Steinberg, M. S. (1963). *Science* **141**, 401–408.

Steinberg, M. S. (1978). *In* "Specificity of Embryological Interactions" (D. Garrod, ed.), pp. 97–130. Chapman and Hall, London.

Stirling, V. (1973). *J. Physiol. (London)* **229**, 657–680.

Stone, L. (1940). *J. Exp. Zool.* **83**, 481–506.

Taylor, A. C. (1943). *Anat. Rec.* **87**, 379–413.

Tennyson, V. (1970). *J. Cell Biol.* **44**, 62–79.

Theriault, E., and Diamond, J. (1981). *Soc. Neurosci. Abstr.* **7**, 179.

Townes, P. L., and Holtfreter, J. (1955). *J. Exp. Zool.* **128**, 53–120.

Tweedle, C. D. (1978). *Neuroscience* **3**, 41–46.

Tyler, A. (1946). *Growth* **10** (Symp. 6), 7–19.

Valverede, F. (1967). *Exp. Brain Res.* **3**, 337–352.

Wall, P. D. (1977). *Philos. Trans. R. Soc. London, Ser. B* **278**, 361–372.

Weber, M. (1979). *In* "The Role of Intercellular Signals: Navigation, Encounter, Outcome" (J. G. Nicholls, ed.), pp. 97–118. Dahlem Konferenzen, Berlin.

Weiss, P. (1941). *Growth* **5**, Suppl., 163–203.

Weiss, P. (1947). *Yale J. Biol. Med.* **19**, 235–278.

Weiss, P. (1955). *In* "Analysis of Development" (B. H. Willier, P. Weiss, and V. Hamburger, eds.), pp. 346–401. Saunders, Philadelphia, Pennsylvania.

Weiss, P., and Edds, M. (1945). *J. Neurophysiol.* **8**, 173–193.

Werner, J. K. (1974). *Am. J. Phys. Med.* **53**, 127–142.

Wigston, D. J. (1980). *J. Physiol. (London)* **207**, 355–366.

Williams, P. L., and Warwick, R. (1975). "Functional Neuroanatomy of Man." Saunders, Philadelphia, Pennsylvania.

Winkelmann, R. K., and Breathnach, A. S. (1973). *J. Invest. Dermatol.* **60**, 2–15.

Wolpert, L. (1971). *Curr. Top. Dev. Biol.* **6**, 183–229.

Young, J. Z. (1951). *Proc. R. Soc. London, Ser. B* **139**, 18–37.

Zelena, J. (1957). *J. Embryol. Exp. Morphol.* **5**, 283–292.

Zelena, J. (1976). *Prog. Brain Res.* **43**, 59–64.

CHAPTER 6

CRITICAL AND SENSITIVE PERIODS IN NEUROBIOLOGY

Reha S. Erzurumlu and Herbert P. Killackey

DEPARTMENT OF PSYCHOBIOLOGY
UNIVERSITY OF CALIFORNIA
IRVINE, CALIFORNIA

I. Introduction

The terms *critical period* and *sensitive period,* along with others such as *vulnerable period, susceptible period,* and *optimal period,* are used interchangeably in developmental neurobiology to refer to a period during which the nervous system is highly malleable. Implicit in the description of such periods are the inferences that specific critical conditions or stimuli are necessary for development and can influence neural development only during that period. These terms have been applied to a wide range of phenomena and, confusingly, the same terms are employed in different contexts while, conversely, different terms have been used in the same context.

A number of investigators have defined these terms in the course of behavioral or neurobiological research (Hinde, 1970; Hess, 1973; Jacobson, 1978; Scott, 1978). For example, Hess (1973) regards the concepts of critical period, susceptible period, and optimal period as

CURRENT TOPICS IN
DEVELOPMENTAL BIOLOGY, VOL. 17

different grades of a sensitive period during behavioral development. Jacobson (1978), on the other hand, differentiates the concepts of critical period and sensitive period on the basis of stimulus or external condition characteristics. In view of these differences, we would like to begin our review of the neurobiological literature related to critical periods and sensitive periods by providing a definition of terms that is based on those provided by Jacobson (1978) and Scott (Scott *et al.*, 1974; Scott, 1978) and that will provide a framework within which various neurobiological studies can be placed.

We define critical period as the time during which the action of a specific external or internal condition or stimulus is required for the normal progress of development (e.g., presence of a normal set of visual stimuli for normal vision to develop, or neonatal secretion of testosterone in order for normal male sexual behavior to develop). The sensitive period is the time period in development during which the nervous system is highly susceptible to the effects of harmful internal or external conditions (e.g., various forms of damage to the nervous system in the neonatal period). Vulnerable and optimal periods can be regarded as special cases of either critical or sensitive periods. During development the organism is vulnerable or susceptible not only to harmful conditions (i.e., sensitive period) but also to the lack or substitution of a different type of condition or stimulus in place of the critical one. Hence, in either of these instances, the term *vulnerable period* implies that the developmental processes will be affected adversely. The term *optimal period,* on the other hand, although referring to the same time periods, implies that the developmental processes will be affected favorably. For example, during optimal periods correction of developmental abnormalities or modification of the nervous system in a favorable direction will be most effective.

It should be emphasized that the conceptual difference between the terms *critical period* and *sensitive period* is a very important one. The first denotes a period during which the presence of certain critical conditions is necessary for the nervous system to develop normally and the other a time during which damage to the nervous system can lead to alterations or reorganization of the system. From a heuristic point of view, defining such periods has immense practical and theoretical value. On one hand, the potential benefits for preventive medicine from the study of various external and internal conditions on the developing nervous system is large, whereas, on the other, these studies may provide valuable insight into mechanisms and principles of neuronal development.

In this chapter we will review studies on critical periods and sensitive periods from a theoretical perspective in an attempt to trace the conceptual origins of these terms, to define the contexts within which they have been used, and to determine the heuristic value of these concepts to modern neurobiology.

II. Historical Background

A. PHILOSOPHICAL ORIGINS

The idea that an organism is highly impressionable during development has been held for a long time. Historically it can be traced to ancient civilizations and related to views of medicine, education, and philosophy (particularly epistemology). The medical writings of ancient civilizations (such as those of Egypt, Mesopotamia, Greece, and Rome) contain isolated observations on the long-term effects of diseases or trauma suffered in childhood. For example, the following Hippocratic aphorisms (Adams, 1950) may serve to illustrate the flavor of some of these observations:

> Such as become hump-backed before puberty from asthma or cough do not recover. . . Those cases of epilepsy which come on before puberty may undergo change; but those which come after twenty-five years of age, for the most part terminate in death.

The idea that a child is more receptive to external influences that have long-lasting effects is central to education. This notion has evolved into the idea that there are differential developmental stages in childhood during which specific capacities are developed (see Hess, 1973, for review). The earliest schoolhouse unearthed is from the time of the Babylonian king Hammurabi (ca. 1955–1913 BC), although education of the young has much earlier roots (Sarton, 1960). The role of exercise in learning and the malleability of the young are fundamental ideas encountered in basically all the ancient civilizations.

A second major and related question that has occupied the minds of many philosophers concerns the acquisition of knowledge. Although a detailed analysis of this issue is outside the scope of this chapter, the current question of the degree to which genetic versus epigenetic factors are involved in the process of development is a continuation of an old philosophical argument into modern neurobiology. In a very simple form, this issue is related to the differential emphasis given to sensory experience as a reliable source for the acquisition of knowledge. The heated argument between the sophists [e.g., Protagoras (481–411 BC), Gorgias (485–380 BC)] and Socrates (469–399 BC) and his followers is perhaps the best example of this issue in ancient philosophy. Following the mystical tradition of Parmenides (sixth century BC), Socrates and later his followers maintained that the senses were deceptive and not to be used as a guide to achieve truth, which could be reliably attained only through the use of reason and logic. On the other hand, sophists accepted sensory experience as the only means of acquiring knowledge and refused any suprasensory reality (Weber, 1925; Durant, 1939). These two opposing views appeared in various forms throughout the history of philosophy.

Several intellectual trends in sixteenth- and seventeenth-century philosophy renewed this controversy. First, the scientific revival during this period and the methods of empirical observations and mathematical reasonings, or in short the "scientific method" as advocated by F. Bacon (1561–1626) and G. Galilei (1564–1642), proved useful in describing naturally occurring phenomena. Second, the intellectual challenge to the dogma and authority of the Christian church and scholasticism opened different avenues in Western philosophy Thus, the issue of acquisition of knowledge and particularly whether sensory experience can be a reliable source of the real world occupied the great philosophers of the seventeenth and eighteenth centuries. The British empiricists (i.e., J. Locke, G. Berkeley, and D. Hume) were strongly influenced by the inductive method as formulated by F. Bacon and strongly rejected the concept of innate ideas. The empiricists advocated that the mind did not contain any ideas at birth; rather, all knowledge and ideas were derived from experience. This is perhaps most succinctly stated by Locke's term *tabula rasa*. This contention was strongly opposed by Descartes, who in parallel with Plato's theory of ideas, defended the thesis that ideas derive not from experience but from the innate properties of the mind. The reasoning followed by Descartes is well known (Weber, 1925); he started by doubting every possible thing and deduced that the only undoubtable thing is the process of doubt itself, i.e., the process of thinking, hence his famous *cogito ergo sum*. Following this line of reasoning, Descartes went one step further by suggesting that the idea of perfection is not present in the world and consequently must be innate. Therefore, there are innate ideas, and reason is the only means to achieve true knowledge. He further emphasized that ideas derived from sensory experience are deceptive and subject to error.

The preceding philosophical views had a strong impact on the emerging field of psychology and were reexpressed as the "nature–nurture" issue. One of the first expressions of this problem at the neuronal level is Hebb's classic book, "The Organization of Behavior"

(1949), in which he attempted a synthesis of the two opposing philosophical trends. He stated that the development of perception involves both innate and acquired perceptual organization. According to Hebb, at birth the neural circuitry possesses basic perceptual processes such as figure–ground relationships but more complex forms of perception such as form perception are acquired through learning and experience. Today this issue is still very much alive, particularly with respect to our ideas on the effects of visual deprivation during development. This will be discussed later; now we will return to the origins during the nineteenth century of the concepts of critical period and sensitive period.

B. Scientific Origins

Clear antecedents of the concepts of critical periods and sensitive periods can be traced to the early nineteenth century. Because these concepts refer to periods of development, it is not unusual that they should have their roots in early experimental embryology. Developmental abnormalities had been noted in ancient civilizations, and the causes of these anomalies were closely related to the prevailing views on embryogenesis. For example, Aristotle held the view that the formation of the embryo is due to the union of the semen with an analogous substance from the female. Furthermore, according to Aristotle the semen represents form, motion, and activity, whereas the female contributes matter and a general substrate (or potentiality) from which the embryo is differentiated by the motion of the semen. Within this view, Aristotle regarded malformations in the newborn as a form of developmental deficiency. The following explanation is given by Aristotle in "De Generatione Animalium" (Platt, 1965).

> If the movements imparted by the semen are resolved and the material contributed by the mother is not controlled by them, at last there remains the most general substratum, that is to say the animal. Then people say that the child has the head of a ram or a bull, and so on with other animals as that a calf has the head of a child or a sheep that of an ox.

The gross differences in embryonic stages of development (i.e., organogenesis) had been described for the chick embryo by Alcmeon (ca. 500 BC; Sarton, 1960; Singer, 1959) and later by Aristotle (384–322 BC) in his "Historia Animalium" (Thompson, 1962). These descriptions are the initial expression of an epigenetic view of development. However, the theory of preformation dominated ideas on embryology until the seventeenth century. This theory held that the process of development is merely an unfolding of the preformed individual and species characteristics (Bodenheimer, 1958). In the seventeenth century this view in embryology was abandoned in favor of the theory of epigenesis, when W. Harvey (1578–1657) showed that the embryo developed from the egg by a successive appearance and development of various structures (Castiglioni, 1958). The idea that certain physical abnormalities (specifically those of various body parts) could be related to the disruption of embryonic stages rather than simple pathology of the previously formed organs was first put forth by E. G. Saint-Hilaire (1772–1844) in 1822. Although Saint-Hilaire deduced that developmental arrest during differentiation of various body parts may be the cause of developmental anomalies, he was not able to experimentally verify his contention. It was only 50 years later, following the introduction of artificial incubation, that another prominent embryologist, C. Dareste (1822–1899), was able to test Saint-Hilaire's ideas. In his account of the production of monstrosities (1891), he has accredited Saint-Hilaire with formulating the idea of developmental arrest, and he demonstrated that changes in incubator temperature altered the rate of growth of chick embryos and was thus able to produce specific physical defects depending on the stage of development when temperature was altered. This finding supported the idea that developmental arrest during organogenesis could produce very specific physical anomalies dependent on the time of insult. Dareste concluded that certain physical conditions, such as changes in incubator temperature or interference with respiration, can modify the development of chick embryos and that this modification was an expression of developmental arrest. Conversely, he suggested that other abnormalities such as hermaphroditism could be an expression of excessive development.

A contemporary of Dareste, C. Féré (1894), extended Dareste's findings to suggest that other harmful conditions, such as mechanical vibration or toxic mercury vapors, could also produce developmental arrest. He further emphasized that the same agent could have totally different effects if applied at different stages of development (such as infertility, monstrosity, abortion, still birth, developmental arrest, and congenital weakness). Féré (1899) drew the important conclusion that the type of anomaly is related to the developmental stage at which disruption has occurred.

The notion of transient periods of susceptibility to harmful external conditions received further support from the studies of the botanist H. De Vries (1848–1935), who reported that in various forms of plants there is a period during which even minimal amounts of disturbance could alter the normal morphology of the plant. He first observed this in a particular variety of poppy in which the number of pistils diminishes irreversibly when subjected to nutritional deficiency before the age of 6 weeks (1905). De Vries referred to this period as the sensitive period and employed this term interchangeably with two others: critical stage and susceptible period. The introduction of this concept of critical periods to biology, particularly embryology, eventually gave rise to teratology.

The concepts of critical periods and sensitive periods were firmly established by Stockard (1921), who studied the effects of various harmful conditions, such as changes in incubation temperature, oxygen deficiency, and toxic substances, on the development of fish eggs. Conceptually, Stockard regarded these periods as critical moments for developmental interruption. He wrote:

> Many of the chief embryonic organs seem also to arise with initial moments of extremely high activity, processes of budding or rapid proliferation and growing out. During these moments a given organ may be thought of as developing at a rate entirely in excess of the general developmental rate of the embryo. Such moments of supremacy for the various organs occur at different times during development. As is well known, a certain organ arises much earlier or later in the embryo than certain others. When these primary developmental changes are on the verge of taking place or when an important organ is entering its initial stage of rapid proliferation or budding, a serious interruption of the developmental progress often causes decided injuries to this particular organ, while only slight or no ill effects may be suffered by the embryo in general. Such particular sensitive periods during development I have termed the "critical moments."

Stockard's experiments in embryology were continued by various investigators who looked at the effects of harmful environmental and genetic conditions on development during critical periods in an attempt to determine the basis for various malformations (Landauer, 1932; Scott, 1937). The results of these experiments added to the notion that developmental acceleration as well as developmental arrest could produce developmental malformations as Dareste (1891) had postulated earlier.

These views were extended by studies on the effects of prenatal malnutrition and toxic substances, which suggested that harmful agents can also alter the course of a particular developmental event even before it has started. In the first experimental study on the effects of malnutrition, Jackson and Stewart (1920) reported reduced brain and body weights in rat pups that were underfed during the preweaning period but not in pups that were malnourished after weaning. However, it was 40 years later before studies on the effects of nutritional deficiency, particularly during prenatal periods, and on the effects of teratogens gained impetus (Werboff *et al.*, 1961; Werboff and Havlena, 1962; Dobbing, 1964, 1968; Winick and Noble, 1966). This line of research changed the predominant view that the embryo leads a parasitic existence *in utero* and that the placenta is a barrier to harmful agents. It also pointed out that various harmful agents could alter the intrauterine environment even before conception and later exert harmful effects on the developing embryo.

The idea of developmental arrest as expressed in the embryology of the late nineteenth and early twentieth centuries was not unique to the field of biology. These biological notions found acceptance in other areas of the study of development, particularly in psychology. For example, around the turn of the century, S. Freud (1856–1939) introduced the developmental aspects of his theory of psychoanalysis. Freud (1938) postulated that the three major stages of psychosexual development (oral, anal, and phallic stages) span the first 5 years of childhood and have profound influences on the development of personality. Furthermore, he related these stages to differential development of the "sexual instinct" with respect to different "erogenous zones" of the body across time. Freud stated that a serious conflict between the sexual instinct and the society at a particular stage (i.e., oral, anal, or phallic) would lead to "fixation" at this stage, in other words, developmental arrest.

The view that development proceeds in stages and that arrest at any given stage could change the course of consequent development has flavored not only theories of personality but also other theories pertaining to various aspects of human development, such as theories of cognitive development that postulate that there are consecutive stages of thought that a child must go through and that each of these stages incorporate the processes of the preceding stage into more complex modes of operation (Piaget, 1926; Bruner, 1968).

The concepts of critical periods and sensitive periods gained wider interest in behavioral biology when Lorenz (1935) emphasized that the process of imprinting (as the basis of social and sexual attachments and of species recognition in birds) is confined to a very definite period of short duration and, furthermore, that this process is irreversible once it has progressed through the period that he called the "receptive period." Lorenz' interpretation of the process of imprinting was much influenced by the work of the embryologist H. Spemann (1869–1941). Spemann had demonstrated the phenomenon of induction. He found that if a piece of tissue from the posterior abdominal region of the frog embryo is transplanted onto the caudal end of the neural groove during a specific period, it becomes part of the new location and differentiates accordingly (for a review of his work, see Spemann, 1938). By analogy, Lorenz suggested that inductive determination could be applied to instinctive behavior whose object is not determined innately:

> the determination of later behaviour by an external influence (derived from a conspecific) during a specific ontogenetic period and the irreversibility of this determination process provide a remarkable analogy between the developmental processes of instinctive behavioural systems and processes which have been identified in morphological development.

Thus, within this context the notion of critical periods refers to differentiation of innately determined mechanisms under environmental influence. Although Lorenz stressed the point that imprinting is a totally different phenomenon than conditioning, various hypotheses as to whether there are specific periods during which learning is enhanced gained importance in behavioral research. These studies were often conducted in conjunction with or in parallel to studies on primary social attachments and development of certain social behaviors in both animals and humans (Harlow and Harlow, 1962; Scott and Fuller, 1965; Hinde, 1970; Hess, 1973).

A conception of critical periods similar to that of Lorenz also emerged from research in the area of neuroendocrinology. The idea that development and neural control of sexual behavior depends on the conditions of the perinatal hormonal environment during a critical period was advanced in the 1950s. This idea evolved from Pfeiffer's (1936) studies on the masculine and feminine patterns of gonadotropin secretion from the pituitary as a consequence of the presence or absence of testicular secretions during the early postnatal period. Subsequently, many researchers studied the effects of pre- and postnatal hormonal treatments following castration or ovariectomy to determine the degree of genetic determination of sexual maturation and behavior (Goy, 1966; Gorski, 1971, 1973; Beach, 1975). In 1959 Phoenix and colleagues demonstrated that prenatal androgen treatment of genetic females changes their sexual behavior to a malelike pattern when supplied with testosterone propionate in adulthood. Later, it was found that in adult males castrated at birth, ovarian transplants form corpora lutea in a cyclic fashion as they do in the female and that when supplied with estrogen these males exhibit female sexual behavior

(Whalen, 1971). Furthermore, it was found that castrated neonatal males exhibit complete female sexual behavior when supplied with low doses of estrogen or ovarian transplants, whereas no such effects were seen in animals castrated in adulthood (Grady *et al.,* 1965). On the basis of these studies, it was suggested that gonadal hormones exert an inductive and organizational action during the development of the central nervous system and an excitatory and activational role in the mature animal (Phoenix *et al.,* 1959; Gorski, 1973). It appears that during a critical period the presence of gonadal hormones alters the sensitivity of the controlling brain mechanisms both with regard to the type and temporal sequence of hormones to which they will respond. Hence, the effect of certain critical factors may not be in altering the rate of development as Stockard had suggested but in altering the organization of the system (depending on the critical factors or conditions involved and the mechanisms affected).

The foregoing lines of research (i.e., the effects of hormones, social stimuli, and learning during specific periods in development) were combined in the study of the ontogeny of bird song. Although acoustic isolation experiments in birds can be traced back to the 1920s, the systematic study of bird song and the emergence of the concepts of critical periods and sensitive periods in this field mainly took place in the late 1950s, after the introduction of the modern methods of recording and analyzing sound (Nottebohm, 1969, 1970; Marler and Mundinger, 1971).

The initial studies on the ontogeny of bird song consisted of raising newly hatched songbirds in various types of acoustic isolation (Thorpe, 1961). In general, most animal calls are predetermined and do not go through an elaborate period of organization. However, in songbirds, acoustic isolation experiments have shown that "learning" from conspecifics, father, mate, and other species, plays a very important role in development of song (Brown, 1975). Studies in this field revealed that most birds learn their songs during a restricted period in their first year of life (Nottebohm, 1970). At the end of the critical period for song learning, the song becomes crystallized and then the song themes or repertoire remain unchanged for the rest of life (Thorpe, 1961; Marler and Mundinger, 1971). Furthermore, the process of crystallization also marks the end of the capacity to imitate and learn new song (Nottebohm, 1969, 1970). It was also found that song learning is dependent not only on social stimuli received from conspecifics or neighbors but also on hormonal factors (Arnold *et al.,* 1976). Furthermore, it is during this time period that central control of song is established on one side of the nervous system (Nottebohm, 1971, 1972). Later studies revealed that instead of one critical period for song learning, there may be two or more (Konishi and Nottebohm, 1969; see Brown, 1975, for a review). Thus, for a given system there may be more than one critical period or sensitive period, and furthermore, these periods may overlap to varying extents in different species.

In a related, but different, line of experimentation, the effects of sensory deprivation during development received wide attention in the 1960s. As early as the end of the nineteenth century, von Gudden (1824–1886) experimented with removal of sensory organs and cranial nerves in young animals and described secondary atrophy of central neural structures (1870, 1874). Later studies on the effects of deafferentation in sensory systems extended von Gudden's observations. In 1958, A. Hess reported dramatic changes in the superior colliculus, lateral geniculate, and the visual cortex following eye removal in fetal guinea pigs, which he interpreted as developmental arrest. Contemporaneously, Riesen and co-workers (Riesen, 1950, 1961; Chow et al., 1957) reported disappearance of retinal ganglion cells and retardation in visual discrimination learning in neonatally light-deprived chimpanzees. However, extensive experimentation on this issue awaited a major breakthrough in the field of sensory physiology, particularly the technical advances that permitted recording of the activity of single neurons in the central nervous system. Hubel and Wiesel demonstrated that neurons in the visual cortex respond to particular spatial attributes of the physical stimuli (1959, 1962). These findings provided the normative data upon which the developmental analysis of visual system functioning could be based. A few years later, Hubel and Wiesel (1963a,b) continued the line of research that was initiated by Riesen and investigated the effects of eye closure in newborn kittens on the receptive field characteristics of the cortical neurons in the adult. They observed that monocular deprivation between 3 weeks and 4 months of age leads to radical changes in the response characteristics of cortical neurons. Normally nearly all cortical neurons are driven binocularly; however, monocular deprivation results in the majority of the neurons responding only to the nondeprived eye. Even more specific effects have been found to result from the exposure of kittens to restricted forms of visual experience such as orientation of the stimulus (Blakemore and Cooper, 1970; Hirsch and Spinelli, 1970, 1971; Spinelli et al., 1972), binocular disparity (Shlaer, 1971), size of the stimulus (Pettigrew and Freeman, 1973), and direction of movement (Daw et al., 1978). Furthermore, all of these changes can be brought about by relatively brief exposure (Blakemore and Mitchell, 1974) and appear to be virtually permanent (Pettigrew et al., 1973).

Morphological effects of visual deprivation can also be demonstrated (Wiesel and Hubel, 1963; Guillery, 1972a,b; Garey *et al.*, 1979; Lin and Kaas, 1980). Other studies confirmed and extended the finding that end organ damage or sensory deprivation in newborn animals produces numerous morphological changes in the nervous system, such as alterations in cell size (Gyllensten *et al.*, 1965), dendritic orientation (Coleman and Riesen, 1968; Valverde, 1968), dendritic spine density (Globus and Scheibel, 1967; Valverde, 1968; Ryugo *et al.*, 1975), synaptic density (Fifková, 1970), cellular organization (Van der Loos and Woolsey, 1973; Weller and Johnson, 1975; Woolsey *et al.*, 1979), and afferent distribution (Belford and Killackey, 1979).

By and large, all of these studies point out that during certain periods in development the nervous system is highly malleable and that intact morphological components as well as a normal range of internal and external stimuli are necessary for development to proceed normally. Hence, the concepts of critical periods and sensitive periods are very important, as they not only delineate times during which the nervous system is prone to irreversible or extensive damage but also contribute to our understanding of the functional organization of the nervous system and the underlying developmental principles. The following account will concentrate on these two major aspects of the concepts of critical periods and sensitive periods. The scope of the discussion will be limited to areas of research directly related to modern developmental neurobiology.

III. Contexts within Which the Terms *Critical Period* and *Sensitive Period* Are Employed

The development of the nervous system is a heterogeneous process that involves complex biochemical mechanisms, operating at different times and rates in different parts of the system. This process can be characterized by well-ordered sequences of organizational mechanisms in which timing is crucial and the operation of which are dependent on the entire sequence of related preceding events (see Fig. 1). Furthermore, various regulatory mechanisms exert their control over this process of organization. Alterations, substitutions, or blocking of these organizational mechanisms can thus have profound effects on the development of the system. Conceptually, critical period and sensitive period denote periods during which organization of the nervous system is taking place; and, furthermore, these organizational processes are vulnerable to factors that disrupt them, leading to abnormal development. The vulnerability of the nervous system to these conditions is not uniform; rather, it is closely tied to the processes of development.

In general, there are no fundamental differences in the develop-

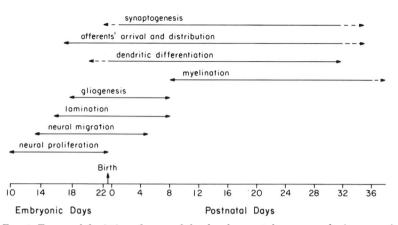

FIG. 1. Temporal depiction of some of the developmental processes during organization of the rat parietal cortex. At a given time during development, various organizational processes (such as migration, differentiation, myelination) that are in progress are vulnerable to interference. (Compiled from Davison and Dobbing, 1968; Berry, 1973.)

mental processes (such as proliferation, migration, cell death, differentiation) and their sequence across different mammalian species. This permits the study of causal relationships between critical factors, harmful conditions, and the alterations in the developing nervous system and making extrapolations across species. However, the state and rate of development with respect to gestational period shows major species differences (Davison and Dobbing, 1968). Accordingly, the temporal limits of a particular critical period or sensitive period exhibit marked species differences. Yet, animal research is invaluable to our understanding of the causal factors behind various developmental defects and to providing protective or corrective measures in medicine.

A. NUTRITIONAL DEFICIENCY AND EXPOSURE TO TOXINS AND OTHER HARMFUL AGENTS DURING DEVELOPMENT

The presence of nutrients, particularly during the synthesis of cellular components of the embryo or fetus, is critical because these factors are essential for cellular proliferation. Thus, prenatal or early postnatal malnutrition can alter the course of development directly or indirectly. In addition, during development the underlying mechanisms are, to varying degrees, vulnerable to the effects of disruptive external agents. Interference with developmental mechanisms at the cellular and/or biochemical levels either by the absence of certain critical components or by the presence of disruptive agents can result in a multitude of effects that may be expressed as abnormal development.

Conversely, different types of interference may have effects that are expressed in a similar fashion. This latter point is illustrated by the paradigm of teratogenic action originally proposed by Wilson (1973) and further discussed by Hutchings (1978). Basically this paradigm states that various disruptive agents may interfere with numerous mechanisms that in the end are expressed by a single effect, for example, cell death (see Fig. 2).

Over the past three decades, animal experimentation has provided data that clearly indicate that nutritional deficiency during development retards or alters various biochemical and morphological features of the nervous system (for reviews, see Morgane *et al.*, 1978; Zamenhof and Van Marthens, 1978). Reports indicate that severe nutritional deficiency during development reduces brain size (Winick and Noble, 1966; Dobbing, 1968; Adlard *et al.*, 1970; Cragg, 1972), neuronal and glial proliferation and myelination (Zamenhof *et al.*, 1968; Clos and Legrand, 1969, 1970; Bass *et al.*, 1970; Winick, 1970, 1971; Davison and Dobbing, 1968), number of axon terminals (Cragg, 1972), and dendritic proliferation and number of spines (Bass *et al.*, 1970; West and Kemper, 1976). Furthermore, transgenerational effects of nutritional deficiency have been reported. Zamenhof and colleagues, 1971, 1972) have found that in rats female offspring of malnourished mothers, even though normally nourished after birth, produce offspring with significantly reduced brain DNA content. Thus, the evidence suggests that the developing nervous system is vulnerable to pre- and/or postnatal nutritional deficiency. In 1968 Dobbin applied the notion of critical periods

```
Cellular Reactions:
    Mitotic interference                    ⎫
    Altered nucleic acid functions          ⎬   Abnormal Development:
    Mutation                                ⎭     *Cell death
    Lack of substrates, precursors                 Failed cellular interaction
    Lack of energy sources                         Reduced biosynthesis
    Enzyme inhibition                              Impaired morphogenic movement
    Changed membrane characteristics               Mechanical disruption
    Osmolar imbalance                              Altered differentiation schedules


                                            End Result:
                                               Intrauterine death
              *Cell death ────▶                Malformation
                                               Growth retardation
                                               Hormonal deficits
                                               Neurobehavioral deficits
```

FIG. 2. Different types of organizational interference can be expressed by a single outcome. (After Wilson, 1973; Hutchings, 1978.)

to the effects of nutritional deficiency and put forth a vulnerable period hypothesis:

> if a developmental process be restricted by any agency at the time of its fastest rate, not only will this delay the process, but will restrict its ultimate extent, even when the restricting influence is removed and the fullest possible rehabilitation obtained (Dobbing, 1968, p. 284).

He further suggested that the time of fastest growth ("the brain growth spurt"), which in the rat takes place between postnatal days 10 and 17 (for other species, see Fig. 6.1 in Davison and Dobbing, 1968), is the period of greatest vulnerability. However, considering the differential effects of the nature of nutritional deficiency (such as the types of nutrients, e.g., carbohydrates, proteins, vitamins, minerals, oxygen), the manner by which it is applied (i.e., maternal malnutrition, placental insufficiency, or neonatal malnutrition), and the precise time and duration of the deficiency on various parameters of the nervous system, the brain growth spurt does not appear to be a good index for the period of maximum vulnerability. In general, various types of nutritional deficiency before or during major developmental events, such as neuronal and glial proliferation, cell migration, differentiation, and myelination, yield different effects. Hence all these developmental events are vulnerable to nutritional deficiency at different times. Consequently, numerous vulnerable periods corresponding to the time of these processes can be demonstrated experimentally. However, when nutritional deficiencies occur outside the laboratory, such as in famine situations, it is not possible to isolate individual vulnerable periods because of the complexities of such circumstances. Nevertheless, there are numerous reports on the effects of nutritional deficiency on the mental development of children (Stoch and Smythe, 1963, 1967, 1976). Usually, these effects are assessed by intelligence quotient; however, they do not provide a full insight into the causal relationships involved, for often socioeconomic variables such as maternal neglect, lack of stimulation and incentive, and poor social environments accompany malnourishment in humans (Jacobson, 1978).

The ill effects of various harmful agents (such as viral and bacterial infections, a wide range of drugs and X irradiation) on developing organisms have been known for a long time. Of particular interest in public health are the effects of various external agents that have little or no harmful effects on the mother during pregnancy but induce structural malformations, and later, behavioral deficits in the offspring. Perhaps the best example of the effects of such an agent is the thalidomide disaster. In the 1960s thalidomide, a nontoxic tranquilizer in the adult, was found to be responsible for severe deformities of the extremities in the newborn (Warkany and Kalter, 1964). Later on, as these affected children matured, it became apparent that along with severe deformities of the extremities and to varying degrees eye, ear, intestinal, and urogenital malformations, the nervous system was also affected (McFie and Robertson, 1973; Sephenson, 1976). In general, drug consumption by pregnant women is fairly common (Peckham and King, 1963; Nora *et al.,* 1967), and studies on this issue have provided ample evidence that exposure of the embryo or fetus to various prescription or nonprescription medications, alcohol, and "recreational drugs" can have deleterious effects on development. The most prominent feature of the effects of teratogens is that they interfere with cell proliferation (i.e., organogenesis), although they may have other specific effects. It has even been suggested that some teratogens have such long-term effects that even preconception exposure may interfere with the development of the embryo (Hutchings, 1978). As in the case of nutritional deficiency, the developing organisms and specifically the nervous system is vulnerable throughout gestation and the early postnatal period.

A range of effects similar to those of teratogens result from exposure to X irradiation during development. The effects of radiation on the nervous system vary, depending on the dose and stage of development at the time of exposure. These effects include changes in differentiation, biochemical characteristics and function, delayed growth, mutations, and cell death (see Hicks and D'Amato, 1978, for review). Mitotic and migratory cells in the nervous system are particularly sensitive to radiation. This has been well demonstrated in the mammalian cerebellum, where the external granule layer has very high proliferative and migratory activity at the time of birth (Hicks, 1958; Hicks *et al.,* 1961; Hicks and D'Amato, 1966; Altman *et al.,* 1968, 1969; Altman and Anderson, 1972). Although the major effect of radiation on the cerebellum is destruction of granule cells, secondary effects that are probably due to loss of granule cells (such as stunting of Purkinje cell dendrites, anomalous synapses between Purkinje cell dendrites and mossy fibers) have been reported (Altman and Anderson, 1972; Llinas *et al.,* 1973).

The effects of radiation on developing human brain have been studied in humans prenatally exposed to radiation as a result of atom bombs during World War II. Wood *et al.* (1967) have reported that such individuals had high incidents of microencephaly and mental retardation. This study along with some clinical observations (Driscoll *et al.,*

1963) suggest that radiation effects in humans are quite similar to those reported for other mammals.

In summary, the studies on the effects of nutritional deficiency and exposure to harmful external agents during development clearly indicate that presence of nutrients is critical, particularly during cellular proliferation in the nervous system, and that throughout pre- and postnatal development the nervous system is highly sensitive to harmful external agents, which have the potential of disrupting one or more developmental mechanisms. As pointed out earlier, it should be borne in mind that during the period of development numerous organizational processes are underway: thus, their course can be most easily altered at this time.

B. HORMONAL DEFICIENCY

The role of hormones during development is quite diverse. First, levels of maternal hormones during gestation can affect the intrauterine environment and placental size and function, which may in turn affect the developing embryo (Zamenhof and Van Marthens, 1971; Zamenhof et al., 1971). Second, endogenous hormones exert their effects in a multitude of ways such as differentiation of target cells, synaptogenesis, myelination, and sexual differentiation of the brain (Whitsett and Vandenbergh, 1978; Gorski, 1973; Toran-Allerand, 1978).

Adequate levels of thyroxine, corticosterone, and somatotropin are very important during dendritic growth and synaptogenesis, as deficiencies of these hormones during these periods produce stunting of dendrites and decreased numbers of dendritic spines and axodendritic synapses in the mammalian cerebral and cerebellar cortices (Balázs, 1974). In mammals, deficiency of thyroid hormone apparently does not affect prenatal brain development (Hamburgh, 1968) but deficiency during postnatal periods results in failure or delayed maturation of the cerebral and cerebellar cortices and, particularly, myelination (Eayrs, 1960; Legrand, 1963; Gomez et al., 1966; Balázs et al., 1969). These effects appear to be widespread.

More specific effects have been attributed to the sex hormones. Important differences in sexual behavior and in sexually dimorphic central nervous structures are, to a great extent, due to the pre- and postnatal hormonal milieu (Gorski, 1973; McEwen, 1978, 1981; Toran-Allerand, 1978; Whitsett and Vandenburgh, 1978; MacLusky and Naftolin, 1981). Although the genetic sex is determined by the presence of X or Y chromosomes, the absence of gonadal hormones during a critical period in development leads mammals of either genetic sex to differentiate as females. In general, in male mammals, testosterone levels in blood increase perinatally, decline after a certain period, and increase again at puberty. It is well accepted that the perinatal increase of blood testosterone levels plays an organizing role on the neural mechanisms underlying later sexual behavior, whereas the increase at puberty activates it (Gorski, 1973; MacLusky and Naftolin, 1981). The critical period for steroid sensitivity in mammals occurs perinatally and extends into the early postnatal period. In the rat, sensitivity of the brain to steroids extends from the eighteenth day of gestation to the eleventh postnatal day (Barraclough, 1971; Lobl and Gorski, 1974). However, the maximal effectiveness of single subcutaneous injections of steroids or of neonatal castration has been delimited to the first 5 postnatal days, a period that some authors consider as the physiological critical period (Toran-Allerand, 1978). The effects of sex steroids during the critical period range from organizing sexually dimorphic brain regions (Raisman and Field, 1971, 1973; Gorski *et al.*, 1978) to establishing various motor behaviors associated with sexual behavior, such as lordosis and mounting behavior (McEwen, 1981).

C. Effects of Neonatal Stimulus Deprivation

The studies on imprinting and various forms of sensory deprivation (where no morphological damage to the sensory system has occurred) have provided ample evidence that the presence of certain stimuli during well-defined periods in development is critical for normal development of the sensory system and species-specific behavior.

The visual system lends itself most readily to the experimental determination of the effects of sensory deprivation during development. Newborn animals can be raised in total darkness or in environments that have only certain types of visual stimuli (e.g., Blakemore tube or Hirsch goggles) or their binocular vision can be disrupted by monocular lid suture and contact lenses or via artificially induced strabismus. In all these cases, no damage to the retina or any other part of the visual pathway occurs. Thus, in the visual system, the effects of sensory deprivation can be isolated with relative ease, whereas in other sensory systems, deprivation studies most often involve concomitant damage to the sensory organ (see Mistretta and Bradley, 1978, for review). In the latter case, it is impossible to isolate the effects of sensory deprivation per se. The lack of sensory stimulation or restricted stimulation studies clearly indicates the presence of critical periods during development. In the visual system, sensory deprivation during critical periods yields both functional and morphological alterations that appear to be irreversible. The majority of the studies on the effects of sensory deprivation during the development of the visual system have focused on the changes in the functional properties of cortical neurons. In the initial studies on this subject it was shown that in normal adult cats most cortical neurons are activated binocularly; however, if kittens had monocular deprivation lasting from about the third week to the third month postnatally, the majority of stimulatable cortical neurons responded to only the nondeprived eye, even though the deprived eye had been opened (Hubel and Wiesel, 1962, 1963a, 1970; Wiesel and Hubel, 1965), thus indicating a critical period for the development of binocular vision. Later studies extended these findings and added that other functional properties of visual cortical neurons, such as orientation specificity, motion or direction specificity, and binocular disparity, also require proper stimulation during critical periods (Blakemore and Cooper, 1970; Hirsch and Spinelli, 1970, 1971; Shlaer, 1971; Pettigrew and Freeman, 1973; Pettigrew and Garey, 1974). A multitude of experiments in this area have provided evidence that abnormal visual experience alters the functional characteristics of cortical neurons. However, the findings by Hubel and Wiesel on the response characteristics of visually inexperienced cats and monkeys have raised another issue. These authors (Hubel and Wiesel, 1963b; Wiesel and Hubel, 1974; LeVay et al., 1980) report that in newborn kittens and monkeys some of the visual cortical neurons display normal adultlike properties prior to opening of the eyes. These observations have raised the issue of the role of experiential and genetic factors in determining the functional integrity of the visual system.

The importance of exposure to various forms of visual stimuli during critical periods has been demonstrated for normal morphology as well as for the functional integrity of the system. For example, it has been reported that monocular deprivation in neonates leads to a decrease in the size of cells in the laminae of the lateral geniculate nucleus normally receiving afferents from the deprived eye (Wiesel and Hubel, 1963, 1965; Guillery, 1972a; Guillery and Stelzner, 1970; Garey et al., 1979; Daniels and Pettigrew, 1976; Lin and Kaas, 1980). Thus, it appears that activity along sensory pathways during critical periods plays an important role in the morphological and functional development of the system.

Despite the temporal differences in critical periods across different species, the practical value of animal research to human health is perhaps best illustrated by the studies on the effects of various forms of visual deprivation in neonatal life. For example, the critical period for binocular vision in the cat extends between 3 and 32 weeks of age (Cynader et al., 1980). The comparable critical period for humans has been determined from studies on the effects of strabismus originating at different ages. The maximum deleterious effect is produced before the age of 3 years (Hohmann and Creutzfeldt, 1975), and the effect may be produced by strabismus up to 6 years of age (Banks *et al.,* 1975). Hence, the optimal period for correcting such conditions in newborns would roughly correspond to the critical period for binocular vision. Additionally, astigmatism not corrected during the first few years of life can cause permanent deficits in terms of reduced acuity in the meridian of impairment even if the refractive error is corrected with lenses later on (Freeman and Thibos, 1973; Pettigrew and Freeman, 1973).

D. The Effects of Early Neural or Sensory End Organ Damage

Damage to either sensory organs or the central nervous system of neonates yields effects that differ from the effects of similar damage to the adult. Numerous studies have led to the conclusion that there is a greater degree of functional sparing if damage to the central nervous system occurs early rather than late in life (Kennard, 1938, 1940; Teuber and Rudel, 1962; Teuber, 1971). This conclusion is most dramatically illustrated by clinical observations on the relocalization of speech and language function that occurs if the left hemisphere is damaged before the age of 12, but rarely afterward (Lenneberg, 1967). On the other hand, other studies clearly indicate that some functions are more impaired in neonatally lesioned animals (Schneider, 1979), and these impairments can probably be related to morphological alterations (Schneider and Jhaveri, 1974; Schneider, 1979). The experimental paradigm used in these studies is to produce specific lesions of the nervous system in newborn animals and study the morphological alterations after having tested the behavior of these animals at maturity. Here it should be noted that in many areas of the nervous system the normal course of development and the events that take place during this time are not well delineated. Thus, it is often difficult to determine whether certain developmental events were subjected to arrest or abnormal morphological changes, particularly connectivity changes initiated as a result of the early damage. For example, studies such as these often conclude that intact axons "sprout" into neonatally denervated zones. However, information on the course of normal developmental events suggests a different interpretation. Land and Lund (1979) have demonstrated that in the newborn rat the retina projects bilaterally across the superficial layers of the superior colliculus but the ipsilateral projection retracts during the first 10 postnatal days, resulting in the normal adult pattern. Furthermore, removal of one eye at birth results in the other retina projecting bilaterally to the superior colliculus in the adult. This evidence would suggest that the abnormal adult pattern is achieved by a failure to retract (e.g., developmental arrest) rather than the "sprouting" of processes. It also points to the danger of drawing interpretations about developmental mechanisms in the absence of normative developmental data. Similar extensive projections during development that are later retracted have been shown for other parts of the nervous system (Innocenti, 1979; Innocenti and Caminiti, 1980; Ivy and Killackey, 1981). Currently, studies on the morphological effects of brain damage during development provide evidence that supports the idea that both developmental arrest and abnormal organization can take place (Land and Lund, 1979; Rhoades and Dellacroce, 1980; Nah et al., 1980).

Research on the rodent trigeminal system serves as an illustrative example of the role of sensory end organs in determining central organization. The orderly arrangement of vibrissae on the face of rodents is represented several times in a precise, topographical fashion along the neuraxis (Belford and Killackey, 1979; Killackey and Belford, 1979). The representation of the vibrissae in the central nervous system is evident in both the cytoarchitectural organization and the distribution of related afferent terminals. Furthermore, damage to the vibrissae during the first few postnatal days alters the organization of the afferent distribution and cellular organization (Van der Loos and Woolsey, 1973; Weller and Johnson, 1975; Jeanmonod et al., 1977; Belford and Killackey, 1979, 1980). In the rat the sensitive period for the afferent organization has been determined to be the first 3 postnatal days (Belford and Killackey, 1980). Some parallels between the effects of damage to the vibrissae and the effects of stimulus deprivation in the visual system have been drawn, and the term *critical period* has been employed in this context (e.g., Woolsey and Wann, 1976; Woolsey et al., 1979). However, there are substantial differences between these experimental manipulations. In one case, there is clear damage to the sensory organ, whereas, in the other, there is no damage. Even a simple depilation of vibrissae cannot be compared to the suturing of an eyelid, in which there is absolutely no damage to the retina.

In summary, the studies on the effects of early injury to the central nervous system and the sense organs collectively indicate that there are sensitive periods during which elimination of a part of the developing system can either arrest development, interfere with its differentiation, or change its course. Furthermore, these sensitive periods correspond to the time period during which the organization of the system, or adultlike connectivity of the system, is underway.

IV. Theoretical Considerations

By and large the majority of critical periods and sensitive periods described in the literature correspond to the period of development. However, there are exceptions. For example, there is evidence that the duration of certain critical periods and sensitive periods can be extended by certain manipulations [e.g., 6-hydroxydopamine treatment for the extension of the critical period for binocular vision (Kasamatsu and Pettigrew, 1979)]; or effects similar to those resulting from various manipulations during the critical periods can in some cases be obtained in the adult [e.g., an increase in the size of certain brain nuclei that normally is characteristic of the male can be brought about in adult female canaries by ovariectomy and androgen treatment (Nottebohm, 1980)]; or some critical periods can be found in the adult [e.g., offspring recognition after parturition in ungulates (Collins, 1956; Cairns, 1966)]. Nevertheless, these exceptions do not pose a major problem to considering critical periods and sensitive periods as highly characteristic of the period of development.

It should be emphasized that determining the temporal boundaries of critical periods and sensitive periods is only a first step and not an end in itself. After this step, various experimental variables can be applied systematically during these periods, as probes to provide insight into the developmental mechanisms and principles that are underway. Hence, aside from the practical value of determining the critical periods and sensitive periods as previously discussed, study of these periods can produce invaluable information as to the nature of mechanisms that underlie the functional organization of the nervous system.

As stated earlier, the development of the nervous system involves the interaction of predetermined (genetic) and environmentally induced (epigenetic) factors. All of these factors are intricately woven into a network of causal relationships. Basically all these events can be considered as an expression of organizational processes. Furthermore, these processes are highly vulnerable to the effects of different disruptive agents or to the absence of critical components at different stages of organization. Viewed from this perspective, the presence of numerous critical periods and sensitive periods is a natural consequence. For example, cellular proliferation is a major developmental event that takes place at different time periods of gestation in different parts of the brain. The presence of certain necessary factors for this event to occur (such as nutrients) is critical (thus, critical period); and, furthermore, the same event is highly sensitive to disruptive agents (such as teratogens and radiation) during a specific period (thus, sensitive period).

Conceptually, the distinctions among critical period, sensitive period, vulnerable period, and optimal period as suggested earlier is very important. In the absence of precision in terminology, totally different events are often compared just because they have been referred to by the same term. For example, there are significant differences in studies on the effects of neonatal sensory manipulation and the related critical periods and sensitive periods. It is inappropriate to compare directly the effects of monocular lid suture to the effects of enucleation and refer to a critical period for binocular vision. Similarly, it is erroneous to compare the effects of end organ damage in other systems to the effects of stimulus deprivation on the developing visual system and to use the term *critical period* in both instances. In cases where sensory deprivation is achieved by damage to a sensory receptor, it is more appropriate to talk about sensitive periods rather than a critical period. Moreover, in such cases one cannot determine if the stimulus is critical or not, as a major aspect of the manipulation is disrupted anatomical organization.

A. Categories of Experimental Paradigms

In view of the distinction between critical period and sensitive period, various experimental studies where these terms have been extensively employed can be divided into four major categories based on the type of experimental manipulation employed:

1. Deprivation of critical factors or conditions without any direct damage to the system (examples: visual deprivation by way of lid suture; dark rearing; induced strabismus; pre- and postnatal nutritional deficiency; hormonal deprivation; sensory isolation in birds).
2. Deprivation of critical factors or conditions and/or direct damage to the system (examples: enucleation; sensory deprivation by end organ damage; X irradiation; teratogens; trauma; deafening and denervation experiments in the ontogeny of bird song).
3. Replacement or alteration of critical factors or conditions without any direct damage to the system (examples: visual form deprivation by use of Blakemore tube or Hirsch goggles; abnormal hormonal treatments; abnormal forms of imprinting).
4. Replacement or alteration of critical factors and conditions plus direct damage to the system (examples: visual deprivation plus retinal damage; hormonal treatments plus castration or ovariectomy).

As pointed out earlier, the majority of research from all of the preceding four categories consists of detailed descriptions of phenomena that involve critical periods and sensitive periods. The common denominator in all of these studies is that there are multiple organizational processes that involve different or overlapping critical periods and sensitive periods.

B. Scott's Theory of Critical Periods

The theory of critical periods proposed by Scott (Scott *et al.*, 1974; Scott, 1978) provides an extremely useful framework within which the diversity of research on critical periods and sensitive periods can be subsumed. The basic premise of Scott's theory is that the organization of a system is most easily modified when it is proceeding most rapidly. Furthermore, Scott ties the existence of critical periods to the nature of the organizational processes involved. He distinguishes four general, theoretical modes of organizational processes (see Fig. 3): (1) If a system organizes uniformly throughout life, it can be modified at any time and there are no associated critical periods. (2) If a system organizes rapidly for a limited period, then stops, there is only one critical period. (3) If a system organizes initially at a slow rate, then more rapidly for a restricted period of time, and then gradually declines but never stops, the system will be affected differentially during a single critical period. (4) If a system has multiple, rapid periods of organization that are distinctly separated over time, there will be many critical periods. Because the development of the nervous system encompasses a multitude of organizational events, Scott's consideration of critical periods is very valuable for neurobiology. For example, in his theoretical analysis of

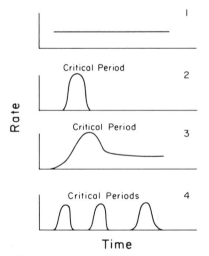

FIG. 3. Diagrammatic illustration of organizational modes according to Scott's theory of critical periods. See text for explanation. (After Scott *et al.*, 1974.)

critical periods, he considers two simple organizational processes, each with a single critical period. Within this framework, these two processes may be either (1) independent or (2) interdependent. Furthermore, in either of these cases the two processes can occur (a) simultaneously, (b) successively, or (c) separately in time. In the first case (1), the critical periods associated with the two processes can be identified no matter when they occur in time because the end result of each process is different. However, in the case of the two processes being interdependent (2), the situation is far more complex (see Fig. 4). If these two processes occur simultaneously (a), the critical periods associated with each process will be the same and one would conclude that a single organizational event is taking place because any modification of one process will inevitably modify the other process. If, on the other hand, these two processes occur successively (b) and if the completion of one process is necessary for the start of the second, the critical period for the system will cover the sum of these two critical periods. If the two processes are affected by similar modifying factors, the individual processes will not be readily detectable, as the main effect of the modifying factors will be to sum the critical periods for both processes. However, if these two processes are modified by different sets of factors, then the

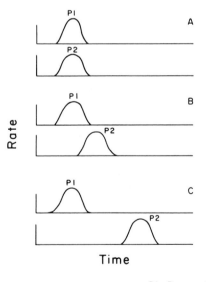

PI: Process 1

P2: Process 2

FIG. 4. Interaction of critical periods of two interdependent organizational processes. See text for explanation. (After Scott *et al.*, 1974.)

underlying two processes will be readily detectable. Finally, if the two organizational processes are separated in time from one another (c), the system will exhibit two separate critical periods, each associated with one process whether the two processes are affected by the same or different modifying factors.

To illustrate, these theoretical considerations can be applied to the organization of the rodent trigeminal system following neonatal damage to the end organ (i.e., the vibrissal pad). As noted earlier, the pattern on the vibrissal pad is anatomically represented along the brain stem, dorsal thalamus, and the somatosensory cortex (Belford and Killackey, 1979; Killackey and Belford, 1979). During normal development, these anatomical representations appear in a sequential fashion; first in the brain stem, next in the thalamus, and finally in the cortex (Killackey and Belford, 1979; Belford and Killackey, 1979). Furthermore, neonatal peripheral manipulations affect the organization at all three levels. The sensitive period during which damage to the periphery alters the central correlates occurs throughout the first 4 postnatal days (Belford and Killackey, 1980). This evidence can be interpreted as suggesting that development of the trigeminal system involves multiple, successively interdependent processes, similar to the organizational mode 2b of Scott's theory. Thus, it is not feasible to differentiate between these processes by means of peripheral manipulations that affect all three of them. Although the research paradigm employed allows us to make inferences about the mechanisms operating at the brain stem level, the organizing processes at the thalamic and cortical levels are not readily detectable as they may be obscured by the organizational processes at the brain stem level. Hence, it is impossible to speak of different sensitive periods for various stations of the trigeminal system as some authors have done (Woolsey et al., 1979). In such studies, description of sensitive periods for the thalamus or cortex must be made cautiously because, by peripheral alterations, we cannot directly infer the nature of the mechanisms operating at a region two or three stations away and ignore intervening events. To deduce the underlying organizational processes during critical periods and sensitive periods, a careful analysis of the experimental conditions is necessary. Studies of visual deprivation provide another example of this. It is often assumed that various forms of visual deprivation exert effects directly on the cortical neurons. Although it is true that cortical neurons exhibit functional changes depending on the type of stimulation during the critical period, these external manipulations also affect the properties of subcortical neurons and what one actually sees is the result of the manipulation on cortical neurons via the effects on the retinal ganglion cells and lateral geniculate cells.

C. PLASTICITY

In the literature pertaining to critical periods and sensitive periods, the term *plasticity* is often encountered. In general, these periods are considered to be times during which the nervous system is highly plastic or malleable (e.g., Daniels and Pettigrew, 1976; Jacobson, 1978; Pettigrew, 1978). Pettigrew (1978) has raised the question of the potential benefit of plasticity during development, as it can be a great handicap and has little apparent value. Perhaps this characteristic—plasticity—is best viewed not as an inherent property of the system but as an epiphenomenon that inextricably accompanies any process of organization. The potential for alteration is a natural property of organizational processes, and, not surprisingly, it occurs when the building blocks and the organizational factors and mechanisms are present. It is this alteration that is often regarded as plasticity. Within this context of plasticity, questions such as why the developing nervous system is highly vulnerable to numerous external factors compared to the adult yet is very resilient to trauma or physical damage in comparison to the adult are raised (Pettigrew, 1978). In view of the foregoing analysis of the concepts of critical periods and sensitive periods, questions like these raise false issues rather than providing insight into the nature of developmental mechanisms. It is not that the developing nervous system is more plastic, but trauma during development either arrests organizational processes and/or alters their course—the net result of this is a different nervous system.

It should also be pointed out that alterations in the timing of sequential phenomena can have profound physiological effects without significantly altering the basic morphology, which can thus appear "normal." In such cases, morphological differences may be expressed in more subtle ways, such as in the pattern of dendritic differentiation and synaptic organization. A good example of this is seen in the effects of hormones on the differentiation of central nervous system structures with respect to sex. For example, Raisman and Field (1971) observed that axons of nonamygdaloid origin synapse upon the shafts of dendrites in the preoptic area of the male rat, whereas in the females they predominantly synapse upon the dendritic spines. Neonatal castration of the males leads to an increase in the number of spine synapses, and androgen treatment of females after birth increases shaft synapses (Raisman and Field, 1973). Such subtle changes in the organization can express themselves in many ways, ranging from small functional modifications to gross behavioral alterations. Furthermore, by changing the course of development, an alternate route of organization may be taken, and the end result may appear to be functionally normal.

D. INTERACTION OF GENETIC AND EPIGENETIC FACTORS DURING NEURAL ORGANIZATION

Another pseudoissue raised in relation to critical periods and sensitive periods is the age-old nature–nurture controversy. The issue of how much of the functional organization of the system is genetically determined and how much of it is experientially determined is usually raised in studies on the effects of visual deprivation during critical periods. Hubel and Wiesel (1970; Wiesel and Hubel, 1963, 1965, 1974; LeVay *et al.*, 1980) have reported that both in the cat and monkey there are some visually inexperienced cells in the visual cortex that show functional organization similar to that of adults. However, there is sufficient evidence pointing out that subsequent experience can either modify this organization or add to it. Modification may involve selective disruption of genetically determined connectivity or alteration of the existing connectivity, so that cortical cells acquire new and different response characteristics. On the other hand, there may be some subpopulation of cells whose connectivity has not been genetically determined so that their response characteristics can be specified in accordance with visual experience (Daniels and Pettigrew, 1976). Rather than segregating genetic and environmental contributions to the development of the nervous system and regarding critical and sensitive periods as times during which this segregation can be inferred from experimental manipulations, it is best to consider the components of the organizational processes, genetic or epigenetic, in a dynamic interplay. As Gottlieb (1973, 1978) has pointed out, the relationship between genetic activity, structural maturation, and function can be regarded as bidirectional. Within this framework the genetic versus environmental, or nature versus nurture, issue becomes a false one.

V. Conclusions

As we have seen, the study of critical periods and sensitive periods involves not only practical aspects but also theoretical inferences into the organizational processes that underlie the functional specificity of the nervous system. Despite the wide variety of mechanisms involved during critical periods and sensitive periods, phenomenologically similar ontogenetic processes, and the ways in which they are influenced by external factors, one may arrive at certain theoretical generalizations. First, the timing and duration of these periods appear to be related to different rates of development across different species; however, the underlying mechanisms are probably similar, hence allowing across-species generalizations. Second, these periods reflect the activity of multiple organizational processes. Furthermore, these processes have different or overlapping critical periods and sensitive periods in a given species. Careful analysis of the experimental conditions and conceptual clarity is necessary in evaluating much of the experimental research. Finally, reversibility of the effects in the adult (i.e., the flexibility versus the rigidity of the organization) seems to be dependent not only on the type of deprivation and the system involved but also on the complexity of the final organization.

ACKNOWLEDGMENTS

We would like to thank Drs. Pauline Yahr and Michael Leon for their helpful comments on the manuscript.

REFERENCES

Adams, F. (trans.) (1950). "The Genuine Works of Hippocrates," Vol. 2. Wm. Wood, New York.

Adlard, B. P. F., Dobbing, J., and Smart, J. L. (1970). *Biochem. J.* **119**, 46P.

Altman, J., and Anderson, W. J. (1972). *J. Comp. Neurol.* **146**, 355–406.

Altman, J., Anderson, W. J., and Wright, K. A. (1968). *Exp. Neurol.* **22**, 52–74.

Altman, J., Anderson, W. J., and Wright, K. A. (1969). *Exp. Neurol.* **24**, 196–216.

Arnold, A. P., Nottebohm, F., and Pfaff, D. W. (1976). *J. Comp. Neurol.* **165**, 487–512.

Balázs, R. (1974). *Br. Med. Bull.* **30**, 126–134.

Balázs, R., Brooksband, B. W. L., Davison, A. N., Eayrs, J. T., and Wilson, D. A. (1969). *Brain Res.* **15**, 219–232.

Banks, M. S., Aslin, R. N., and Letson, R. D. (1975). *Science* **190**, 675–677.

Barraclough, C. A. (1971). *In* "Steroid Hormones and Brain Function" (C. H. Sawyer and R. A. Gorski, eds.), pp. 149–159. Univ. of California Press, Berkeley.

Bass, N. H., Netsky, M. G., and Young, E. (1970). *Arch. Neurol. (Chicago)* **23**, 289–302.

Beach, F. A. (1975). *Psychoneuroendocrinology* **1**, 3–23.

Belford, G. R., and Killackey, H. P. (1979). *J. Comp. Neurol.* **188**, 63–74.

Belford, G. R., and Killackey, H. P. (1980). *J. Comp. Neurol.* **193**, 335–350.

Berry, M. (1973). *In* "Studies on the Development of Behavior and the Nervous System" (G. Gottlieb, ed.), Vol. 2, pp. 7–67. Academic Press, New York.

Blakemore, C., and Cooper, G. F. (1970). *Nature (London)* **228**, 477–478.

Blakemore, C., and Mitchell, D. E. (1974). *Nature (London)* **241**, 467–468.

Bodenheimer, F. S. (1958). "The History of Biology: An Introduction." Dawson, London.

Brown, J. L. (1975). "The Evolution of Behavior." Norton, New York.

Bruner, J. S. (1968). *In* "Contemporary Issues in Developmental Psychology" (N. S. Endler, L. R. Boulter, and H. Osser, eds.), pp. 476–494. Holt, New York.

Cairns, R. B. (1966). *J. Comp. Physiol. Psychol.* **62**, 298–306.

Castiglioni, A. (1958). "A History of Medicine." Alfred A. Knopf, New York.

Chow, K. L., Riesen, A. H., and Newell, F. W. (1957). *J. Comp. Neurol.* **107**, 27–42.

Clos, J., and Legrand, J. (1969). *Arch. Anat. Microsc. Morphol. Exp.* **58**, 339–354.

Clos, J., and Legrand, J. (1970). *Brain Res.* **22**, 285–297.

Coleman, P. D. and Riesen, A. H. (1968). *J. Anat. (London)* **102**, 363–374.

Collins, N. E. (1956). *Ecology* **37**, 228–239.

Cragg, B. G. (1972). *Brain* **95**, 143–150.

Cynader, M., Timney, B. N., and Mitchell, D. E. (1980). *Brain Res.* **191**, 545–550.

Daniels, J. D., and Pettigrew, J. D. (1976). *In* "Studies on the Development of Behavior and the Nervous System" (G. Gottlieb, ed.), Vol. 3, pp. 196–232. Academic Press, New York.

Dareste, C. (1891). "Recherches sur la production artificielle des monstruosités." Reinwald, Paris.

Davison, A. N., and Dobbing, J. (1968). *In* "Applied Neurochemistry" (A. N. Davison and J. Dobbing, eds.), pp. 253–286. Blackwell, Oxford.

Daw, N. W., Berman, N. E. J., and Ariel, M. (1978). *Science* **199**, 565–566.

De Vries, H. (1905). "Species and Varieties." Open Court Publishing Co., Chicago, Illinois.

Dobbing, J. (1964). *Proc. R. Soc. London, Ser. B* **159**, 503–509.

Dobbing, J. (1968). *In* "Applied Neurochemistry" (A. N. Davison and J. Dobbing, eds.), pp. 287–316. Blackwell, Oxford.

Driscoll, S. G., Hicks, S. P., Copenhaver, E. H., and Easterday, C. L. (1963). *Arch. Pathol.* **76**, 113–119.

Durant, W. (1939). "The Story of Civilization," Parts II and III. Simon & Schuster, New York.

Eayrs, J. T. (1960). *Br. Med. Bull.* **16**, 122–126.

Féré, C. (1894). *C. R. Seances Soc. Biol. Ses Fil.* **46**, 462–465.

Féré, C. (1899). *In* "Cinquantenaire de la société de biologie," Vol. jubilaire, pp. 363–369. Masson Cie., Paris.

Fifková, E. (1970). *J. Neurobiol.* **1**, 285–295.

Freeman, R. D., and Thibos, L. N. (1973). *Science* **180**, 876–878.

Freud, S. (1938). "The Basic Writings of Sigmund Freud" (A. A. Brill, trans. and ed.). Modern Library, New York.

Garey, L. J., Blakemore, C., and Vital-Durand, F. (1979). *Prog. Brain Res.* **51**, 445–456.

Globus, A., and Scheibel, A. (1967). *Exp. Neurol.* **18**, 116–131.

Gomez, C. J., Ghittoni, N. E., and Dellacha, J. M. (1966). *Life Sci.* **5**, 243–246.

Gorski, R. A. (1971). *In* "Frontiers in Neuroendocrinology" (L. Martin and W. F. Ganong, eds.), pp. 237–290. Oxford Univ. Press, London and New York.

Gorski, R. A. (1973). *Prog. Brain Res.* **39**, 149–162.

Gorski, R. A., Gordon, J. H., Shryne, J. E., and Southam, A. M. (1978). *Brain Res.* **148**, 333–346.

Gottlieb, G. (1973). *In* "Studies on the Development of Behavior and the Nervous System" (G. Gottlieb, ed.), Vol. 1, pp. 3–45. Academic Press, New York.

Gottlieb, G. (1978). *In* "Studies on the Development of Behavior and the Nervous System" (G. Gottlieb, ed.), Vol. 4, pp. 3–5, 331–335. Academic Press, New York.

Goy, R. W. (1966). *J. Anim. Sci.* **25**, 21–35.

Grady, K. L., Phoenix, C. H., and Young, W. C. (1965). *J. Comp. Physiol. Psychol.* **59**, 176–182.

Guillery, R. W. (1972a). *J. Comp. Neurol.* **144**, 117–130.

Guillery, R. W. (1972b). *J. Comp. Neurol.* **146**, 407–420.

Guillery, R. W., and Stelzner, D. J. (1970). *J. Comp. Neurol.* **139**, 413–422.

Gyllensten, L., Malmfors, I., and Norrlin, M. (1965). *J. Comp. Neurol.* **124**, 149–160.

Hamburgh, M. (1968). *Gen. Comp. Endocrinol.* **10**, 198–213.

Harlow, H. F., and Harlow, M. K. (1962). *Sci. Am.* **207,** 136–146.

Hebb, D. O. (1949). "The Organization of Behavior." Wiley, New York.

Hess, A. (1958). *J. Comp. Neurol.* **109,** 91–115.

Hess, E. H. (1973). "Imprinting." Van Nostrand-Reinhold, Princeton, New Jersey.

Hicks, S. P. (1958). *Physiol. Rev.* **38,** 337–356.

Hicks, S. P., and D'Amato, C. J. (1966). *Adv. Teratol.* **1,** 196–250.

Hicks, S. P., and D'Amato, C. J. (1978). *In* "Studies on the Development of Behavior and the Nervous System" (G. Gottlieb, ed.), Vol. 4, pp. 36–72. Academic Press, New York.

Hicks, S. P., D'Amato, C. J., Coy, M. C., O'Brian, E. D., Thurston, J. M., and Joftes, D. L. (1961). *Brookhaven Symp. Biol.* **14,** 246–261.

Hinde, R. A. (1970). "Animal Behavior," 2nd ed. McGraw-Hill, New York.

Hirsch, H. V. B., and Spinelli, D. N. (1970). *Science* **168,** 869–871.

Hirsch, H. V. B., and Spinelli, D. N. (1971). *Exp. Brain Res.* **12,** 509–527.

Hohmann, A., and Creutzfeldt, O. D. (1975). *Nature (London)* **254,** 613–614.

Hubel, D. H., and Wiesel, T. N. (1959). *J. Physiol. (London)* **148,** 574–591.

Hubel, D. H., and Wiesel, T. N. (1962). *J. Physiol. (London)* **160,** 106–154.

Hubel, D. H., and Wiesel, T. N. (1963a). *J. Physiol. (London)* **165,** 559–568.

Hubel, D. H., and Wiesel, T. N. (1963b). *J. Neurophysiol.* **26,** 993–1002.

Hubel, D. H., and Wiesel, T. N. (1970). *J. Physiol. (London)* **206,** 419–436.

Hutchings, D. E. (1978). *In* "Studies on the Development of Behavior and the Nervous System" (G. Gottlieb, ed.), Vol. 4, pp. 7–34. Academic Press, New York.

Innocenti, G. M. (1979). *Prog. Brain Res.* **51,** 479–487.

Innocenti, G. M., and Caminiti, R. (1980). *Exp. Brain Res.* **38,** 381–394.

Ivy, G. O., and Killackey, H. P. (1981). *J. Comp. Neurol.* **195,** 367–390.

Jackson, C. M., and Stewart, C. A. (1920). *J. Exp. Zool.* **30,** 97–106.

Jacobson, M. (1978). "Developmental Neurobiology," 2nd ed. Plenum, New York.

Jeanmonod, D., Rice, F. L., and Van der Loos, H. (1977). *Neurosci. Lett.* **6,** 151–156.

Kasamatsu, T., and Pettigrew, J. D. (1979). *J. Comp. Neurol.* **185,** 139–162.

Kennard, M. A. (1938). *J. Neurophysiol.* **1,** 477–496.

Kennard, M. A. (1940). *Arch. Neurol. Psychiatry* **44,** 377–397.

Killackey, H. P., and Belford, G. R. (1979). *J. Comp. Neurol.* **183,** 285–304.

Konishi, M., and Nottebohm, F. (1969). *In* "Bird Vocalizations" (R. A. Hinde, ed.), pp. 29–48. Cambridge Univ. Press, London and New York.

Land, P. W., and Lund, R. D. (1979). *Science* **205,** 698–700.

Landauer, W. (1932). *J. Genet.* **25,** 367–394.

Legrand, T. (1963). *Arch. Anat. Microsc. Morphol. Exp.* **52,** 205–214.

Lenneberg, E. H. (1967). "The Biological Foundations of Language." Wiley, New York.

LeVay, S., Wiesel, T. N., and Hubel, D. H. (1980). *J. Comp. Neurol.* **191,** 1–51.

Lin, C.-S., and Kaas, J. H. (1980). *Neurosci. Lett.* **18,** 267–273.

Llinas, R., Hillman, D. E., and Precht, W. (1973). *J. Neurobiol.* **4,** 69–94.

Lobl, R. T., and Gorski, R. A. (1974). *Endocrinology* **94,** 1325–1330.

Lorenz, K. (1935). *In* "Studies in Animal and Human Behaviour" (R. Martin, trans.), Vol. I. Harvard Univ. Press, Cambridge, Massachusetts (1970 ed.).

MacLusky, N. J., and Naftolin, F. (1981). *Science* **211,** 1294–1303.

McEwen, B. S. (1978). *Prog. Brain Res.* **48,** 291–307.

McEwen, B. S. (1981). *Science* **211,** 1303–1311.

McFie, J., and Robertson, J. (1973). *Dev. Med. Child Neurol.* **15,** 719–727.

Marler, P., and Mundinger, P. (1971). *In* "The Ontogeny of Vertebrate Behavior" (H. Moltz, ed.), pp. 389–450. Academic Press, New York.

Mistretta, C. M., and Bradley, R. M. (1978). *In* "Studies on the Development of Behavior and the Nervous System" (G. Gottlieb, ed.), Vol. 4, pp. 215–247. Academic Press, New York.

Morgane, P. J., Miller, M., Kemper, T., Stern, W., Forbes, W., Hall, R., Bronzino, J., Kissane, J., Hawrylewcz, E., and Resnick, O. (1978). *Neurosci. Biobehav. Rev.* **2,** 137–230.

Nah, S. H., Ong, L. S., and Leong, S. K. (1980). *Neurosci. Lett.* **19,** 39–44.

Nora, J. J., Nora, A. H., Sommerville, R. J., Hill, R. N., and McNamara, D. G. (1967). *JAMA, J. Am. Med. Assoc.* **202,** 1065–1069.

Nottebohm, F. (1969). *Ibis* **111,** 36–387.

Nottebohm, F. (1970). *Science* **167,** 950–956.

Nottebohm, F. (1971). *J. Exp. Zool.* **177,** 229–262.

Nottebohm, F. (1972). *J. Exp. Zool.* **179,** 35–50.

Nottebohm, F. (1980). *Brain Res.* **189,** 429–436.

Peckham, C. H., and King, R. W. (1963). *Am. J. Obstet. Gynecol.* **87,** 609–624.

Pettigrew, J. D. (1978). *In* "Neuronal Plasticity" (C. W. Cotman, ed.), pp. 311–330. Raven, New York.

Pettigrew, J. D., and Freeman, R. D. (1973). *Science* **182,** 599–601.

Pettigrew, J. D., and Garey, L. J. (1974). *Brain Res.* **66,** 160–164.

Pettigrew, J. D., Olson, C., and Hirsch, H. V. B. (1973). *Brain Res.* **51,** 345–351.

Pfeiffer, C. A. (1936). *Am. J. Anat.* **58,** 195–225.

Phoenix, C. H., Goy, R. W., Gerall, A. A., and Young, W. C. (1959). *Endocrinology* **65,** 369–382.

Piaget, J. (1926). "The Language and Thought of the Child." Routledge & Kegan Paul, London.

Platt, A. (trans.) (1965). "The Works of Aristotle," Vol. V. Oxford Univ. Press, London and New York.

Raisman, G., and Field, P. M. (1971). *Science* **173,** 731–733.

Raisman, G., and Field, P. M. (1973). *Brain Res.* **54,** 1–29.

Rhoades, R. W., and Dellacroce, D. D. (1980). *Brain Res.* **90,** 248–254.

Riesen, A. H. (1950). *Sci. Am.* **183,** 16–19.

Riesen, A. H. (1961). *In* "Functions of Varied Experience" (D. W. Fiske and S. R. Maddi, eds.), pp. 57–80. Dorsey Press, Homewood, Illinois.

Ryugo, D. K., Ryugo, R., and Killackey, H. P. (1975). *Brain Res.* **96,** 82–87.

Saint-Hilaire, E. G. (1822). "Philosophie anatomique des monstruosités humains." Bailliere, Paris.

Sarton, G. (1960). "A History of Science," Vol. 1. Harvard Univ. Press, Cambridge, Massachusetts.

Schneider, G. (1979). *Neuropsychologia* **17,** 557–583.

Schneider, G. E., and Jhaveri, S. R. (1974). *In* "Plasticity and Recovery of Function in the Central Nervous System" (D. G. Stein, J. J. Rosen, and N. Butters, eds.), pp. 65–109. Academic Press, New York.

Scott, J. P. (1937). *J. Exp. Zool.* **77,** 123–157.

Scott, J. P., ed. (1978). "Critical Periods." Dowden, Hutchingon, & Ross, Stroudsburg, Pennsylvania.

Scott, J. P., and Fuller, J. L. (1965). "Genetics and the Social Behavior of the Dog." Univ. of Chicago Press, Chicago, Illinois.

Scott, J. P., Stewart, J. M., and De Ghett, V. J. (1974). *Dev. Psychobiol.* **7,** 489–513.

Shlaer, R. (1971). *Science* **173**, 638–641.

Singer, C. (1959). "A History of Biology." Abelard-Schuman, London.

Spemann, H. (1938). "Embryonic Development and Induction." Yale Univ. Press, New Haven, Connecticut.

Spinelli, D. N., Hirsch, H. V. B., Phelps, R. W., and Metzler, J. (1972). *Exp. Brain Res.* **15**, 289–304.

Stephenson, J. B. P. (1976). *Dev. Med. Child Neurol.* **18**, 189–197.

Stoch, M. B., and Smythe, P. M. (1963). *Arch. Dis. Child.* **38**, 546–552.

Stoch, M. B., and Smythe, P. M. (1967). *S. Afr. Med. J.* **41**, 1027–1030.

Stoch, M. B., and Smythe, P. M. (1976). *Arch. Dis. Child.* **51**, 327–336.

Stockard, C. R. (1921). *Am. J. Anat.* **28**, 115–266.

Teuber, H.-L. (1971). *In* "Physical Trauma as an Etiological Agent in Mental Retardation" (C. R. Angle and E. A. Bering, Jr., eds.), pp. 7–28. Natl. Inst. Health, Bethesda, Maryland.

Teuber, H.-L., and Rudel, R. G. (1962). *Dev. Med. Child Neurol.* **4**, 3–20.

Thompson, D. W. (trans.) (1962). "The Works of Aristotle," Vol. IV. Oxford Univ. Press, London and New York.

Thorpe, W. H. (1961). "Bird-song. The Biology of Vocal Communication and Expression in Birds." Cambridge Univ. Press, London and New York.

Toran-Allerand, C. D. (1978). *Am. Zool.* **18**, 553–565.

Valverde, F. (1968). *Exp. Brain Res.* **5**, 274–292.

Van der Loos, H., and Dorfl, J. (1978). *Neurosci. Lett.* **7**, 23–30.

Van der Loos, H., and Woolsey, T. A. (1973). *Science* **179**, 395–398.

von Gudden, B. A. (1870). *Arch. Psychiatr. Nervenkr.* **2**, 693–723.

von Gudden, B. A. (1874). *Arch. Ophthalmol.* **2** (Abt. 20), 249–268.

Warkany, J., and Kalter, H. (1964). "Proceedings of the Biregional Institute on Maternity Care-Primary Prevention," pp. 102–121.

Weber, A. (1925). "Histroy of Philosophy" (F. Thilly, trans.). Scribner's, New York.

Weller, W. L., and Johnson, J. I. (1975). *Brain Res.* **83**, 504–508.

Werboff, J., and Havlena, J. (1962). *Exp. Neurol.* **6**, 263–269.

Werboff, J., Gottlieb, J. S., Havlena, J., and Word, T. J. (1961). *Pediatrics* **27**, 318–324.

West, C. D., and Kemper, T. C. (1976). *Brain Res.* **107**, 221–237.

Whalen, R. E. (1971). *In* "Ontogeny of Vertebrate Behavior" (H. Moltz, ed.), pp. 229–261. Academic Press, New York.

Whitsett, J. M., and Vandenbergh, J. G. (1978). *In* "Studies on the Development of Behavior and the Nervous System" (G. Gottlieb, ed.), Vol. 4, pp. 73–106. Academic Press, New York.

Wiesel, T. N., and Hubel, D. H. (1963). *J. Neurophysiol.* **26**, 978–993.

Wiesel, T. N., and Hubel, D. H. (1965). *J. Neurophysiol.* **28**, 1060–1072.

Wiesel, T. N., and Hubel, D. H. (1974). *J. Comp. Neurol.* **158**, 307–318.

Wilson, J. G. (1973). "Environment and Birth Defects." Academic Press, New York.

Winick, M. (1970). *Pediatr. Clin. North Am.* **17**, 69–78.

Winick, M. (1971). *Pediatrics* **47**, 969–978.

Winick, M., and Noble, A. (1966). *J. Nutr.* **89**, 300–306.

Wood, J. W., Johnson, K. G., and Omori, Y. (1967). *Pediatrics* **39**, 385–392.

Woolsey, T. A., and Wann, J. R. (1976). *J. Comp. Neurol.* **170**, 53–66.

Woolsey, T. A., Anderson, J. R., Wann, J. R., and Stanfield, B. B. (1979). *J. Comp. Neurol.* **184**, 363–380.

Zamenhof, S., and Van Marthens, E. (1971). *In* "Cellular Aspects of Neural Growth and Differentiation" (D. C. Pease, ed.), pp. 329–359. Univ. of California Press, Berkeley.

Zamenhof, S., and Van Marthens, E. (1978). *In* "Studies on the Development of Behavior and the Nervous System" (G. Gottlieb, ed.), Vol. 4, pp. 149–186. Academic Press, New York.

Zamenhof, S., Van Marthens, E., and Margolis, F. L. (1968). *Science* **160**, 322–323.

Zamenhof, S., Van Marthens, E., and Gravel, L. (1971). *Science* **172**, 850–851.

Zamenhof, S., Van Marthens, E., and Gravel, L. (1972). *Nutr. Metab.* **14**, 262–270.

CHAPTER 7

FROM CAT TO CRICKET: THE GENESIS OF RESPONSE SELECTIVITY OF INTERNEURONS

R. K. Murphey and H. V. B. Hirsch

NEUROBIOLOGY RESEARCH CENTER
STATE UNIVERSITY OF NEW YORK
ALBANY, NEW YORK

I. Assembly of the Brain

Brains are characterized by specific patterns of synaptic connections between neurons. For developmental neurobiologists there are two issues: First, how is the basic wiring diagram of the brain assembled? Second, can the wiring diagram be modified in response to environmental demands? These questions cannot always be answered independently of one another because in some nervous systems the environment helps "guide" assembly of the wiring diagram.

We have been studying these questions in sensory systems of two dissimilar species: the visual system of the cat and a mechanoreceptor system of the cricket. Results of our investigations reveal fundamental similarities between these two systems, suggesting that development of vertebrate and invertebrate brains is governed by the same mechanisms. These similarities, and possible underlying mechanisms, are described in this chapter. Specifically, we have compared the genesis of response selectivity and receptive field characteristics of interneurons in the cat visual system and in the cricket mechanoreceptor system.

II. Somatotopy and Receptive Fields

The basic response capabilities of central neurons in most sensory systems are provided by excitatory afferent connections (Shepherd, 1979). These afferents are often topographically organized with reference to the body surface or the world surveyed by the sensory surface.

CURRENT TOPICS IN
DEVELOPMENTAL BIOLOGY, VOL. 17

Thus, an important determinant of an interneuron's receptive field is the organization of its excitatory afferent input. The development of this organization has been intensively examined in the visual system of the vertebrates because visual space is mapped in a very direct manner from eye to brain. Evidence indicates that the axons of retinal ganglion cells are able to sort themselves into orderly groupings as they grow, retaining the near-neighbor relations of their cell bodies both in the optic nerve and as they grow over their target area (Scholes, 1979). In the mammalian visual system, axons of the retinal ganglion cells grow to the lateral geniculate nucleus (LGN), as well as to other targets, where their terminals form retinotopic arrays. Relay cells in the LGN preserve the topography of the visual projection in mapping the visual world onto various areas in the cortex. Thus, the location of a cortical neuron's receptive field is largely determined by the general location of its dendritic field and cell body relative to the afferent array.

The selectivity of receptive fields of interneurons may also be determined by the organization of dendritic fields with respect to the afferent array. To study this requires an analysis at a more microscopic level. This is made possible by the use of revolutionary dye injection technologies, which are now being routinely applied to vertebrates. For example, Sherman and his co-workers have applied these techniques to the study of the LGN (Friedlander *et al.,* 1981; Stanford *et al.,* 1981). Injection of neurons in the LGN demonstrates that the physiologically measured size and circular organization of receptive fields of at least some cell types is likely to be a consequence of the relationship of their radially symmetric dendritic fields to the distribution of the retinal afferents.

At the next level of the visual system, in the visual cortex of the cat, Colonnier (1964) [following Young's (1960) work on the octopus] observed that the dendritic fields of layer IVB stellate cells are elongated and that their long axes occur in different orientations. He suggested (1964, 1966) that cortical cells with elongated dendritic fields might have elongated receptive fields on the retina and therefore respond optimally to appropriately oriented lines or edges. Just such elongated receptive fields and a preferential response to one orientation characterizes the simple cells in the visual cortex of the cat (Hubel and Wiesel, 1959, 1962). Colonnier hypothesized that the elongated dendritic fields of layer IVB stellate cells provide the anatomical basis for the orientation selectivity of simple cells. He thus suggested that the topography of excitatory inputs may account for some of the response properties of cells in the visual cortex of the cat. Although there have been studies of the morphology of physiologically identified cells in the visual cortex (Kelly and VanEssen, 1974; Gilbert and Wiesel, 1979), apparently none have generated sufficient information to test the functional significance, if any, of the branching pattern of the dendrites. There is thus relatively little direct evidence from study of the vertebrate visual system for the role of dendritic morphology in determining response selectivity of interneurons.

The relevance of dendritic morphology in determining receptive field properties has been studied more intensively in the cricket and in other insect nervous systems. The insect nervous system is assembled from two separate pools of neurons. Central neurons (interneurons and motoneurons) are generated in a well-defined way from neuroblasts (Bate, 1976; Goodman and Spitzer, 1979) and peripheral (sensory) neurons are generated in an equally well-defined manner from epidermal precursors. Because peripheral neurons are spatially separate as well as distinctive in time of production and clonal origin from their central targets, an important step in constructing the insect CNS is the growth of the processes of the afferent neurons from periphery to center.

One system being exploited for correlated studies of the role of somatotopy in receptive field properties is the escape circuit of crickets and other insects. Many authors, beginning with Roeder (1948), have studied the abdominal giant neurons of these insects. A pair of cone-shaped appendages called cerci bear mechanoreceptive hairs that detect low-velocity air movements. Each hair is innervated by the dendrite of a single, peripherally derived sensory neuron, which projects an axon to the CNS where it makes synapses along the dendrites of the giant interneurons. Relevant stimuli such as the air movements produced by an approaching predator are analyzed and converted to an appropriate escape maneuver—usually turning and running or jumping away from the stimulus (Camhi *et al.*, 1978).

Edwards and Palka (1974) were the first to study the receptive fields in detail and demonstrated that the interneurons were directionally sensitive. They then determined that the mechanoreceptive hairs were also directionally sensitive; each hair was hinged in such a way that it was mechanically constrained to vibrate in a single plane. Many different planes of hair movement were detected and initially they were divided into two classes: those hairs vibrating in a plane transverse to the long axis of the cercus and those vibrating in a plane parallel to the cercal axis (Fig. 1). This division now seems artificial, as there appears to be a continuum of directional sensitivities representing all possible directionalities (R. K. Murphey, unpublished; Palka *et al.*, 1977). However, this exaggeration focused attention on the very important fact that the transversely vibrating hairs matched the directional sensitiv-

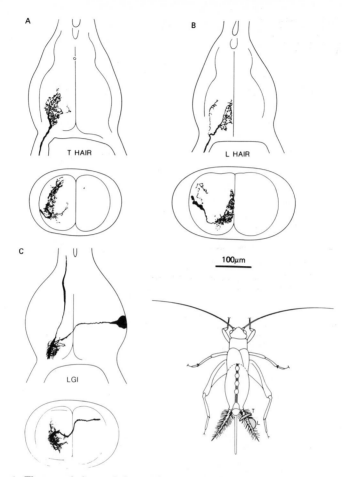

FIG. 1. The morphology of the cricket cercal to giant neuron system. (A) and (B) Terminal arborizations of a T and an L hair in the last abdominal ganglion of crickets. The upper figure in each panel is a dorsal view; the lower panel a transverse section reconstructed from 20-μm sections of the ganglia. (C) The morphology of the lateral giant interneuron. The cell receives excitatory input exclusively from T hairs, and its dendrites are located in areas innervated by T hairs but not by L hairs. The inset shows the structure and location of the abdominal nervous system and cercal sensory system. The planes of movement of one T hair and one L hair are indicated by the stippled areas. (From J. P. Bacon and R. K. Murphey, unpublished.)

ity of the interneurons. They proposed that these T hairs, as they called them, which were confined to the dorsal and ventral regions of the cercus, excited the interneurons, whereas hairs with other directionalities did not. These results provided the basic circuit diagram and were confirmed later by intracellular methods (e.g., Levine and Murphey, 1980b). As with all "circuit breaking," the work did not explain how an apparently uniform array of sensory neurons could be selectively connected with another array of interneurons.

A first step toward answering this question emerged when the afferent terminal arborizations were stained individually (Murphey *et al.*, 1980; Bacon and Murphey, 1981; Murphey, 1981). The terminal arborizations form a somatotopic map of the cercal surface within the CNS. The mechanoreceptive afferents terminate in a glomerulus, which is a hollow, club-shaped structure formed by the afferent terminals. Each neuron on the cercal surface projects to a particular region of the glomerulus, and adjacent neurons on the cercus project to adjacent regions in the CNS, thereby forming a somatotopic map. What this means is that interneuron dendrites located in the proper places could receive monosynaptic input from a subset of afferents, thus providing a mechanism for selective responsiveness. For example, the largest interneuron in the escape system, the lateral giant interneuron (LGI), has its main dendrites embedded in parts of the map corresponding to the dorsal and ventral surface of the cercus (Fig. 1). Because the transversely vibrating hairs are located dorsally and ventrally, the directionality of this neuron can be explained, at least in part, by assuming it receives monosynaptic excitatory input only from those areas of the somatotopic map in which its dendrites are embedded.

These results add significantly to our understanding of the wiring diagram, but they do not answer the basic mechanistic question: How is the somatotopic organization formed in the first place? The question is identical to one that developmental biologists have come to call the problem of pattern formation, namely, how is a uniform array of cells such as the cricket sensory cells converted to a pattern that distinguishes cells one from the other? Developmental biologists have constructed a hypothesis known as the positional information (PI) theory (Sperry, 1963; Wolpert, 1969), which suggests that cells are "labeled" according to their position on the body surface. The essence of the theory as it has been applied to the insect CNS is that all sensory neurons on the insect body are "labeled" according to their position on the body surface and that this labeling directs their growth to topographically corresponding places in the CNS (Bate, 1978; Palka, 1980).

The data reported earlier for the cricket CNS are consistent with these ideas and in fact are some of the first morphological evidence in support of the theory. These ideas can be tested further by transplanting the cercus, and thereby the sensory neurons, to various locations on the body and examining the terminal arborizations regenerated after the transplant. The logic of such experiments focuses on two issues:

First, if position is important, then a change in position might alter the neuron, causing it to project to a new area of the CNS consistent with its new location on the body. Second, such transplants often induce a process of regeneration, which can lead, for example, to the generation of supernumerary structures.

We have carried out a number of experiments that attempt to alter identified sensory neurons by transplantation, and they have all failed spectacularly. Virtually all transplant maneuvers we have tried lead to the same result—if a regenerating neuron gets to the cercal glomerulus, 90% of the time it finds its way to the target region appropriate to its *original* position. Thus, we have been unable to test the PI theory in this way. However, some transplants produce supernumerary structures, including neurons, and these structures can serve as a test of the PI theory. The theory predicts their form and symmetry and consequently should, if applicable to neurons, predict the target areas innervated by supernumerary neurons. We have examined the terminal arborization of supernumerary neurons and find that they project to parts of the CNS appropriate to their position on the supernumerary cercus. Thus, although direct proof of the PI theory is difficult to obtain, this result supports the idea.

Logically similar hypotheses have been proposed in numerous sensory systems, for example, the vibrissal–barrel system of rodents (Killackey, 1981) and the visual system of lower vertebrates (Conway *et al.*, 1981, Vol. 1), which cannot be reviewed adequately here. The point is that an important step in the production of a nervous system is the mapping of sensory neurons onto the target area, and it is clear that the afferent neurons must receive a signal corresponding to their location at the body surface, which is converted to a particular program of growth and synaptogenesis. Attempts to examine the mechanisms for response selectivity will be obliged to determine the nature of this positional information and its translation by the sensory neurons into specific patterns of neuronal connectivity.

In summary, the distribution of excitatory afferents and the shape of the postsynaptic cell's dendritic field can partially account for the selective response properties of some cat and cricket neurons. However, there is evidence of a second mechanism that is involved in the genesis of the selectivity of neurons in sensory systems.

III. Inhibitory Tuning of the Basic Selectivity

The basic selectivity generated by the excitatory connections is sharpened by inhibition in both cats and crickets, and in some cases inhibition may even be preeminent for producing selectivity. Sillito (1975a,b, 1977, 1979) and co-workers (Sillito *et al.*, 1980) have provided evidence for this in the visual cortex. They have shown that some of the selectivity of neurons in the visual cortex of the cat is dependent upon intracortical inhibitory processes. They injected bicuculline iontophoretically into the region near the neuron being recorded from. Bicuculline blocks γ-aminobutyric acid (GABA), a putative inhibitory transmitter in the visual cortex, and would release the neuron from GABA-mediated inhibition (Sillito, 1975a). A range of effects that appeared to depend upon the type of cell being studied were obtained. For example, application of bicuculline abolished or reduced orientation selectivity of complex cells while having relatively little affect on orientation selectivity of simple cells (Fig. 2) (Sillito, 1979). Also, application of bicuculline transformed many cells that responded only to

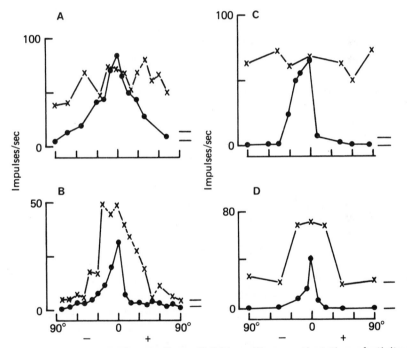

FIG. 2. Effect of iontophoretically applied bicuculline on orientation selectivity of complex cells in the visual cortex of the cat. For each of the four cells, the solid line shows the relationship between the response of the cell and the orientation of the stimulus before application of bicuculline and the interrupted line shows the relationship between these variables during iontophoretic application of bicuculline. Note that cells A and C show a loss of orientation selectivity during application of bicuculline, whereas cells B and D show a reduction, but not a loss, of orientation selectivity. [Each point shows the average response per trial (impulses/second) for 25 trials. Bars to the right of the curves indicate spontaneous activity level. Bicuculline application currents were between 70 and 100 nA.] (Reprinted with permission from Sillito, 1979, Fig. 2.)

stimulation of one eye into cells that responded to stimulation of either eye (Sillito *et al.,* 1980). Thus, the selectivity of cortical cells appears to be sharpened by intracortical inhibitory connections.

As with the excitatory connections, these inhibitory connections may be determined by the morphology of the dendritic fields of cells in the visual cortex. Schiller *et al.* (1976) have proposed a role for elongated dendritic fields that may complement that suggested by Colonnier (1964, 1966). They suggested that, in the monkey, simple cells are pyramidal cells and that it is the topographical organization of inhibitory inputs to the cortical dendrites of these pyramidal cells that determines their orientation preferences. Thus, it is possible that there are two classes of cells and that the distribution of inhibitory inputs determines the orientation selectivity of one class, as Schiller *et al.* (1976) suggested, whereas the distribution of excitatory inputs determines the orientation selectivity of the other class, as Colonnier (1964, 1966) suggested. Studies of the physiological and anatomical effects of altered postnatal visual exposure described later provide some indirect evidence in support of this suggestion.

We turn now to the cricket cercal system for another look at the joint role of excitatory and inhibitory inputs in the genesis of the selectivity of neurons. As we have already discussed, the basic directionality of giant interneurons is due to excitatory input from a subset of the available receptors. The remaining receptors (L hairs, Fig. 1) activate an inhibitory pathway that sharpens the basic directional sensitivity of the interneurons (Levine and Murphey, 1980b). The role of inhibition in refining direction selectivity can be demonstrated by blocking the movement of those hairs that activate the inhibition. When this is done, the central sensory interneurons are more broadly tuned than normal. Thus, the basic directional sensitivity conferred by the specificity of the excitatory afferent input is enhanced by inhibitory inputs, which themselves are oriented orthogonal to the excitatory input. This mechanism probably extends the range over which the system is directionally selective because the receptors begin to lose their directional sensitivity at modest stimulus intensities (Levine and Murphey, 1980a,b). Although none have been located, it is easy to imagine that the inhibitory interneurons could have dendritic fields in those parts of the afferent projection corresponding to L hairs (Fig. 1).

As in the mammalian visual system, the role inhibition plays in shaping the stimulus selectivity of a given neuron varies from neuron to neuron. In the cricket the stimulus selectivity of one interneuron, LGI (see Fig. 1), is primarily determined by its excitatory input, which is unilateral, and inhibition plays a minor role in determining its selec-

tivity. At the other extreme, the directional response of sensory interneuron 10-2 is almost completely determined by inhibition driven
bilaterally (Levine and Murphey, 1980b).

In summary, interneurons in the cricket cercal system and the cat
visual system have several basic features in common. Some neurons in
both systems are selective because of the pattern of excitatory inputs
they receive, whereas other neurons are selective because of the inhibitory input they receive.

These results suggest to us that although some of the changes in
response properties of neurons that are observed when the development
of the nervous system is manipulated experimentally may reflect
changes in excitatory connections, others are likely to reflect alterations in the balance between local inhibitory interactions and the dominant excitatory inputs. We will consider these possibilities in the
remainder of this chapter.

IV. The Development of Excitatory Connections

The basic excitatory inputs to neurons in the cat visual cortex may
be altered by experience. Depriving a cat of normal patterned visual
stimulation by suturing one eye shut at birth (Wiesel and Hubel, 1963,
1965) leads to spreading of thalamic afferents from the nondeprived
eye beyond their normal range in the cortex, whereas those from the
deprived eye project to a reduced portion of cortex (Shatz and Stryker,
1978). Thus, it appears that neuronal activity guides the production of
the afferent array. The consequences of this morphological change are
probably as well known as any result in neurobiology—many fewer
cortical neurons are activated by deprived eye afferents (Wiesel and
Hubel, 1963, 1965). This physiological result is consistent with the
morphological finding of a smaller spread of deprived afferents and, we
presume, their ability to synapse with cortical neurons. By opening the
deprived eye and suturing the nondeprived eye at the appropriate
stage of development, it is even possible to reverse this process
(Blakemore and Van Sluyters, 1974).

In addition to these gross changes in the distribution of presynaptic
elements, there are, at least in the cat, also alterations of the synapses
themselves. Tieman (1979) injected radioactive amino acids into
thalamic layers of the deprived eye or into thalamic layers of the nondeprived eye and examined labeled terminals in the cortex. She found
that afferents from the deprived eye make relatively fewer synaptic
contacts per axon than do afferents from the nondeprived eye. Furthermore, the terminals from the deprived eye contain fewer mitochondria than do terminals from the nondeprived eye. In summary, both the

extent of cortical space occupied by excitatory afferents and the synapses they make can be altered by deprivation.

The first indication that development of connections in the cricket CNS could be influenced by the environment was the observation that cricket neurons became more responsive to remaining excitatory synaptic inputs when they were partially deafferented during development (Palka and Edwards, 1974). A specimen's right cercus was removed at hatching, partially deafferenting many of the giant neurons, and the animal was reared throughout postembryonic development with only one cercus. When the response properties of interneurons deafferented throughout early development were examined in adults and compared with acutely deafferented control specimens, it was observed that long-term deafferented interneurons responded better to the remaining cercus than did the controls. This result was initially attributed to a change in the excitatory input to these cells. It was hypothesized that the remaining afferents had sprouted, crossed the midline, and innervated denervated dendrites (Palka and Edwards, 1974; Murphey et al., 1975). However, the ability of cercal afferents to sprout and cross the midline has since been examined with higher resolution techniques, and there is no convincing evidence for it. None of the mechanosensory neurons studied could be encouraged to sprout across the midline by removing the cercus contralateral to the afferent being studied (R. K. Murphey, unpublished). Those reports in the literature claiming to demonstrate such crossing of the midline can be attributed to chemosensory cells that normally cross the midline. Another possibility for enhancing the basic excitation would be for the deafferented target neurons to grow new dendrites into normally afferented regions. A careful test of this hypothesis also turned up a negative result (Murphey and Levine, 1980). Thus, in the cricket escape system, the excitatory connections are impressively stable in the face of experimental perturbations.

In summary, results from the cat visual system, but not from the cricket cercal system, show that experience can exert an effect on the basic excitatory input from sensory end organs to target cells in the central nervous system. Superimposed upon these changes in excitatory connectivity are further changes in the type and distribution of inhibitory interconnections.

V. The Development of Inhibitory Connections

In the visual cortex of the normal adult cat, each eye can activate 80–90% of the cells. After one eye is deprived of normal patterned stimulation, virtually all of the cells in the visual cortex respond only to stimulation of the deprived eye (Wiesel and Hubel, 1963, 1965). The few cells activated by the deprived eye are abnormal and most seem to be restricted to layer IV (Shatz and Stryker, 1978). The situation is reported to change drastically when the nondeprived eye is removed (Kratz et al., 1976). Significant numbers of neurons can then be activated by the animal's remaining, deprived eye. Many of these neurons display abnormal response properties. Kratz et al. (1976) suggest that monocular deprivation produces a tonic inhibitory imbalance, which ultimately results in suppression of the response to stimulation of the deprived eye. This suppression is removed when the nondeprived eye is taken out.

Consistent with the hypothesis of an inhibitory imbalance following monocular deprivation is the report that the binocular response of cortical neurons can be restored in cats deprived of vision in one eye by the intravenous administration of bicuculline. This effect of bicuculline may involve elimination of the inhibitory suppression of afferents from the deprived eye (Duffy et al., 1976).

The results of studies in which cats are deprived of all normal visual stimulation, either by suturing both eyes shut at birth (Wiesel and Hubel, 1965; Watkins et al., 1978) or by raising the animal in total darkness (Buisseret and Imbert, 1976; Leventhal and Hirsch, 1977, 1980; Fregnac and Imbert, 1978) provide additional, indirect evidence that visual stimulation is required for the development of intracortical inhibitory connections. First, neurons in the visual cortex of cats that have had both eyelids sutured shut lack the inhibitory sidebands that normally contribute to the orientation selectivity of simple cells (Watkins et al., 1978). Second, there are numerous nonselective cells in the visual cortex of cats reared in total darkness (Leventhal and Hirsch, 1977, 1980). These neurons have a number of common features that make it possible to compare them with similar, but selective, cells found in normal cats. Results of this comparison suggest that nonselective cells in dark-reared cats receive excitatory inputs but lack the inhibitory inputs that normally confer selectivity (Leventhal and Hirsch, 1980) (Fig. 3). Third, those cells in dark-reared cats that do have an orientation preference are less selective than comparable neurons found in normal cats (Leventhal and Hirsch, 1980). Most noticeable is the absence of a preference for one direction of stimulus motion, a response property that is thought to reflect intracortical inhibition (Goodwin and Henry, 1975; Sillito, 1975b, 1977). Thus, several lines of evidence indicate that deprivation affects the development and/or maintenance of inhibitory connections between cells in the visual cortex of the cat. We suggest, therefore, that only those cells whose

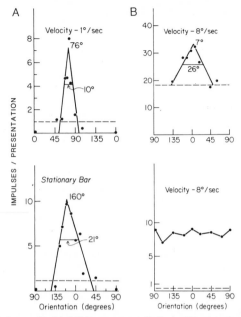

FIG. 3. Loss of intracortical inhibitory connectivity in the visual cortex of dark-reared cats. In normal adult cats, the response of cortical cells varies with stimulus orientation. There is an optimal stimulus orientation that evokes a maximal response; the response of the cell decreases as the stimulus orientation deviates from the optimal value. This is illustrated by plotting the cell's response (discharges/stimulus presentation: y axis) as a function of stimulus orientation (orientation in degrees: x axis). Two cortical cell types are illustrated. (A) Cells that probably receive excitatory input from X-type cells in the LGN; (B) cells that probably receive excitatory input from Y-type cells in the LGN. Cells illustrated in the top row were recorded from normal adult cats; cells illustrated in the bottom row were recorded from dark-reared cats. Note that cortical cells receiving excitatory input from X-type LGN cells are selective for stimulus orientation in both normal and dark-reared adults. Cortical cells receiving excitatory input from Y-type LGN cells are orientation selective in normal adult cats but not in dark-reared cats. (The dashed line indicates the spontaneous discharge rate for the cell. The solid lines were fitted to the data points by computer. The short horizontal line is the width of the turning curve at half of its peak value; the number to the right indicates the length of this width at half height. The number next to the peak indicates the preferred orientation for the cell as determined by the intersection of the lines that were fit to the data points.) (Redrawn from Hirsch and Leventhal, 1978.)

orientation preferences are determined primarily by excitatory afferents remain orientation selective in dark-reared cats, whereas those
cells whose orientation preferences are determined primarily by inhibitory inputs will not be orientation selective. For the moment, however,
we have only indirect evidence that binocular deprivation leads to a
failure in the development of inhibitory connections whereas monocu-

lar deprivation leads to an imbalance in inhibitory connections. To explore these mechanisms at the cellular level, we return to the cricket cercal system.

Murphey and Levine (1980) repeated the early experiments of Palka and Edwards on long-term partial deafferentation of cricket giant interneurons by removal of one cercus from developing crickets. Using intracellular recording methods, they confirmed the basic result that unilateral deafferentation increased responsiveness of interneurons to the remaining afferents. They focused their attention on one sensory interneuron that was amenable to standard electrophysiological tests, which helped to determine the mechanism for the observed changes (Levine and Murphey, 1980a). The most important feature of this neuron (sensory interneuron 10-3 = SI 10-3) was that its inhibitory inputs could be measured quantitatively. By depolarizing the neuron to a potential above threshold with current injection and then pitting this against the hyperpolarization produced by postsynaptic inhibitory inputs, a quantitative index of the strength of these inhibitory inputs could be obtained and compared between specimens.

When the strength of postsynaptic inhibition impinging on SI 10-3 in long-term and short-term deafferented specimens was quantified, it was found that inhibition was virtually absent in animals that had been reared with only one cercus throughout their early life (Fig. 4). Because blockage of inhibition in control specimens will cause SI 10-3 to respond more vigorously to excitatory inputs (Levine and Murphey, 1980b), it was concluded that the prolonged imbalance in input produced by unilateral deafferentation led to a loss of inhibition, resulting in the observed difference between acute and long-term treated animals. The functional consequence of this loss of inhibition was a loss of stimulus selectivity (Levine and Murphey, 1980a), as it is in the cat (Watkins *et al.*, 1978; Leventhal and Hirsch, 1980). This example represents direct proof for a novel mechanism for changing a neuron's selectivity—loss of inhibition—and all three of the identified cricket neurons that were examined exhibited this effect.

Deafferentation did not produce a uniform effect on cricket interneurons. The lateral giant interneuron (Fig. 1), for example, is the only neuron studied that receives excitation only from one cercus. Thus, removing one cercus can remove all of its monosynaptic excitatory drive. When the LGI was deafferented throughout life, no change in excitatory input could be detected; it was *not* more excitable. However, when the inhibitory inputs were tested, they were weaker. Unlike other bilaterally excited neurons (e.g., SI 10-3), LGI receives no contralateral excitatory input. Thus, none could be revealed by a loss of inhibition. Levine and Murphey (1980a) concluded that all of the ob-

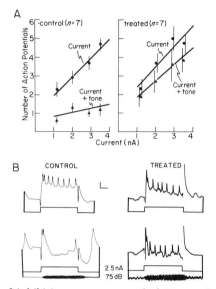

FIG. 4. Influence of inhibition on current-evoked trains of action potentials. (A) Graphic summary of the effect of inhibition on trains of current-evoked action potentials. The number of action potentials ($\pm$ SE) evoked during the 20–70 milliseconds following the onset of a 100-millisecond current pulse. Abscissa: current intensity. Each point represents the mean from 10 trials on each of seven animals. In controls, the number of action potentials evoked by a current pulse (circles) was reduced by concurrent sound activation of the inhibitory pathway (triangles). In treated siblings, the reduction was negligible, indicating that inhibition was less effective. (B) Examples of current-evoked trains of action potentials, recorded via a bridge circuit, in the absence (upper) and presence (lower) of an inhibitory tone. Calibration: vertical, 5 mV; horizontal, milliseconds. (From Levine and Murphey, 1980a, with permission.)

served changes in responsiveness in the cricket nervous system were due to the disrupted inhibitory pathway rather than enhanced excitation. The observed change, at least for some interneurons, is an increase in responsiveness produced by a reduction in inhibitory connectivity. This is very much like the increase in the range of stimuli to which some cells in the visual cortex of dark-reared cats respond (Leventhal and Hirsch, 1980) (Fig. 3). In the absence of appropriate input, these inhibitory connections may fail to develop—or regress through disuse in both the cat and cricket nervous system.

VI. Conclusion

There is evidence, indirect for the cat and direct for the cricket, that the selectivity of neurons in sensory systems depends to a varying degree upon a framework established by the excitatory input to the cell. These excitatory connections determine the universe of stimuli to which the neurons may respond. Probably the most important next step in the analysis of this aspect of neuronal organization is the determination of the mechanisms by which the orderly arrays of afferents are produced. Until we understand this, it seems unlikely that we will understand the changes in connectivity produced by experimental manipulation of the developing nervous system.

To the extent that response selectivity depends on inhibitory inputs to interneurons, modulation of inhibitory connections provides a means for controlling response selectivity. The discovery, based on indirect evidence in the cat and direct evidence in the cricket, that experimental interruption of inputs can affect the formation of inhibitory connections suggests that this may be an important mechanism responsible for much of the sensitivity to experience of the developing nervous system.

It seems to us that the major advance in the analysis of the neural circuits of invertebrate animals in the past decade emphasizes the technical ease with which invertebrate nervous systems can be utilized to solve neurobiological problems. When these simple systems are used, conceptual problems are susceptible to a focal, rapid attack. It is our observation that common mechanisms guide neuronal development in animals as diverse as the cat and the cricket; and just as *Escherichia coli* and λ phage have served molecular biology as models of the genome, so will simple nervous systems of invertebrates serve neurobiology as models of the more sophisticated vertebrate brain. The breakthroughs of the past decade will look modest in the face of ideas now emerging for the analysis of the assembly of the nervous system. Central to this will be the underlying assumption that the function of nervous systems of all organisms, from cats to crickets, is based on a few basic rules.

REFERENCES

Bacon, J. P., and Murphey, R. K. (1981). *Neurosci. Abstr.* (in press).
Bate, C. M. (1976). *J. Embryol. Exp. Morphol.* **35,** 107–123.
Bate, C. M. (1978). *In* "Handbook of Sensory Physiology" (M. Jacobson, ed.), Vol. 9, pp. 2–53. Springer-Verlag, Berlin and New York.
Blakemore, C., and Van Sluyters, R. C. (1974). *J. Physiol. (London)* **237,** 195–216.
Buisseret, P., and Imbert, M. (1976). *J. Physiol. (London)* **255,** 511–525.
Camhi, J. M., Tom, W., and Volman, S. (1978). *J. Comp. Physiol.* **128,** 203–212.
Colonnier, M. L. (1964). *J. Anat.* **98,** 327–344.
Colonnier, M. L. (1966). *In* "Brain and Conscious Experience" (J. C. Eccles, ed.), pp. 1–23. Springer-Verlag, Berlin and New York.
Conway, K., Feiock, K., and Hunt, R. K. (1980). *Curr. Top. Dev. Biol.* **15,** 217–212.
Duffy, F. L., Snodgrass, S. R., Burchfiel, J. L., and Conway, J. L. (1976). *Nature (London)* **260,** 256–257.

Edwards, J. S., and Palka, J. (1974). *Proc. R. Soc. London, Ser. B* **185**, 83–103.
Fregnac, Y., and Imbert, M. (1978). *J. Physiol. (London)* **278**, 27–44.
Friedlander, M. J., Lin, C. S., Stanford, L. R., and Sherman, S. M. (1981). *J. Neurophysiol.* **46**, 80–129.
Gilbert, C. D., and Wiesel, T. N. (1979). *Nature (London)* **280**, 120–125.
Goodman, C. S., and Spitzer, N. C. (1979). *Nature (London)* **280**, 208–214.
Goodwin, A. W., and Henry, G. H. (1975). *J. Neurophysiol.* **38**, 1524–1540.
Hirsch, H. V. B., and Leventhal, A. G. (1978). *In* "Frontiers in Visual Science" (S. J. Cool and E. L. Smith III, eds.), pp. 660–673. Springer-Verlag, Berlin and New York.
Hubel, D. H., and Wiesel, T. N. (1959). *J. Physiol. (London)* **148**, 574–591.
Hubel, D. H., and Wiesel, T. N. (1962). *J. Physiol. (London)* **160**, 106–154.
Kelly, J. P., and VanEssen, D. C. (1974). *J. Physiol. (London)* **238**, 515–547.
Killackey, H. P. (1980). *Trends Neurosci. (Pers. Ed.)* **3**, 303–305.
Kratz, K. E., Spear, P. D., and Smith, D. C. (1976). *J. Neurophysiol.* **39**, 501–511.
Leventhal, A. G., and Hirsch, H. V. B. (1977). *Proc. Natl. Acad. Sci. U.S.A.* **74**, 1272–1276.
Leventhal, A. G., and Hirsch, R. K. (1980). *J. Neurophysiol.* **43**, 1111–1132.
Levine, R. B., and Murphey, R. K. (1980a). *J. Neurophysiol.* **43**, 383–394.
Levine, R. B., and Murphey, R. K. (1980b). *J. Comp. Physiol.* **135**, 269–282.
Murphey, R. K. (1981). *Dev. Biol.* **88**, 236–246.
Murphey, R. K., and Levine, R. B. (1980). *J. Neurophysiol.* **43**, 367–382.
Murphey, R. K., Mendenhall, B., Palka, J., and Edwards, J. S. (1975). *J. Comp. Neurol.* **159**, 407–418.
Murphey, R. K., Jacklet, A., and Schuster, L. (1980). *J. Comp. Neurol.* **191**, 53–64.
Palka, J. (1980). *Adv. Insect Physiol.* **14**, 256–349.
Palka, J., and Edwards, J. S. (1974). *Proc. R. Soc. London, Ser. B* **185**, 105–121.
Palka, J., Levine, R. B., and Schubiger, M. (1977). *J. Comp. Physiol.* **119**, 267–283.
Roeder, K. D. (1948). *J. Exp. Zool.* **76**, 353–374.
Schiller, P. H., Finlay, B. L., and Volman, S. F. (1976). *J. Neurophysiol.* **39**, 1362–1374.
Scholes, J. H. (1979). *Nature (London)* **278**, 620–624.
Shatz, C. J., and Stryker, M. P. (1978). *J. Physiol. (London)* **281**, 267–283.
Shepherd, G. M. (1979). "The Synaptic Organization of the Brain," 2nd ed. Oxford Univ. Press, London and New York.
Sillito, A. M. (1975a). *J. Physiol. (London)* **250**, 287–304.
Sillito, A. M. (1975b). *J. Physiol. (London)* **250**, 305–329.
Sillito, A. M. (1977). *J. Physiol. (London)* **271**, 699–720.
Sillito, A. M. (1979). *J. Physiol. (London)* **289**, 33–53.
Sillito, A. M., Kemp, J. A., and Patel, H. (1980). *Exp. Brain Res.* **41**, 1–10.
Sperry, R. W. (1963). *Proc. Natl. Acad. Sci. U.S.A.* **50**, 703–710.
Stanford, L. R., Friedlander, M. J., and Sherman, S. M. (1981). *J. Neurosci.* **1**, 578–584.
Tieman, S. B. (1979). *Soc. Neurosci. Abstr.* **5**, 810.
Watkins, D. W., Wilson, J. R., and Sherman, S. M. (1978). *J. Neurophysiol.* **41**, 322–337.
Wiesel, T. N., and Hubel, D. H. (1963). *J. Neurophysiol.* **26**, 1003–1017.
Wiesel, T. N., and Hubel, D. H. (1965). *J. Neurophysiol.* **28**, 1029–1040.
Wolpert, L. (1969). *J. Theor. Biol.* **25**, 1–47.
Young, J. Z. (1960). *Nature (London)* **186**, 836–839.

CHAPTER 8

THE NEUROEMBRYOLOGICAL STUDY OF BEHAVIOR: PROGRESS, PROBLEMS, PERSPECTIVES

Ronald W. Oppenheim

NEUROEMBRYOLOGY LABORATORY
DIVISION OF MENTAL HEALTH RESEARCH
DOROTHEA DIX HOSPITAL
RALEIGH, NORTH CAROLINA

I. Introduction

In his presidential address to the American Association of Anatomists in 1933, the pioneer neuroembryologist and psychobiologist G. E. Coghill spoke on the "neuro-embryologic study of behavior," by which he meant the attempt to understand behavioral development and its neural correlates, especially (but not exclusively) during embryonic, larval, and fetal stages. It has sometimes been suggested that the central focus of the neuroembryological approach to behavior is neuroanatomical or neurobiological development and that behavior merely serves as a convenient index or bioassay—in lieu of, or in addition to, neurophysiology—of structural maturation (e.g., Gottlieb, 1970). If this were, in fact, the case, then the subject matter would constitute little more than a minor subfield of developmental neurobiology and thus would be of only slight interest to students of developmental psychobiology, as it would have little obvious significance for a comprehensive conceptualization of behavioral development. Although space does not permit a detailed rebuttal to this argument, the evidence does not appear to support such a position (see

257

CURRENT TOPICS IN
DEVELOPMENTAL BIOLOGY, VOL. 17

Oppenheim, 1978). Coghill, for instance, undertook his neuroembryological studies of the salamander embryo with a clear recognition of psychological problems and with a firm commitment to behavioral development. As his colleague, friend, and biographer, C. Judson Herrick, has commented, "The mastery of anatomical detail to which most of his time and attention was directed was not the objective—his primary interest was in the activity itself" (1949, p. 147).

Although some neuroembryologists both past and present obviously do not fully share Coghill's devotion to behavioral development, his interests in this respect were by no means novel. Virtually all of the early neuroembryologists recognized that embryos are behaviorally active, that the nervous system functions before birth, and thus that behavior is a legitimate and potentially important subject for investigation by embryologists interested in neurogenesis. As the founder of the study of embryonic behavior, Preyer noted almost 100 years ago, "A science of biochemical and physiological embryology is . . . necessary for the understanding of men and animals when they are born. In the same way in which we can only understand organs, tissues, and cells following the investigation of their genesis, *function* can also only be understood through its history" (1885, p. 3; italics added). Many subsequent neuroembryologists, such as R. Harrison, S. Detwiler, P. Weiss, W. Windle, R. Levi-Montalcini, A. Hughes, H. Tracy, R. Sperry, J. Piatt, and V. Hamburger (and the list could be continued), have, in fact, carried out investigations of the behavior of embryos and larvae. In his 1949 Silliman lectures on the organization and development of the embryo, Ross Harrison underlined the significance of this approach when he commented, "By ablation and transplantation of parts of the embryo it is possible to fashion almost any kind of nervous system desired and subsequently *to study its function* without the disturbing effects of trauma. *This is one of the most promising lines of investigation* leading off from the field covered in the present lecture" (1969, p. 163; italics added). And in his textbook of experimental embryology, published in 1939, Weiss included a chapter on the development of behavior, for, as he later put it, "No account of neurogenesis can be complete without relating itself to the problem of behavior" (1955, p. 390).

By drawing on Coghill's characterization of this area of investigation as the *neuro*embryological study of behavior—in contrast, for example, to the embryology of behavior, psychoembryology, or behavioral embryology—I do not wish to imply that the behavior of embryos, larvae, and fetuses cannot be profitably studied for its own sake, independent of any putative neural correlates. The work of Carmichael, Kuo, Gottlieb, Impekoven, Vince, and many others, which I discuss later, is ample proof of the value of a strict emphasis on behavioral or psychological development in the embryo. Nevertheless, it is a historical reality that for most students of behavioral embryology the nervous system has appeared as such an alluring attraction that few have been able to resist the temptation to study (or at least to draw parallels and suggest correlations between) behavioral and neural changes. Perhaps this is not surprising. By virtue of its relative simplicity, the embryo provides a rare opportunity—equaled perhaps only by adult invertebrate and tissue culture preparations—for understanding the neural basis of behavior, the central focus of neurobiology.

In contrast to a purely behavioral approach, however—and without claiming that behavioral development can be explained solely by the emergence of specific neural mechanisms—the neuroembryological study of behavior has always had as a central tenet the contention that the early differentiation of the nervous system establishes the framework and determines the potentialities for behavioral performance. As Paul Weiss has put it:

> the nervous system emerges from its embryonic phase well patterned and nothing could be more misleading than the impression that embryonic neurogenesis merely fabricates blank sheets on which experiential input from the outer world is then to inscribe operative patterns (1970, p. 60).

Two major assumptions of such a position are, first, that the *structure* of the nervous system is a fundamental determining factor in the organization of specific behavior patterns and, second, that many aspects of this structural organization are present even before the system begins to function. It is primarily this last contention that in the past has served to distinguish the neuroembryological approach from certain other conceptions of behavioral embryology (e.g., see Holt, 1931; Kuo, 1967).

A major goal of the neuroembryological study of behavior is to examine the emergence of behavioral patterns from their origin in the less complex activities of the embryo and fetus and to relate these, insofar as possible, to neural changes, to environmental (including "ecological") factors, and to the maturation of other emerging organ systems (e.g., cardiovascular, hormonal, muscular). The framework is biological by virtue of an abiding interest in the evolutionary significance and adaptive value of embryonic behavior; more specifically, the approach is both comparative (in that all invertebrate and vertebrate forms, not just man, provide the material for study) and embryological (to the extent that the principles and concepts of developmental biology are considered fundamental for understanding behavioral ontogeny throughout the life cycle from conception to aging) (see Oppenheim, 1981a). [Regrettably, space does not permit me to review the still small, but growing, number of studies on invertebrate behavioral development. Suffice it to say that it seems likely that in the very near future studies on invertebrate preparations will make increasing contributions to this field and thus will have to be seriously considered in any valid and credible theory of behavioral development. The interested reader should consult several of the chapters in Fentress (1976) and Young (1973).]

It is implicit that the neuroembryological study of behavior includes a firm commitment to studying and understanding the fundamental features of neurogenesis, from primary induction to synaptogenesis. For, as just noted, it is a central tenet of this approach that behavioral development is dependent, to one degree or another, on the orderly emergence of neurobiological mechanisms in the embryo. [In this regard, one investigator has suggested that "in tracing the connections between genes and behavior . . . one is aware of the possibility that understanding this might well involve solving all of the outstanding questions in biology" (Brenner, 1973, p. 269).] However, it is not my intention to review this information in this chapter, in which the major focus is behavioral development. The interested reader is referred to the excellent and comprehensive treatment of neurogenesis by Jacobson (1978). (For a recent insightful review of related neurophysiological issues in the embryology of behavior, see Bekoff, 1981.) Rather my inquiry begins at a relatively advanced stage of neurogenesis, when nerve cells first begin to secrete minute amounts of neurotransmitter substances sufficient to activate neurons and muscles, resulting ultimately in overt behavior.

The neuroembryological approach to behavioral development was initiated by G. E. Coghill in his now classic studies of the salamander embryo and larva, which began in 1906 and continued until shortly before his death in 1941. For much of this period Coghill's work was ignored by neurologists, embryologists, and psychologists, presumably because the conclusions he was reaching concerning behavioral and neural development in the salamander were so different from the prevailing views, in which reflexes and simple stimulus–response associations were thought to be the major guiding force in the organization of the vertebrate nervous system and behavior. Despite the unorthodoxy of his views, it was apparently felt by many that as long as Coghill restricted his attention to salamanders his work posed no immediate threat to the prevailing reflexology, which was derived mainly from work on higher vertebrates, especially mammals.

By the late 1920s, however, Coghill (1929a,b) became more bold and began to argue that the salamander pattern was neither unique nor aberrant, but rather that it reflected a general vertebrate plan and was representative of all members of the group from fishes to humans. Coghill's fundamental position was, "The behavior pattern from the beginning expands throughout the growing normal animal as a perfectly integrated unit, whereas partial patterns arise within the total patterns, and by a process of individuation acquire secondarily varying degrees of independence—partial patterns emerge and tend toward independence, but under normal conditions always remain under the supremacy of the individual as a whole" (1929a, p. 989).

In contrast to Coghill's contention that the salamander pattern represented a general vertebrate principle of behavioral development, Coghill's chief opponent, the developmental neuroanatomist W. Windle, argued that in fetal mammals the earliest movements were always independent reflexes. He maintained that "the basic elements in the genesis of mammalian behavior are relatively simple reflex responses—the more complex reactions of older fetuses are formed by progressive neuronal integration of the less complicated activities (1940, p. 141)." In other words, according to this view, not only were the earliest movements thought to be reflexogenic (and thus local) in nature, but the entire subsequent emergence of complex behavior was thought to be guided by the principle that there is inexorable concatenation of simpler behavioral units (i.e., reflexes) into complex, integrated behavior patterns. As the lines of battle could hardly have been drawn more clearly, one might expect that the proof or disproof of one or the other view would be a simple matter of turning to the embryo and fetus for an unequivocal answer. Unfortunately, as is often the case, the apparent simplicity of the issue (and its resolution) proved deceptive. In fact, a final answer has been so elusive that even today, almost 40 years after the original debate reached an impasse, many questions still remain unanswered.

In fishes and amphibians, the development of behavior appears to adhere rather closely to Coghill's original observations, although even here there is not a consensus (e.g., see Faber, 1956; Holtzer and Kamrin, 1956; Macklin and Wotjkowski, 1973). In birds, as was first pointed out by Windle and Orr (1934) and later by Kuo (1939) and Hamburger (1963), the initial stages follow a rather stereotyped "Coghill-like" sequence, but by the beginning of the second "trimester" of incubation (e.g., by day 7 in the chick), this pattern breaks down as the embryo begins to perform irregular, seemingly uncoordinated jerks, twitches, startles, and convulsions in no discernible pattern or order (see Bekoff, 1976; Provine, 1980, for exceptions). Thus, during the major portion of incubation, the chick embryo behaves in a manner that defies explanation in Coghill's terms (but see Crain, 1974, for a different view). Moreover, as I discuss in greater detail later, because much of the activity of the chick is spontaneous (i.e., it is nonreflexogenic or generated endogenously), it is also irreconcilable with Windle's scheme, which depends upon a behavioral process whose guiding force is a stimulus–response reflex mechanism.

In view of the fact that to a very great extent the resolution of the Coghill–Windle debate hinged upon the degree to which behavioral development in mammals conforms to the salamander pattern, it might be expected that the issue would have been settled on the basis of what the mammalian fetus actually does. Despite the fact that many studies were carried out on mammalian embryos during this period (see Hooker, 1952, for a review), however, a consensus was never reached. Part of the reason for this impasse stems from the fact that until quite recently it has been technically impossible to carry out systematic normative studies on mammalian fetuses *in utero* under sufficiently undisturbed conditions such that one could rule out such things as asphyxiation, surgical trauma, and drug effects. This is not to say that available studies tell us nothing about the behavioral capacities of mammalian embryos and fetuses or that they shed no light on the organization of the developing nervous system. Rather, it is only meant to draw attention to the fact that these data, by their very nature, tell us little, if anything, about the normal progression of mammalian behavioral development *in utero*. In addition to the serious physiological shortcomings noted earlier, there is also the procedural consideration, stemming from the prevailing conceptual emphasis on reflexology, that virtually all of these early studies used stimulation to evoke behavioral activity. As one of the foremost investigators of human fetuses, Hooker admitted, "The whole problem of spontaneous movements has been ignored" (1944, p. 36). This bias could not have been stated more clearly than when Windle commented, "With the advent of conduction from afferent to efferent neurons through synaptic centers, reflex responses are manifested. *At this point in development, behavior may be said to have its genesis*" (1940, p. 139; italics added). Thus, behavior was arbitrarily dated from the inception of reflexes, with little, if any, consideration for the role of spontaneous movements. Because the technology now exists for observing the behavior of mammalian fetuses *in utero* under undisturbed conditions, the time is ripe for a renewed attack on this and other issues.[1]

It has long been recognized that the theoretical views of both Coghill and Windle concerning the *reflexogenic* versus *integrated* nature of behavioral development are untenable as general principles for understanding the emergence of embryonic behavior in most vertebrates. Although both theories can explain some of the facts, at some stages, and in some forms (e.g., see Hamburger, 1963), their usefulness as global conceptual leitmotivs has been seriously questioned and thus must be rejected.

In contrast, many of the specific empirical questions addressed by Coghill, Windle, and their supporters still remain as fundamental concerns for the present-day neuroembryological study of behavior. For instance, although some progress has been made (Oppenheim, 1975; Bekoff, 1976; Provine and Rogers, 1977; Singer *et al.*, 1978; Nornes *et al.*, 1980), we still do not know most of the details regarding the formation of anatomical or functional reflex circuits or of longitudinal integrating tracts in most vertebrates; nor do we understand when and how (or even if) specific parts of the brain are important for the prenatal emergence of behavior. As alluded to earlier, in the case of mammals we know surprisingly little about the normal sequence of behavioral events *in utero* and we have only the dimmest idea of what the various differences and similarities in behavior between the embryos of different species mean (if anything) in terms of adaptation. These are some of the questions that might give valuable information if pursued within the broad framework of the neuroembryology of behavior.

Historically, the period between 1925 and 1940 can now be seen to have been, in many respects, a halcyon era for the neuroembryological study of behavior. Motivated by the theoretical challenge of Coghill, investigators throughout the world studied the embryos and fetuses of a wide variety of vertebrate species ranging from the primitive dogfish to humans. Then, for a number of reasons (Oppenheim, 1978), but mainly because of a technical and theoretical impasse over the issue of the reflexogenic versus integrative nature of embryonic behavior in higher vertebrates, the golden age of behavioral embryology came to a

---

[1] Techniques involving ultrasonic radiography (Chef, 1979) and intrauterine fetal visualization with fiber optics (Kaback and Valenti, 1976) are just beginning to be used in normative studies of fetal behavior. Nevertheless, they offer the hope for a successful, renewed attack on the behavioral embryology of mammals. And when coupled with the increasing feasibility of fetal neurosurgery (Miller and Lund, 1975; Taub, 1976; Goldman and Galkin, 1978), the neuroembryological study of behavior in mammals may not be an unrealistic prospect for the near future.

rather abrupt halt in the early 1940s. The entire field languished for over 20 years until, in the early 1960s, Viktor Hamburger, a neuroembryologist, started a research program on the neuroembryology of behavior in the chick. With a few notable exceptions, much of the work in this field since that time has been carried out by Hamburger and his students or else was directly inspired by their efforts.

Following the important review paper by Hamburger, which appeared in 1963, the only other comprehensive attempts to chart the recent progress of this field were written in 1972 and appeared in a book edited by Gottlieb (1973). Thus, it would seem to be a propitious time to once again take stock of the status of the field and perhaps attempt to foresee in what direction(s) future work should proceed in order to answer the many questions that still remain unresolved. In doing this, I have made no attempt to focus on only the progress that has been made. In fact, if anything, I have made a deliberate effort to emphasize the lack of progress in order to highlight our ignorance about many—indeed, about most—aspects of the neuroembryology of behavior. My rationale in this regard has been guided by the words of the philosopher Alfred North Whitehead, who once commented, "Not ignorance but ignorance of ignorance is the death of knowledge."

II. Are Embryonic and Fetal Behaviors Spontaneous or Stimulus Induced?

In all vertebrate embryos that have so far been examined, overt movements occur that are not the result of deliberate stimulation and that often appear to be unreleated to environmental perturbations. In teleosts and amphibians, such activity may be infrequent, whereas avian embryos are almost continuously active at some stages (Oppenheim, 1974). Mammals appear to assume an intermediate position in this respect. Preyer, in his book "Specielle Physiologie des Embryo" (1885), was the first person to draw attention to this form of activity. As he pointed out, "The embryos of all animal classes show characteristic movements which are of the highest physiological interest because some of them came about without a noticeable outside stimulus. These movements, which I have observed in the embryos of fish, amphibia, reptiles, birds, and mammals, I have called impulsive to distinguish them from all other movements of the pre- and postnatal periods. They precede all other movements and form the starting point for the development of the mind" (1885, p. 540).

Although virtually all subsequent investigators have supported Preyer's observations on the ubiquity of spontaneous movements, there has been considerable debate over the extent to which such activity is, strictly speaking, spontaneous (i.e., endogenously generated). A number of early investigators of embryonic behavior (e.g., Kuo, 1932; Barcroft and Barron, 1939; Windle, 1940) have implied that spontaneous movements are virtually always induced by sensory input (but input that may be of such a subtle nature that it is easily overlooked).

It should be pointed out that there are actually two related, but separate, questions at issue here. First is the question of the necessity for external stimuli in eliciting embryonic movements (one extreme view, for instance, would contend that were it possible to exclude all forms of sensory input, the embryo would be totally quiescent). Second is the question of whether the *form* of the spontaneous activity (i.e., the behavior *pattern*) is the result of previous sensory experience. According to this view, it is possible that at some stage movements can occur spontaneously, but, as the argument goes, the type, form, or pattern of such behavior is largely a result of prior patterns of sensory input (presumably acting by the "functional validation" of a specific neural organization); change the prior input (or eliminate it altogether) and a different behavior (and neural) pattern will result. In practice these two questions are often linked, in that the occurrence and the form of the movements are both considered to be either dependent on or independent of sensory input. At the moment I only wish to deal with the first question; the second will be considered in a subsequent section.

The earliest evidence for the existence of truly spontaneous activity was the observation made by Preyer (1885), showing that the chick embryo exhibits overt motor activity several days before sensory stimulation is effective in evoking reflexes, thereby implying that the efferent or motor system begins to function prior to the sensory system (i.e., that there is a *nonreflexogenic* period). Not only was Preyer the first to describe this phenomenon, but more importantly he clearly recognized the general, theoretical implications of his observation for the development of behavior, commenting, "I consider this fact as one of the most important in the entire field of the physiology of the embryo" (1885, p. 470.[2] In the chick, this nonreflexogenic phase appears to be explained by the fact that, in the spinal cord, the motor system differentiates in advance of sensory structures (Hamburger, 1948; Bekoff, 1976; Hollyday and Hamburger, 1977; Carr and Simpson, 1978). Con-

---

[2] The notion that spontaneous activity constitutes an important feature of embryonic development was not entirely original with Preyer, although he does appear to have been the first person to provide scientific documentation for its existence. In his famous book, "Elements of Physiology" (1843), the German physiologist Johannes Mueller argued that voluntary behavior develops out of the spontaneous movements of the embryo, fetus, and infant.

sequently, when the motoneurons have already innervated the musculature, the sensory reflex arc has apparently not yet been completed (Windle and Orr, 1934; Visintini and Levi-Montalcini, 1939).

A prereflexogenic period of motility also exists in the altricial pigeon embryo (Harth, 1974) and in the precocial duck (R. W. Oppenheim, unpublished observations), and, as in the chick, it is probably related to the retrograde sequence of maturation of spinal neurons in these forms.[3] This retrograde sequence of development, which starts with the "target" organ and ends with the sensory afferents, may be a general feature of neural development. In addition to the spinal motor action system, it also occurs in the autonomic nervous system (Slotkin et al., 1980), the superior colliculus (Stein et al., 1980), and the cerebral cortex (Bruce and Tatton, 1980). I hasten to add, however, that such a retrograde morphological sequence can occur even in the absence of a behavioral prereflexogenic period. For instance, in the human fetal spinal cord, there is also a clear retrograde sequence of neurogenesis, with the motor system preceding the sensory system (Okado et al., 1979, and personal communication). Yet, according to the best evidence available (Hooker, 1952), spontaneous motility does not precede the earliest reflexes; both apparently begin at about the same time. (It is even conceivable that one could have a prereflexogenic period in the absence of a retrograde sequence of neurogenesis; it would only be necessary that the sensory system remain functionally inactive until after the motor system matures and begins to function.)

The presence of a prereflexogenic period of motility illustrates most clearly that the neural mechanisms that mediate specific behavior patterns can develop many of their major features prior to the onset of a sensory mechanism. Presumably, the early neurogenetic processes involving proliferation, migration, growth, and differentiation are sufficiently canalized in such cases so that the system is prepared to function more or less appropriately from the beginning.

Besides its existence in avian embryos, the only other clear example of a prereflexogenic period was described by Tracy (1926) in his comprehensive and classic monograph on the development of behavior in

---

[3] It should be noted that my use of the term spontaneous activity does not imply that such activity is generated independently of a stimulus. All that is meant is that there is a motor output to the muscles that can occur in the absence of environmental input from sensory receptors. Although it is possible that such activity may result from intracellular changes in the motoneurons, it is considerably more likely that afferent input from interneurons is a key feature of such activity. Furthermore, in agreement with Hamburger (1963), I distinguish spontaneous embryonic activity from myogenic activity by virtue of there being a neural involvement in the former (i.e., it is neurogenic).

the teleost embryo *Opsanus tau* (the toadfish). Tracy noted that there is a span of approximately 2 weeks between the onset of motility and the first appearance of reflex responses. In addition to its longer duration, motor primacy in the toadfish is also more striking than in the chick, in that adultlike swimming activity is perfected before the newly hatched larva can respond to stimulation. Tracy also anticipated the later suggestion of Hamburger (1963) concerning the possible postnatal persistence of spontaneous activity, when he commented that "movements of endogenous origin continue through the whole life of the animal and determine its habitual activity" (1926, p. 275). In this regard, Tracy made the interesting observation that some species of fishes that are highly active as adults (e.g., *Fundulus*) are also more active in embryonic and larval stages compared to species such as the toadfish, which is rather sluggish in both adult and embryonic stages. Unfortunately, with the possible exception of sleep mechanisms and related activity rhythms (Hamburger, 1975; Corner, 1977), there is little else known about the persistence of endogenous motor activity in postnatal life (but see Taub, 1976). By saying this, I do not mean to deny the fact that endogenous or centrally generated behavior patterns are a common occurrence in all vertebrates and invertebrates that have been examined postnatally (for a review, see Delcomyn, 1980). I only wish to emphasize the fact that in most of these cases specific prenatal antecedents have not been identified, and thus any possible relationship between centrally generated patterns of behavior in the adult and endogenous embryonic activity, which is also quite often patterned in nature, must remain speculative.

Although spontaneous activity occurs in amphibians and mammals, the fact that these forms appear to exhibit reflexes at least as early as spontaneous motility (i.e., a nonreflexogenic period is apparently absent) raises the question of whether the spontaneous activity in these forms might, in fact, be due to undetected sensory input. In this context it is pertinent to note that in several older studies it was shown that amphibian larvae that have been surgically deafferented as embryos exhibit complex and often highly coordinated spontaneous locomotor activity (Weiss, 1941; Yntema, 1943; Taylor, 1944; Detwiler, 1947). Furthermore, in a more recent, elegant study, combining behavioral and electrophysiological techniques, it has been shown unequivocally that bullfrog embryos and larvae exhibit endogenously generated, patterned, motoneuronal discharges that are the specific neural correlate of swimming (Stehouwer and Farel, 1980); both the motoneuronal discharges and the swimming pattern persist following sensory deafferentation. Similarly, newborn monkeys whose limbs

were deafferented *in utero* display a remarkable repertoire of spontaneous motor activity (Taub, 1976), although definite abnormalities in
voluntary fine motor movements were also found.

Unfortunately, no detailed observations have yet been made of the
behavior of deafferented mammalian preparations during fetal stages.
However, Dawes *et al.* (1972) have reported that rhythmic thoracic
trunk movements in the sheep fetus appear unaffected by sensory deafferentation. And Narayanan *et al.* (1971) reported that certain features
of motility in rat fetuses appear to be independent of extrinsic stimuli
and thus may be spontaneous (also see Barcroft, 1938). It should be
noted, however, that the absence of a prereflexogenic period does not
rule out the possibility that endogenously generated movements occur.
In many animals, centrally generated movements occur in the presence
of fully functional and mature sensory mechanisms (Delcomyn, 1980).

Perhaps the most convincing evidence that spontaneous embryonic
activity is not eliminated, suppressed, or masked following the onset of
reflex responses comes from the work of Hamburger and his associates.
In a series of studies involving sensory deafferentation in the chick
embryo, they found that for at least 10 days after the onset of reflexes,
the rate and temporal pattern of spontaneous motility continued virtually normally in the absence of any sensory input (Hamburger *et al.*,
1966; Hamburger and Narayanan, 1969; Decker, 1970; Narayanan and
Malloy, 1974a,b). Other experiments in the chick, involving sensory
deprivation and "enrichment" (Oppenheim, 1966, 1972a,b), electrophysiological recordings of spontaneous spinal cord activity in
curarized and in deafferented embryos (Provine, 1973; Sharma *et al.*,
1970), and electromyographic recordings (Bekoff, 1976), also support
these findings. Thus, even following the emergence of somatosensory
responsiveness, much of the behavior of the embryo continues to be
endogenously generated.

To briefly summarize: Although all vertebrate embryos, larvae,
and fetuses appear to display movements in the absence of any obvious
external stimuli, with the exception of the chick and frog, the question
remains whether such activity is strictly spontaneous and thus initiated and maintained by endogenous pattern generators in the central
nervous system. Because spontaneous activity apparently is not related to any unique species-specific function in the frog or chick, however, and because it also occurs in many developing invertebrates, it is
difficult not to believe that it is a widespread and phylogenetically
persistent feature, even though its significance still remains somewhat
of an enigma (but see later). As noted earlier, endogenous pattern
generators within the central nervous system are now thought to be an
important mechanism in the organization of complex motor patterns in

postnatal animals (Fentress, 1976; Delcomyn, 1980; Bekoff, 1981). Although direct evidence is lacking, it seems likely that in many cases the endogenous or spontaneous motility of embryos represents a developmental antecedent of these postnatal mechanisms.

Finally, in view of the paramount role of sensory mechanisms in the development and control of many types of motor behavior in postnatal and adult animals, the question arises as to why, despite being "functionally" connected with the motor system, the somatosensory system of the early embryo appears to be almost entirely "irrelevant" for normally observed motor output. It could be argued that the endogenous generation of motility in the embryo reflects a means of assuring a specific amount, frequency, or pattern of movement and neural function that is necessary, for example, for normal neuromuscular development (see following and footnote 8). Although the embryo, if stimulated, can respond, it is conceivable that if specific motility parameters are critically involved in neuromuscular or other aspects of development, they could less reliably be based upon reflexogenic behavior. The embryo develops in an environment that, although not one of total isolation, nonetheless, appears to be monotonously regular and lacking in the kinds of changes that may be necessary for the initiation and maintenance of adequate rates or patterns of motility. Spontaneous or endogenously generated motility may have evolved as an adaptation to serve this role in the embryo.

It has also been suggested that one mechanism by which the embryo's motor activity is relatively isolated from naturally occurring sensory input may be active neural inhibition (Crain, 1973, 1976; Oppenheim and Reitzel, 1975) (i.e., by blocking the transmission of sensory signals to the motor system). There is, in fact, evidence that inhibitory mechanisms develop relatively early during neurogenesis (see later).

It is also entirely plausible that in some cases the early appearance of function in the somatosensory system of the embryo is merely another example of the common embryological rule that cells, tissues, and organs appear to mature morphologically, and even become functional, prior to their need by the organism for survival. From studies of prematurely born human infants, for example, it is known that the visual system can respond to light several months prior to normal birth (see review by Gottlieb, 1971a); it seems exceedingly unlikely, however, that this capacity plays any role in the prenatal development of the visual system. Carmichael (1954) has referred to such phenomena as examples of the "law of anticipatory morphological maturation" (it is also known eponymously as "Carmichael's law"). Anokhin's theory of "systemogenesis" is an elaboration on this idea (Anokhin, 1964), as is

Weiss' concept of "forward reference," in which it is held that the embryonic nervous system develops in forward reference to, and without the benefit of, function (see Weiss, 1939).[4] Finally, it is possible that the early onset of sensory activity may serve some unique and transient adaptive function in the embryo, unrelated to its later role as a monitor and transducer of environmental change. For instance, the phasic and/or tonic input from sensory receptors during embryogenesis could, in some cases, serve as a kind of "trophic" agent for the normal development of the neurons and neural circuits mediating later adaptive motor or other behavior patterns: There is a growing recognition that many features of the developing nervous system are transient or provisional (see below; Purves and Lichtman, 1980; and Oppenheim, 1981a,b; Innocenti, 1982). In the final analysis, however, we simply do not know why such phenomena exist. Between observation and understanding lies a vast and alluring expanse of conjecture which only future studies can hope to reduce.

III. Are There Functionally Mediated Adaptations in the Neuroembryological Development of Behavior?

This question, which is related to the issue discussed in the previous section, is one of the central problems in the study of both neural and

---

[4] This so-called law was actually appropriated, not invented, by Carmichael. As Harrison (1904) recognized, this principle had its origin, in part, in the conceptualizations of the pioneer embryologist Wilhelm Roux. In 1881 Roux proposed that all cells, tissues, and organs go through two general phases of development: an initial period when cells, tissues, and organs are formed in anticipation of, and without the participation of, function; and a later period in which the final stages of differentiation require functional activation. Although Roux believed that some cells, tissues, or organs might depend upon the later (functional) stage to a greater extent than others, he nonetheless argued that to one degree or another the functional stage was always involved. It was Harrison's critical experiments in 1904, showing that skeletal muscle and swimming behavior developed normally in frog tadpoles that were chronically paralyzed as embryos, that first indicated that Roux might be wrong, thereby providing the earliest evidence for what later came to be variably known as the concept of "forward reference" (Weiss, 1939), "anticipation" (Barcroft, 1938), or "anticipatory morphological maturation" (Carmichael, 1954). As Harrison put it, "the production of the specific structure and arrangement of the muscle fibers takes place independently of the functional activity of the muscles. . . . While the above fact cannot but strike one as remarkable it is, nevertheless, . . . in accordance with what should in reality be expected, for such complex mechanisms as, for example that used in respiration, develop in the mammalian embryo during intrauterine life without ever having been brought into action" (1904, pp. 216, 218). Thus, as with several other concepts and principles in embryology, Harrison also deserves credit not only as the originator of "Carmichael's law," but also, together with Roux, as the first proponent of the related notion of "forward reference."

behavioral development.[5] Even in prehistoric times, women were certainly aware of the movements of the fetus during pregnancy and they may have contemplated the significance of such movements for mental development. [Preyer (1885) has documented the existence of similar concerns in historic times.] Empiricist views on this question arose out of the psychological tenets of the eighteenth- and nineteenth-century British and French associationists and the "psychobiological" environmentalist views of Lamarck. The nativist position, in contrast, stems in part from embryological theories of preformationism and from the philosophical views of Kant and other so-called nativist philosophers in the eighteenth and nineteenth centuries (Oppenheim, 1982).

Although the early associationists expressed only the vaguest speculations about the anatomical–physiological basis for the development of knowledge and perception, they left little doubt that their views were based on the assumption that the nervous system begins as a rather uniform, undifferentiated structure in which function, experience, and associations gradually inscribe more definite patterns. As the nineteenth-century British physiologist (and associationist) Carpenter aptly expressed it, "the nervous system grows to the manner in which it is habitually exercised" (1874, p. 106). Among the early adherents of this view who expressed an explicit belief in the existence of prenatal "learning" mechanisms were Helvitius (1769, see Cumming, 1955), Darwin (1796), Lamarck (1809), Mueller (1843), Bain (1855), Roux (1881), and Luys (1882); and this list is by no means exhaustive.[6]

In a few instances, suggestions about the anatomical or physiological basis of such functional adaptations were also proposed, although in

---

[5] I use the term *functionally mediated adaptations* here to draw attention to the need to make a distinction between putative adaptations that result from function, use, or experience versus merely demonstrating that some aspect of neural or behavioral development can be modified by such factors. Although it is sometimes assumed that functionally mediated modifications always reflect adaptations, this is often not the case. In a large measure, environmental factors frequently encountered in the natural habitat of a species evoke adaptive reactions, whereas rare and unusual environments often fail to do so.

[6] By mentioning all of these individuals in the same context, I do not mean to imply that there were no differences among them on this issue. There certainly were. What they had in common, however, was the belief that to one degree or another behavior patterns and their neural correlates are acquired prior to birth by learning or learning-like associative mechanisms. The pioneer English psychiatrist H. Maudsley, for instance, held a rather modern position on this issue. Although Maudsley expressed the belief that "the nervous system of man and animals is moulded structurally according to the modes of its functional exercise," he went on to point out that "The mind is not like a sheet of white paper which receives just what is written upon it, nor like a mirror which simply reflects more or less faithfully every object, but by it is connoted a plastic power minister-

no case was any scientific evidence presented to support the belief that brain structure can be "molded" or modified by experience. Nevertheless, these early theoretical proposals were often ingenious and in some instances anticipated modern views of neural plasticity (Oppenheim, 1979). Despite this rich intellectual history, contemporary discussions of this issue often give the impression that such ideas first appeared in the present century in the writings of S. Ramón y Cajál, C. Ariens-Kappers, E. B. Holt, D. O. Hebb, and others.

Berger (1900), discoverer of the EEG, appears to have been the first to attack this problem experimentally (but see Gates, 1895). Berger sutured the eyelids of neonatal dogs and cats (thereby depriving them of patterned light) and later examined the histology of the visual cortex. His findings, which were later confirmed and expanded by Riesen and others (see Riesen, 1975, for a review), showed that there were marked anatomical deficits in the visual system as a result of the early deprivation. Berger's study, as well as many subsequent investigations, failed, however, to distinguish clearly between the role of visual experience in the *initial* development of the brain and its role in *maintaining* already developed or mature structures against later atrophy. Despite an increasingly sophisticated technology and the appearance of a plethora of related studies in the last 10 years, this still remains as a difficult and nagging problem in the interpretation of many of the studies of the role of early experience in brain development (e.g., Gottlieb, 1976a). Moreover, virtually all of the studies done in this tradition, beginning with Berger's pioneer work, have focused on the role of experience in *postnatal* life. Problems of interpretation notwithstanding, the present consensus on this issue is that experience is important for certain aspects of normal brain and behavioral development postnatally (as well as being necessary for their maintenance), although there is still vigorous debate over many important details (see Barlow, 1975; Grobstein and Chow, 1975; Greenough, 1976; Jacobson, 1978). With this in mind, one may then ask whether *prenatal* function or experience also plays a role in embryonic or fetal neurogenesis and behavior. In contrast to the major focus of attention in the previous section (i.e., are embryonic and fetal movements dependent upon sensory input for their manifestation and pattern), the central issue here is whether sensory input, experi-

---

ing to a complex process or organization, in which what is suitable to development is assimilated, what is unsuitable is rejected" (1876, p. 220). It is worth noting that such views were motivated in a large measure by Lamarckism. It was widely believed that in order for behavioral evolution to occur there must be *acquired* ontogenic modifications in the structure of the nervous system that could be *inherited* (Oppenheim, 1979, 1982).

ence, or use during embryonic stages is necessary for the differentiation and manifestation of adaptive behavior patterns seen at *later* stages of development.

As with so many other fundamental issues in neuroembryology, the first experimental study of this question was performed by Harrison (1904). In a classic experiment, Harrison placed the developing eggs of two species of frogs into an anesthetic solution (chloretone) prior to the earliest signs of motility (and before any significant muscle development had begun). After 7 days in this condition—and at a time when control animals were free-swimming and feeding on their own—the immobilized tadpoles were placed in tap-water; after as little as 20 minutes, some animals were reported to swim normally. Harrison concluded that "It is clear then that the mechanism requisite for carrying out the complex muscular movements of locomotion and respiration develops normally without ever having functioned, although in the normal development of the embryo, the acquirement of this power is a gradual one, being accompanied by the frequent activity of the parts" (1904, p. 214). Several years later, these findings were replicated (and extended to salamanders) by a student of Harrison, S. R. Detwiler (see Matthews and Detwiler, 1926), and by the psychologist Leonard Carmichael (1926, 1927).[7]

Despite the fact that these older studies have been widely heralded as supporting the belief that, for the embryo and fetus, the nervous system develops in forward reference to and without benefit of function, because of a number of shortcomings they should not be considered as the final word on the subject. First, because all of the studies used the anesthetic chloretone, one would like to know the extent to which this drug blocks *all* neural transmission. Although it is generally assumed that it does, it is also possible that, like many of the general anesthetics (e.g., the barbiturates), it preferentially affects selected portions of the CNS at specific doses, developmental stages, etc. Regrettably, there appears to be no definitive information on this point.

Furthermore, although there can be no question that the tadpoles in these early studies could swim following release from the drug solu-

---

[7] It is of some considerable historical interest that despite the apparent clarity of his findings in arguing against the role of function, Carmichael, who was apparently greatly influenced by the antiinstinct movement of the 1920s and 1930s, nonetheless argued that his findings supported the notion that stimulus–response mechanisms were involved in the ontogeny of swimming (see Carmichael, 1927, p. 47). Later he modified his views for, as he confessed, "I was so under the domination of a universal conditioned reflex theory of the development of adaptive responses that I denied categorically the role of maturation" (1941, p. 17).

tion, more rigorous and quantitative information on the details of swimming or other behaviors are needed before one can accept the conclusiveness of the findings concerning the endogenous nature of behavioral development. For instance, in a widely cited study by Fromme (1941), it was found that, in agreement with the studies of Harrison, Carmichael, and others, frog tadpoles exhibited a normal pattern of swimming after being immobilized during the entire embryonic period. However, through the use of quantitative testing procedures, Fromme discovered that the experimental tadpoles swam considerably slower than controls of the same age. Because neither Fromme nor Harrison or Carmichael followed the experimental embryos in their studies beyond 1–2 hours after their release from the drug treatment, it is also conceivable that the animals might behave differently after longer periods (e.g., perhaps the deficits in swimming speed reported by Fromme only reflect transient, residual drug effects, which subsequently disappear). Apparent evidence against this possibility was reported by Matthews and Detwiler (1926) in a study involving salamander embryos. They reported that under certain conditions following embryonic immobilization with chloretone, the tadpoles were later found to exhibit deficits in swimming that persisted for up to 2 weeks following the cessation of drug treatment.

As part of a more general effort to study early behavioral development in amphibians, Mr. Lanny Haverkamp, working in my laboratory, has attempted to replicate and extend some of these early studies involving embryonic immobilization. With regard to the specific effects reported by Matthews and Detwiler, Haverkamp has found that chronic immobilization of axolotl salamander embryos with chloretone for even shorter periods than were used by Matthews and Detwiler almost always results in gross morphological abnormalities involving rigidity, asymmetry, and malformations of the trunk. Although these animals exhibit persistent behavioral deficits in swimming, which resemble those described by Matthews and Detwiler, we assume that these deficits are an indirect outcome of the trunk rigidity and contortions and thus do not reflect a direct effect of the immobilization on the ontogeny of swimming mechanisms in the CNS.

In related experiments with frog embryos (*Xenopus laevis*) in which it is possible to avoid the morphological abnormalities seen in salamanders, Haverkamp has found that chronic embryonic immobilization with chloretone, α-bungarotoxin, or xylocaine has no measurable effect on the swimming pattern of the tadpoles after they are removed from the drug. However, in tests designed to measure swimming speed, he has found that the frog tadpoles do, in fact, swim significantly slower than controls, at least for the first 2 hours after the cessation of treatment. In contrast to the study of Fromme, however, Haverkamp continued to test his animals beyond 2 hours after release from the drug, and, in fact, some animals have been followed for 9–10 days. Beginning at least as soon as 24 hours after being removed from the drug, the swimming speed of all of the experimental groups was indistinguishable from controls (Haverkamp and Oppenheim, 1981).

Although these studies are still in progress, it seems likely that the initial deficits in swimming speed in *Xenopus* may only reflect the residual systemic effects of unmetabolized drug. An alternative explanation is that the deficit is real but that it is eliminated following only a few hours of normal swimming experience. At present we cannot decide between these two possibilities. Despite the uncertainty concerning its cause, however, it is quite clear that even this quantitative deficit is exceedingly short-lived. Although much more work needs to be done, it appears that the original conclusions reached by Harrison and Carmichael may be basically correct in that the elimination of all embryonic motility appears to have little, if any, effect on the normal ontogeny of swimming.

It is obviously of great interest to determine whether the immobility of other vertebrate embryos can also be eliminated without any long-lasting effects on subsequent behavior. Unfortunately, it is not nearly as easy to carry out similar studies with birds or mammals. In both groups, long-term immobilization invariably results in skeletal and other gross defects similar to those seen in salamanders (e.g., Murray and Drachman, 1969). In birds, at least, it has proved possible, however, to immobilize embryos pharmacologically for shorter periods of time (i.e., from 2 to 4 days) at different stages during development without any serious morphological defects.

Using drugs that suppress embryonic motility by blocking physiological activity at the neuromuscular junction, my colleagues and I have been carrying out such studies with chick embryos during the past several years. We have found that in spite of eliminating virtually all embryonic movements—thereby also eliminating any sensory feedback that might normally accompany such activity—during either early (e.g., days 6–9) or late (e.g., days 10–13) stages of incubation, the subsequent behavioral development of such embryos is remarkably normal. Depending upon the specific drug used, within a few hours or days following such treatment, the rates of motility appear normal, sensory responsiveness is similar to controls, and in many cases the embryos are able to hatch (Oppenheim *et al.*, 1978; Pittman and Oppenheim, 1979). Although we have not carried out any systematic studies of behavioral development following hatching, it is our impression that the animals are normal. Thus, although we have not been able to suppress the behavior of these animals during the entire, or even during a major, part of the 21-day incubation period, to the extent that this has been possible, these data are consistent with the findings from amphibians. I hasten to add, however, that there is one other very important difference between the two sets of studies. Although it is not known for certain, it is generally assumed that the drug chloretone, which was used in all of the amphibian studies, acts as a general anesthetic that suppresses all electrophysiological activity in the peripheral and central nervous system, whereas the drugs used in the chick studies (i.e., curare-like agents) act exclusively at the neuromuscular junction. Thus, in the chick studies, endogenously generated CNS activity has not been eliminated, but rather only the manifestation of this activity in overt behavior has been suppressed. Regrettably, all general anesthetics or other agents that block all neural activity and that have been tried in the chick are lethal because of the blockade of cardiac activity.[8]

At present, the only solution to this dilemma in avian or mammalian embryos is to grow pieces of isolated embryonic neural tissue in culture dishes and expose them to agents that block all neural activity. It is now well established that isolated fragments of nervous tissue from mouse, rat, and chick spinal cord or brain, if removed prior to synapse formation and before the onset of function, can develop apparently normal organotypic and tissue-specific structures and functions when grown in tissue culture dishes (Nelson, 1975; Crain, 1973, 1976). Furthermore, in some studies it has been shown that when these neural fragments are allowed to develop in a tissue culture medium containing anesthetics or neurotoxins that suppress all bioelectric activity, they also appear to develop normally. When the pharmacological

---

[8] Part of the reason for this difference between chick and amphibian embryos may lie in the fact that in the chick the heart forms and begins to function at a relatively earlier stage than in frogs, in which the general features of the embryo are already determined before the heart is formed. This heterochronic acceleration of heart development in the chick is probably related to the fact that, in contrast to the frog, the chick requires a functional vascular system for the transport of food from the yolk. Thus, in relation to the emergence of behavioral activity, the cardiovascular system becomes sensitive to pharmacological disruption considerably earlier in the chick than in the frog.

agents are washed out at a stage when control cultures have formed functional synapses, the deprived cultures begin to function almost immediately and synaptogenesis is similar to that found in control tissue (Crain *et al.,* 1968; Model *et al.,* 1971; Obata, 1977; Bird, 1980; Crain, 1980).

Thus, because these *in vitro* results using neural tissue from mammalian embryos are consistent with the data from the amphibian studies done with intact embryos, collectively these data provide convincing evidence for the contention that early neurogenesis, including synapse formation and the onset of neural function, can take place in the absence of either endogenous neural activity or sensory input. In apparent contrast to these findings, however, Bergey *et al.* (1978) found that when fetal mouse spinal cord and dorsal root ganglia are grown in tissue culture in the presence of the neurotoxin tetrodotoxin, which blocks all electrical activity, the synapses that were initially present in the tissues were lost and a large percentage of postsynaptic cells either failed to develop or later died. Because *mature* neural cultures exposed to the same tetrodotoxin regimen appeared normal, the authors concluded that synaptic activity is necessary for the initial development and survival of neurons but not for the subsequent maintenance of neurons following the major period of neural maturation. It is not yet clear whether the apparent difference between these results and those from the earlier tissue culture studies are due to the use of different cell types, to a longer duration of treatment, to use of different chemical agents, or to other factors (but see Crain, 1980). In any event, the results of Bergey *et al.* (1978) suggest that we may still have a lot to learn about the role of function in early neurogenesis.[9] Tissue culture preparations, as Crain has often argued (e.g., Crain, 1973, 1976), may provide an especially powerful technique for answering these questions.

Despite the attractiveness of *in vitro* preparations, however, there is always the nagging question of their relevance for events occurring in the intact organism, especially regarding behavioral development. For instance, physiological and anatomical deficits analogous to those reported by Bergey *et al.* (1978), if they were to occur *in vivo,* may be prevented from manifesting themselves in behavioral deficits by self-

[9] It is of interest in this regard that a virtually total suppression of behavioral activity in the chick embryo during those stages when normal cell death of spinal motoneurons occurs results in the prevention of cell death (Pittman and Oppenheim, 1979; Oppenheim, 1981b, for review). Thus, early embryonic neural and behavioral activity may serve a variety of functions (Harris, 1981, review).

regulatory, compensatory mechanisms that may not be present in isolated fragments of neural tissue.[10] Similarly, the apparent lack of physiological or morphological deficits as reported by Crain *et al.* (1968) and other may be more apparent than real. If it were possible to repeat those experiments *in vivo*, it is conceivable that one might find behavioral deficits. Although there does not appear to be any immediate solution to this dilemma, recent developments appear to offer some hope for the future. For instance, in a recent report it was shown that pieces of normal fetal brain tissue can be transplanted into the brain of adult rats who have had the related brain region destroyed (Perlow *et al.*, 1979). What was remarkable about these experiments was that such transplants were able to compensate for or alleviate the behavioral dysfunctions resulting from the earlier brain lesions. Thus, it is conceivable that it may soon be possible to grow pieces of brain tissue in tissue culture in the presence of drugs or other agents that block function and then transplant them back into the brain of an intact organism and assess their functional capacities. For the present, however, and despite their many drawbacks, studies of intact embryos in which neural function is suppressed or otherwise altered provide the only available alternative to the more traditional tissue culture studies (but see Harris, 1980).

Although immobilization studies similar to those described earlier with amphibian and avian embryos have not been carried out with fetal mammals, there is an interesting analogous situation involving cases of premature human infants born with respiratory distress, such as hyaline membrane disease. This pathological condition is most common in 32- to 35-week-old premature infants and requires the use of an artificial respirator. The use of this device necessitates that the infant be rather rigidly restrained so that motor activity will not mechanically dislodge the respirator. As a result of the restraint, there is a gradual loss of all motility, reactivity, and sensitivity, including a failure of the infant to respond to painful stimuli and the loss of all primary reflexes. Despite the total immobilization and inactivity of these premature infants, which may last for up to 1 week, such treat-

---

[10] It may be possible to partly obviate this and other problems associated with tissue culture by using organ or "systems" preparations in which larger pieces of embryonic or fetal neural tissue are placed in an organ bath medium where they can be maintained for many hours or even days. In such an environment, the organ or "system" (e.g., the lumbar spinal cord along with the limb or select limb muscles) is amenable to physiological and pharmacological studies not possible *in vivo*. (For some recent attempts to use this approach, see Landmesser, 1978; Saito, 1979; Stehouwer and Farel, 1980; Ziskind-Conhaim, 1981.)

ment apparently has no long-term ill effects. In the words of one investigator, "when the lung disease abates . . . he (the infant) almost instantly assumes his own breathing rate, and his neurological level once again approximates the norm according to his maturative state" (Saint-Anne Dargassies, 1977, p. 255).

As this example illustrates, the study of behavioral development in prematurely born infants offers a potentially rich source of information concerning several issues in the neuroembryology of behavior. Because these exteriorized "fetuses" may spend up to the last 13 weeks of gestation in an abnormal environment, they provide a unique opportunity to evaluate the role of altered sensory input and neural function on subsequent behavioral development.

The pioneer child psychologist A. Gesell was one of the first to recognize the value of premature infants for studies of behavioral development. In his 1928 book summarizing the scanty evidence then available on this issue, he asked, "is this precocious entrance into the world but an incident and will the nervous system proceed unperturbed in its growth, punctual to the usual program [or] . . . will this confer upon him a precocious adjustment and carry him hurriedly through his own infancy?" (1928, p. 299). Gesell's conclusion was that, in large measure, "the preterm infant grows much like a fetus even though he is out of the womb" (1928, p. 318).

Subsequent studies have tended to corroborate Gesell's early interpretation on this point. Nerve conduction velocity (Thibeault et al., 1975), electrophysiological (EEG) development (Dreyfus-Brisac, 1975; Parmelee, 1975), visual perception (Sigman and Parmelee, 1974; Fantz and Fagan, 1975), visually evoked EEG activity (Watanabe et al., 1972), motor behavior, postures and reflexes (Dubignon et al., 1969; Prechtl et al., 1975; Gardner et al., 1976; Saint-Anne Dargassies, 1977), intelligence, academic achievement, and language (Dweck et al., 1973; Francis-Williams and Davies, 1974; Fitzhardinge, 1975; Douglas and Gear, 1976; Kelsey and Barrie-Blackley, 1976) all appear to develop relatively normally (and at the appropriate rates) in those prematurely born infants lacking any obvious or suspected signs of neurological damage upon delivery (i.e., "nonrisk" premature infants). What is particularly remarkable is that despite 1 to 2 months of additional respiratory "experience," the neural mechanisms controlling breathing and related activities are similar in premature and full-term babies of the same postconceptional age (Parmelee et al., 1971).

One of the leading investigators in this field Saint-Anne Dargassies, has concluded, "Foetal maturation is independent of the living conditions imposed upon the infant since it takes place in the incubator just as it does *in utero*" (1977, p. 299). And she adds, "Despite the afferent stimuli and experience due to precocious extra-uterine life, the environment has not hastened the appearance of the visual and auditory functions: the former premature behaves . . . as does a term neonate only a few days old" (1977, p. 244). This is not to deny that in some respects premature infants do differ from normal term infants (see Wolff and Berber, 1979 review). Weight gain, for instance, is often retarded, and there are reports that certain types of stimulation may mitigate these and other deficits in premature infants (e.g., Campbell, 1982; Powell, 1974; White and Labarba, 1975; Rice, 1977). Neural and behavioral differences may also exist, which either are too subtle to be detected by current techniques or which simply have not yet been investigated (e.g., see Dubowitz *et al.*, 1980). Despite such caveats, the extent to which neural and behavioral development appears normal following such a drastic environmental perturbation is still rather remarkable and implies that there is a substantial degree of intrinsic regulation in the prenatal development of the nervous system and behavior in humans.

In apparent contrast to these studies with premature infants is the report of Prechtl (1965) concerning the role of the intrauterine position of the human fetus on postnatal reflexes. Prechtl found that infants born in the breech position ("feet first") tended to have abnormal resting postures of the legs as well as reversed leg reflexes following stimulation of the foot (i.e., they exhibited extension whereas normal babies typically flex, and vice versa). The most obvious explanation of these results, namely, that the fetal breech position and the abnormal leg posture and leg reflexes all reflect a single underlying neurological defect, and not a cause and effect relationship, has apparently been ruled out. Prechtl and his colleagues found that fetuses who are in the breech position but who spontaneously assume the normal vertex (i.e., head-first) position before the seventh or eighth month of gestation have normal leg reflexes and posture after birth. Unfortunately, these interesting findings on breech-born babies have not yet been replicated, nor have there been any long-term follow-up studies on the persistence of the abnormal reflexes or on the possible occurrence of other abnormal behaviors following infancy. Obviously such studies are urgently needed if we are to understand the significance of such observations for normal behavioral development in humans.

The studies on embryonic or larval sensory deafferentation (reviewed earlier) are also pertinent to the question of whether functionally mediated adaptations occur prenatally, in that, despite being deprived of one aspect of neural function (i.e., sensory input), the animals in those studies were able to develop relatively complex motor patterns. This is especially striking in the case of the chick, as the embryos were deprived of sensory input prior to the first signs of overt motor behavior and thus never had an opportunity to respond to stimulation. Notwithstanding the rather remarkable performance of these animals, I do not wish to imply that such findings indicate that prenatal sensory input is either irrelevant for neurogenesis or unimportant for the emergence of normal behavior patterns. As was the case for the embryonic, fetal, and premature infant immobilization studies reviewed earlier, there remain a number of important unanswered questions, including the generality of these findings and the effects of such treatments on behavior patterns more complex than motility, reflexes, hatching, or locomotion. For instance, in a replication of the earlier deafferentiation study by Hamburger *et al.* (1966), Narayanan and Malloy (1974a,b) found that although chick embryos with deafferented fore- or hindlimbs were behaviorally normal before hatching, they were unable to stand, walk, or make wing-flapping movements after hatching (but see Provine, 1979). In the future, more rigorous and detailed studies involving a combination of behavioral and neurophysiological techniques will be required in order to answer many of the outstanding questions concerning the role of somatosensory input in the embryo. Moreover, it is important to emphasize that even less is known about the role of prenatal experience or function in most other sensory modalities, such as taste, smell, audition, vision, or vestibular, for behavioral development. A notable exception concerns the role of prenatal auditory input in birds. There is now rather convincing evidence that prenatal auditory experience is a normal and necessary feature in the ontogeny of a number of auditory-related behaviors in birds. Although these studies have been ably summarized and discussed in several reviews (e.g., Vince, 1973; Gottlieb, 1976b; Impekoven, 1976), there has been substantial progress made recently in the understanding of some of these phenomena such that they deserve to be discussed in the present context.

In one of the earliest reports, Margaret Vince observed that in quail—and probably in many other precocial birds—hatching time among a large clutch of eggs is rather precisely synchronized to within a few hours. The mechanism mediating this remarkable behavioral synchronization involves auditory sensory input. During the last few days before hatching, the embryos of many avian species emit audible clicking sounds produced in association with pulmonary respiration (i.e., they are not vocalizations in the strict sense). Vince has shown that when eggs are in physical contact, these sounds are transmitted from one egg to another. Depending on the rate of the emitted clicks, adjacent embryos are either accelerated or retarded in their development such that the entire clutch hatches together. Although the specific processes by which the clicking produces this effect are still not entirely clear, it has been demonstrated that the onset of particular coordinated motor patterns involved in prehatching and hatching behavior (see Hamburger and Oppenheim, 1967) are preferentially enhanced and their rate is increased during exposure to such clicks (Vince *et al.*, 1976). Although groups of embryos that are deprived of hearing these clicks can eventually initiate these movements and can hatch normally, their hatching time as a group is random and asynchronous.

In the case of quail, at least, the adaptive function of synchronized hatching is apparently related to the early exodus of the mother and young from the nest. If hatching of the embryos were drawn out over 1 to 2 days, as is the case in the deprived embryos, the retarded embryos would be abandoned and probably could not hatch or, even if they hatched, could not survive on their own. [Another even more remarkable instance of sensory-induced, synchronized hatching occurs in cockroach embryos. In this case, the synchronization apparently is triggered by tactile stimulation between the embryos within a single eggcase, leading to cooperative "hatching" activity (Provine, 1976). As a result of this behavioral cooperation, the embryos are able to open the eggcase and emerge, whereas the activity of only one or even a few of the embryos is apparently unable to accomplish this feat.]

Embryonic clicking is not the only factor involved in the synchronization or regulation of hatching time in birds. In addition to the role of intrinsic factors, involving specific neuroanatomical and neurotransmitter systems, in these phenomena (Oppenheim, 1973; Pittman *et al.*, 1978), it has also been reported that specific vocalizations of the incubating parent during the last few days of hatching may regulate the prenatal behavior and hatching time of the embryo (Tschanz, 1968; Hess, 1973).

The work of Tschanz (1968) and Tschanz and Hirsbrunner-Scharf (1975) is of particular interest in this regard. They have demonstrated that the prenatal experience of hearing parental vocalizations is not only involved in hatching, but that it is also a crucial component for the development of individual recognition of the parent by the chick following hatching. In guillemots (*Uria aalge*), which breed in dense colonies on cliff ledges—conditions which could lead to a considerable confusion and chick mortality if no personal bonds existed immediately after hatching—this propensity is especially well developed. In contrast, in the closely related razorbill (*Alca torda*), whose nests are isolated in caves and crevices and thus are spatially distinct, prenatal auditory experience is not an important feature of individual recognition. Apparently, in the razorbill, the evolutionary selection pressures for the occurrence of immediate posthatching recognition have not been as strong as in the guillemots, in which the developmental process of individual recognition has been pushed back to prenatal stages.

Similarly, laughing gulls (*Larus atricilla*) and herring gulls (*Larus argentatus*), which also live in spatially well-defined nesting areas (like the razorbill), do not require prenatal experience for the development of *individual* recognition (Impekoven, 1976), although such experience may be an important component of *species* recognition after hatching (Evans, 1973). In the same manner, Impekoven (1976) has also shown that prenatal exposure to specific parental vocalizations that later aid the hatchling in feeding and predator avoidance may also enhance or suppress the postnatal responsiveness of gull chicks to these calls.

The most detailed and comprehensive examination of the role of prenatal auditory experience in birds is the important work of the psychologist Gilbert Gottlieb. Over a period of many years, Gottlieb has doggedly examined the role of prenatal auditory experience in an attempt to illustrate that certain behaviors that appear to be more or less perfect shortly after hatching (and on that basis have traditionally been called innate) are not devoid of an important experiential history.

Gottlieb's work was inspired by the naturalistic observation that in many species of precocial ducks, the newly hatched young appear to approach and form attachments with their mother in large measure on the basis of vocalizations emitted by the mother. Gottlieb then went on to demonstrate that under experimental conditions in the laboratory the young would preferentially approach and stay close to surrogate mothers emitting the species-typical maternal call (as opposed to the maternal call of a different species) and that, in the case of the mallard (*Anas platyrhynchos*), certain aspects of this perceptual discrimination depend upon prenatal auditory experience (Gottlieb, 1971b, 1976b). Contrary to expectations, however, Gottlieb found that it is not the prenatal experience of the *maternal* call—to which the embryos are, in fact, normally exposed—that is crucial. Rather, it is the experience of the embryo with its own prenatal vocalizations (most precocial avian embryos begin to vocalize a few days before hatching). When deprived by embryonic muting of hearing its own vocalizations, the hatchling experiences difficulty in making a choice between its species-typical maternal call and that of certain other species (e.g., the chicken maternal call).

Furthermore, Gottlieb has found that particular acoustic features of the maternal call (i.e., frequency and repetition rate) are used by the young in making the postnatal discrimination. Similarly, it is a specific aspect of the embryo's own vocalizations (i.e., high-frequency acoustic energy) that represents the critical sensory cue involved in sharpening the posthatching frequency preference. He showed that muted embryos exposed to tapes of high-frequency sounds perform similarly to normal ducklings after hatching, whereas embryonic exposure to low-frequency sounds fails to have this meliorating effect. Interestingly, prenatal experience is apparently not necessary for the initial development of the preference for a specific repetition rate. However, once developed, prenatal auditory experience of a highly specific nature becomes necessary for maintaining repetition rate preferences in the embryo (see Gottlieb, 1978, 1979, 1980). In the absence of such experience, the preference for a specific repetition rate is eventually lost.

It is important to note, however, that despite these subtle experiential effects, the prenatally deprived ducklings, which are unable to discriminate a mallard from a chicken maternal call after hatching, have neither completely lost nor failed to develop a preference for the mallard call; when tested against certain other sounds or vocalizations, the ducklings typically prefer their own species call (Gottlieb, 1976b). Thus, the perceptual deficits resulting from prenatal auditory deprivation are not general but rather are manifested only under rather specific conditions.

Although it remains to be critically tested, it seems probable that in cooperation with visual and perhaps other cues, the auditory identification of the mother by the duckling serves to initiate and promote early social interactions that are important to the maintenance of the parent–young bond. In this way the "poikilothermic" and vulnerable young are prevented from becoming lost and thus being exposed to potentially lethal temperature fluctuations and to predation. This early bond may also represent a first crucial component of species recognition that is later used in a number of social contexts (e.g., breeding, group feeding, flock formation, migration) by mature birds.

Despite the obvious advantages accruing to an early establishment of species recognition, it is still not clear why, in the case of ducks, it is important for the whole process to begin prior to hatching. The obvious selection pressures for the role of prenatal experience that are present in the avian species studied by Tschanz (1968) do not appear to exist in these ducks. Moreover, the young of many precocial species of ducks may require 12–24 hours, or even longer, before they are capable of sustained locomotion, and in the mallard the mother does not leave the nest with the young until approximately 16–20 hours after hatching (Bjärvall, 1967; Miller and Gottlieb, 1978). Therefore, one might expect that there would be ample opportunity after hatching for the establishment of species recognition. Comparative studies of the ecology of precocial avian species that do not depend upon embryonic experience (if any exist) for the development of these capacities may be able to shed some light on this question.

Another question of even deeper developmental significance concerning these findings is whether prenatal auditory deprivation per se has a detrimental effect on posthatching survival in nature. Despite the carefully documented presence of perceptual deficits following embryonic deprivation, one would like to know how successfully the deprived hatchlings could compete with normal siblings in a natural environment where they may never be required to make crucial discriminations based solely on sounds with a particular frequency or repetition rate (e.g., as in the case of the chicken versus mallard calls). The observation that a preference for the mallard maternal call can develop even in the absence of virtually all auditory experience (Gottlieb, 1976b) suggests that the mechanisms involved may be sufficiently canalized so as to insure adaptive responses despite experiential perturbations. That is, it is conceivable that, in a natural context, a sufficiently large number of deprived animals might be able to develop social attachments such that their perceptual deficits would never be manifest in lower survivorship or lowered reproductive success. It is easy to see that, in principle at least, such a mechanism, involving this kind of a developmental homeostasis or self-righting tendency, could be as important as the role of experiential mechanisms, because it is obvious that animals that can survive and reproduce even in nonoptimal conditions would have a selective advantage over animals tightly locked into specific determinants, be they maturational or experiential (for a more complete discussion of these and other issues concerning species identity, see Roy, 1980).

Notwithstanding such nagging questions, these experiments of Gottlieb (and the related studies of Vince, Impekoven, Tschanz, and others on prenatal auditory experience) are of fundamental importance in that they constitute a beginning effort to evaluate the role of embryonic or prenatal experience within the context of the natural history and ecology of the species. When considered within an evolutionary framework, in which questions of natural selection and adaptation are central, it simply is not sufficient to demonstrate that an embryo or fetus is responsive to a given stimulus or that its pre- or posthatching behavior can be modified by experience (although this obviously represents a necessary first step). A number of studies, for instance, have reported prenatal conditioning (Gos, 1935; Spelt, 1948; Hunt, 1949; Sedlacek, 1964; Fried and Gluck, 1966), and others have found that various atypical or unnatural kinds of prenatal stimulation or manipulation can affect postnatal behavior (Grier et al., 1967; Adam and Dimond, 1971; Dimond and Adam, 1972; Rajecki, 1974). Beyond demonstrating that the neural mechanisms involved in learning, conditioning, or related modifications are present prenatally, however, it remains to be seen what relevance such observations have for understanding normal mechanisms of ontogenesis. It seems exceedingly unlikely that traditional mechanisms of learning or conditioning play any significant role in the organization of the nervous system or behavior during the prenatal period.

Although claims have also been made for prenatal auditory learning in humans (Salk, 1962, 1973; Macfarlane, 1977; Rosen, 1978; Goodlin, 1979; Rosner and Doherty, 1979), the data on this issue are still equivocal. It is worth noting, however, that it has been shown that external sounds can be detected by microphones placed within the amnion of the sheep fetus (Armitage et al., 1980) and, moreover, sounds played to the guinea pig fetus in utero are selectively responded to after birth (Vince, 1979). Thus, it is conceivable that, as appears to be the case with birds, certain aspects of auditory-mediated behavior after birth in mammals, such as the infant–maternal bond in sheep or the selective auditory preferences of human neonates for the maternal voice (DeCasper and Fifer, 1980), may also depend, in part at least, on auditory experience in utero. This is an exciting prospect, and there will undoubtedly be much work in the future devoted to such questions.

Finally, despite the demonstration that for some avian (and, perhaps other) embryos prenatal auditory experience is apparently an essential feature for normal behavioral development, it is conceivable that these examples are not entirely representative of prenatal development. For example, sensory systems may be considerably more dependent on prenatal function or use than motor systems. Or, in the examples involving prenatal auditory experience in birds it may be that because of the particular ecological conditions in which these forms live, they have been forced to evolve unique adaptations that utilize prenatal sensory input, in much the same way that the early embryo utilizes specific inductive stimuli and other cellular interactions for triggering or channeling development along a specific pathway. As August Weismann pointed out already in the last century,

"When it is important to regulate different possibilities of development, nature makes use of external influences as stimuli" (1894, p. 49). This represents an early recognition of the point later stressed by Waddington (1957) and others that natural selection acts on the entire epigenetic system and not just on the genome.

To avoid any misunderstanding, however, I feel compelled to repeat that my intention in making these comments is not to denigrate or diminish the likelihood that prenatal function or sensory input may play an essential role in the neuroembryology of behavior or that in other cases it may normally exert significant modifying effects. Nor do I mean to undermine an interest in pursuing this problem experimentally. I only wish to emphasize that, considering our present lack of understanding of the neuroembryology of behavior, any attempt to fashion theories based largely or exclusively on the effects of early experience and function on behavioral ontogeny is both premature and unrealistic in that it ignores a growing literature in embryology, neuroembryology, and developmental psychobiology that argues against such a narrow approach. There appear to be powerful homeostatic or stabilizing mechanisms, largely endogenous in nature, which play fundamental roles in development, including the ontogeny of the nervous system and behavior (Waddington, 1957; Jacobson, 1978). Because these mechanisms almost always work in cooperation with extrinsic factors, both processes must be considered in any credible theory of behavioral development.

IV. The Development of Inhibition and the Neuroembryology of Behavior

In view of his pioneering role in the establishment of a neuroembryology of behavior, it is perhaps not surprising that Coghill was the first to consider inhibition as a fundamental feature of behavioral development (Oppenheim, 1978). Coghill's conceptual scheme (in which behaviors are initially integrated or global in nature, followed by a later individuation of localized movements and reflexes) depended upon the orderly emergence of inhibition to account for the emancipation of local or reflex movements out of the more global pattern. Because large parts of the body had to be inactivated in specific spatiotemporal patterns, Coghill was convinced that active inhibitory mechanisms were involved, similar to those that had so recently been identified in the adult mammalian spinal cord by C. Sherrington and other neurophysiologists early in this century; fatigue or the passive inactivity of excitatory mechanisms appeared to be inadequate to explain his behavioral observations of the developing salamander. It is important to note, however, that Coghill always maintained that all behavior, even local movements and reflexes, constituted a total "behavior" pattern, in that some parts of the nervous system were actively excited, whereas others were actively inhibited (see Crain, 1973).

Although Coghill never carried out any experiments on the actual neural mechanisms involved in inhibition, he did make predictions from his anatomical data about the putative inhibitory role of Mauthner's neurons in the development of swimming (Coghill, 1934), a suggestion that was later confirmed in experiments by Detwiler (1936). Interestingly, in an all too brief report, Koppanyi and Karczmar (1947) have claimed that the stage of development at which salamander larvae (*Ambystoma*) first respond to the drug strychnine (which is a specific pharmacological antagonist of the inhibitory neurotransmitter glycine) corresponds to Coghill's data on the time of emergence of local movements (Coghill, 1929b). In view of the remarkable increase in knowledge since Coghill's time concerning the physiology, biochemistry, and pharmacology of inhibitory mechanisms, it would seem to be an especially propitious time to attempt to relate behavioral development in the salamander (and in other species) with the development of inhibition.

Because most of the early studies on the ontogeny of inhibition in vertebrates have only dealt with reflexogenic behavior or with electrophysiological mechanisms involving antagonistic or similar types of inhibition (see Oppenheim and Reitzel, 1975), they shed little light on the role of inhibition in spontaneous embryonic activity. Neither do these studies say anything about the possible role of inhibitory mechanisms whose effects may not be immediately apparent in overt behavior. Moreover, as Crain (1976) has pointed out, studies of inhibition carried out in tissue culture preparations suffer from the disadvantage that synaptic structures and functions involved in inhibition may develop abnormally in response to physicochemical factors in the culture environment or in association with increased collateral sprouting (or other factors) peculiar to the surgically isolated nerve cells. Notwithstanding these caveats, at present, the evidence from both *in vivo* and *in vitro* studies strongly implies that behaviorally relevant inhibitory mechanisms appear very early in all vertebrate embryos that have been examined, ranging from fishes to humans (Hamburger, 1973; Oppenheim and Reitzel, 1975; Crain, 1976). Although most of these data still support the old idea that excitation precedes inhibition during ontogeny, more recent evidence suggests that in some cases this sequence may be more apparent than real. Tissue culture studies of fetal rodent CNS tissue (e.g., Crain, 1976; Nelson *et al.*, 1977) indicate a remarkably early onset of synaptic inhibitory mechanisms. Work from my own laboratory is also consistent with this view.

In an early publication on the ontogeny of glycine-mediated behavioral inhibition in the chick embryo (Oppenheim and Reitzel, 1975), we reported that day 8 or 9 of incubation was the earliest age at which we could detect the expected behavioral activation following administration of the glycine antagonist strychnine. More recently, however, using a more direct method of drug application involving intravenous injection, we have been able to demonstrate a behavioral effect of strychnine at day 7 and of glycine at day 6 (Reitzel and Oppenheim, 1980). These newer data are in good agreement with the first appearance of antagonistic muscle activity involving spinal inhibitory circuits (Bekoff, 1976) and thus support the idea that glycine-mediated inhibitory circuits are functional by at least day 7 in the chick embryo.

Furthermore, experiments with another inhibitory neurotransmitter γ-aminobutyric acid (GABA) and its pharmacological antagonists picrotoxin and bicuculline suggest an even earlier onset, compared to glycine, of GABA-mediated spinal inhibition in the chick embryo (Reitzel et al., 1979). Exogenous application of GABA as early as 4 to 5 days of incubation was found to decrease spontaneous motility, whereas picrotoxin and bicuculline first increased motility on day 6. This latter observation implies that there is a sufficient amount of endogenous GABA in the day 6 embryo to functionally stimulate GABA receptors and that bicuculline or picrotoxin blocks this effect. Consistent with these behavioral effects is our observation that the enzymes necessary for the synthesis and degradation of GABA appear to be present in the spinal cord at (or before) the onset of spontaneous motility at 3.5 days of incubation (Reitzel et al., 1979). Experiments currently in progress in my laboratory on other putative inhibitory neurotransmitters (e.g., noradrenaline, dopamine, and the opiates) indicate that these systems may also be functional at early embryonic stages.

If we are correct in ascribing such an early onset of inhibition to the vertebrate embryo—and I hasten to add that even for the chick the data are not entirely consistent on this point (e.g., Stokes and Bignall, 1974; Sedlacek, 1976, 1977a,b; Fein et al., 1978)—the question arises as to the functional relevance of this phenomenon for neural or behavioral ontogeny. The least interesting possibility is that it merely represents still another instance of the "law" of anticipatory functional maturation (i.e., that early inhibition is merely another functional expression of neural maturation that does not become behaviorally adaptive until later in development; see footnote 4).

Alternatively, early inhibition may subserve some unique ontogenetic function unrelated to, or at least independent of, its role in the mature organism. Crain (1976), for instance, has suggested that in the chick, inhibition may act to dampen the amplitude of spontaneous motility, thereby preventing the tearing of membranes or the premature breaking of the shell. This possibility seems unlikely, however, in light of our observations that massive, strychnine- or picrotoxin-induced convulsions in the chick embryo failed to cause any such damage (Oppenheim and Reitzel, 1975, and unpublished observations). Early embryonic inhibition may also be responsible for regulating the periodic or rhythmic nature of spontaneous motility in the chick (Hamburger, 1963, 1973) and thus could be a precursor of mechanisms underlying rhythmic activities postnatally (Wolff, 1966; Oppenheim, 1974; Corner, 1977). Similarly, if endogenous motility is a necessary factor in the normal ontogeny of the neuromuscular system (see earlier), the rhythmic or periodic nature of such motility may be a crucial aspect of this relationship.

A number of workers have also suggested that the early appearance of inhibition may serve to suppress complex behavioral patterns whose neural mechanisms are developing or are already mature in the embryo but that are not needed until after birth or hatching (e.g., pecking, locomotion). Evidence from the chick embryo (Oppenheim and Reitzel, 1975) and from the newborn rat (Plykkö and Woodward, 1961), however, argues against this possibility, in that, following the release of inhibition by strychnine or picrotoxin injection, no new, different, or more organized behavior patterns appear, other than convulsions. It is conceivable, however, that this negative result could merely be an inadvertent artifact of the systemic administration of the drug; localized application of the drugs directly into specific regions of the central nervous system of the chick might release specific complex motor patterns. Evidence from fish embryos supports the suggestion that inhibition may suppress complex behavior as it has been reported that strychnine treatment can elicit complex behavior patterns (i.e., swimming) considerably earlier than they would normally appear spontaneously (Pollack and Crain, 1972; Pollack and Kuwada, 1976). Similarly, silk moth pupae treated with the GABA antagonist picotoxin (Truman, 1976) and tobacco hornworm pupae given the inhibitory antagonist L-canaline (Kammer et al., 1978) exhibit adultlike behavior patterns earlier than normal. In these cases, it would appear that, although the neural mechanisms for the behavior are functionally competent, they are maintained in a latent state by active inhibition. In light of these studies with developing fishes and insects, it seems likely that additional studies on higher vertebrates would be illuminating concerning the role of inhibition in suppressing functionally competent behavioral patterns in the embryo.

It is also conceivable that the early appearance of inhibitory neurotransmitters may subserve a transient, neurogenetic trophic function similar to that suggested for the biogenic amines (e.g., Lauder and Krebs, 1978). That is, the release of inhibitory neurotransmitters may regulate proliferation, migration, cell death, or differentiation in "postsynaptic" cells, independent of their later role in impulse transmission (see Chronwall and Wolff, 1980).

Finally, in light of the reports that GABA-mediated inhibition may be involved in some of the neurophysiological effects of early visual deprivation (Duffy et al., 1976; Burchfiel and Duffy, 1981) and in the normal mediation of orientation selectivity (Tsumoto et al., 1979), it may not be farfetched to suggest that the early appearance of inhibition in the embryo represents a mechanism whereby functional adaptations could be incorporated into the developing nervous system during prenatal development. Roberts (1976) has offered a similar proposal for the role of inhibition in postnatal and adult stages. Although it is presently unclear how such a mechanism would work in practice, in theory, at least, it is conceivable that the induction and/or validation of selective inhibitory circuits by endogenous or sensory-evoked neural function could serve to suppress certain neurogenetic alternatives and thereby guide neural and behavioral development along specific, adaptive pathways. I realize that this as well as some of the other suggested functions for embryonic inhibition are speculative and vague. But I have chosen to express them, nonetheless, in the hope that they may lead to further interest and experimentation. For, as the physicist Max Planck has pointed out, "We must never forget that ideas devoid of a clear meaning frequently give the strongest impulse to the further development of science" (1936, p. 112).

V. Continuity and Discontinuity in the Neuroembryology of Behavior: Is the Embryo a Prophet or an Adaptation?

Perhaps the central concern of the neuroembryological study of behavior, as with any developmental discipline, is to understand the processes and mechanisms of organismic change and transformation. This was as true for the earliest supporters of epigenesis in the eighteenth century as it is for today's adherents of the casual–analytic, epigenetic approach to development. Despite this early recognition that change over time is the main focus of an epigenetic approach to biological development, the modern view that each stage of development is the cause of the next is of relatively recent origin, having first been clearly expressed by W. His in the last century. He wrote,

> Embryology is, in essence, a physiological science; it has not only to describe the building of every single form from the egg, according to its different phases, but to trace it back in such a way that every stage of development with all its pecularities appears as the necessary result of those immediately preceding it (1874, p. 2).

This view is often taken to mean that all development is a continuous epigenetic progression in which the organism gradually takes on more of the features of the adult. In other words, that developmental characteristics are merely incomplete or quantitatively less perfect versions of adult features. This is the prevailing belief in developmental biology, developmental psychology, and behavioral embryology. An underlying assumption of this view is that the major, if not the sole, focus of evolution, regarding ontogeny, has been the formation of a reproductively mature organism. It is sometimes forgotten, however, that life histories are often complex and frequently involve fundamental transitions, deletions, regressions, additions, and other apparent discontinuities, in which some stages may only be fully understood when considered as transitory or provisional adaptations in their own right and not solely as nascent preparations for the adult situation.

Embryos, fetuses, larvae, and postnatal animals frequently inhabit environments that are markedly different from those of the adult. Consequently, each of these stages may have required the evolution of adaptations involving unique morphological, biochemical, physiological, and behavioral mechanisms that differ from the adult, and that may, in fact, require modification, reorganization, inhibition, or even total destruction before the adult stage can be attained. Although I realize that the ontogenetic adaptations exhibited by many invertebrates and amphibians (Etkin and Gilbert, 1968), as well as the familiar adaptations of vertebrate embryos (e.g., the protective membranes enclosing the embryo), have long been recognized, I would argue that with a few notable exceptions (see Oppenheim, 1981a) the possible significance of ontogenetic adaptations for other, less obvious, cases in higher vertebrates has been consistently ignored, especially with regard to behavioral development.

To avoid any misunderstanding, I should add that I am not denying that at some level of organization from the genome to the organism, all steps or stages in ontogeny can in a sense be considered as a preparation for, or necessary to, adulthood. Continuity and ontogenetic adaptations are not necessarily mutually exclusive processes. I recognize that in many, perhaps in most, cases ontogeny can serve immediate needs as well as influence future goals. I do wish, however, to emphasize what I perceive to be a neglected feature of behavioral development, namely, that the "immediate" function of certain neurobehavioral events in ontogeny may have been the major consideration in their adoption by natural selection.

With this in mind, it is obvious that the question posed in the heading for this section—Is the embryo a prophet or an adaptation—is considerably more complex and should read: To what extent is the embryo a preparation for the adult and to what extent do embryonic traits reflect embryonic (i.e., ontogenetic) adaptations? A related question is: How do these two processes interact during ontogeny and together explain the developmental history of any particular animal? Because space does not permit me to present a detailed examination of all the implications of these questions, I will only attempt to discuss them within the framework of the neuroembryology of behavior (for a more thorough coverage, see Oppenheim, 1981a).

Although it is self-evident that for any behavior to function adaptively there must have been a prior history during which the neural and other mechanisms—if not the behavior itself—were being constructed or prepared, the fact is that we have remarkably little information on specific early antecedents or of how they become integrated into later behavior, and this is especially true for prenatal antecedents of postnatal behavior. Part of the reason for this stems from the fact that the only means for obtaining unequivocal evidence for an antecedent involves the demonstration that experimental manipulation of a putative antecedent modifies some subsequent events; and for the prenatal period especially, this is often a technically difficult, and sometimes impossible, task.

If it were possible, for example, to describe the gradual building-up of complex postnatal motor patterns from "simpler" prenatal movements, then such a sequence might tentatively be thought of as a temporal sequence of antecedents in which each step is dependent upon previous steps. Perhaps the clearest and most detailed example of just such a *temporal* process comes from the observations of Coghill (1929b) and others on the emergence of swimming in the amphibian embryo.

Swimming emerges by the gradual appearance of increasingly complex movements, starting with the early, simple, unilateral, trunk flexures and ending with S-waves and sustained swimming. One gets the clear impression from such observations that swimming is perfected by the overt performance of the simpler antecedents and their sequential integration into progressively more complex patterns. In other words, it appears that the earlier events reflect necessary steps or antecedents for the later patterns. Yet, even here appearances can be deceptive. For as I have already discussed, there are data showing that the prevention of the "preparatory" steps (trunk flexure, S wave, etc.) by immobilization of the embryo does not lead to deficits in the swimming pattern of the tadpole. Consequently, performance of the "earlier" behaviors in this case cannot be a necessary antecedent of swimming, but rather appears to merely represent an overt manifestation of the development of the underlying neuronal connectivity subserving swimming. Although such "antecedent" movements may serve some provisional function for the embryo (e.g., insuring normal bone, joint, or muscle development), they apparently are not necessary for the emergence of the neurogenetic mechanisms involved in swimming.

Although many similar behavioral sequences have been described for the ontogeny of motor patterns (both prenatally and postnatally) in a number of vertebrates (e.g., McGraw, 1942; Hooker, 1952; Narayanan et al., 1971), in none of these cases do we know the extent to which early steps may be skipped or prevented without disrupting later behavior. Furthermore, in the case of birds and mammals, the data are even less clear because in these forms many of the spontaneous embryonic activities consist of irregular, unorganized movements of all parts of the embryo that occur in no apparent or meaningful order or pattern with regard to later more complex behavior patterns (Hamburger, 1963, 1973; Oppenheim, 1974). I hasten to add, however, that because detailed, quantitative techniques such as those used to study postnatal motor patterns (Hinde, 1970; Fentress, 1976) have seldom been applied to the avian or mammalian embryo, there may, in fact, be more order in this apparent chaos than we recognize or are willing to admit (e.g., Bekoff, 1976; Provine, 1980). Such studies, although tedious, time-consuming, and technically difficult, might yield valuable information that would justify the efforts involved. Although it has often been implied that because of the apparent random nature of embryonic motility in birds their activity cannot be considered as a direct precursor of complex postnatal patterns (e.g., Hamburger and Oppenheim, 1967), it is obvious that in the absence of detailed quantitative information our understanding of the significance of many aspects of embryonic behavior will remain in the realm of conjecture, in which case our conclusions will probably often be wrong. The embryo is a scrupulous guardian of its secrets, often revealing to the casual observer only what is superficial, irrelevant, or trivial, while hiding its most fundamental characteristics.

Despite the general lack of information on the issue of prenatal antecedents to postnatal behavior, however, there is now abundant evidence showing that overt embryonic and fetal activity in birds and mammals regulates certain aspects of muscle differentiation, prevents the atrophy of differentiated muscle (Jacobson, 1978, review), and is important for the normal formation of joints and bone articulations (Murray and Drachman, 1969). In the chick, the prevention of activity by pharmacological agents like curare for even relatively short periods of time (i.e., 48 hours) can lead to long-lasting and perhaps permanent muscle and joint abnormalities (Oppenheim et al., 1978). Although relatively trivial from a behavioral or neuroembryological point of view, these mechanisms are of obvious importance for the embryo and newborn animal; based on observations of newly hatched chicks that were immobilized prenatally, it is clear that such animals could not survive in nature (Oppenheim et al., 1978).[11]

For the avian embryo, hatching constitutes an obvious example of a behavior serving a transient adaptive function. During the last few days prior to its escape from the shell, much of the behavior of the embryo is devoted to "prehatching" activities that are involved in attaining a position within the shell necessary for successful hatching (Hamburger and Oppenheim, 1967; Oppenheim, 1973). And, of course, the hatching act itself is a unique, complex, and stereotyped motor pattern that is used only once in the life of the bird. Similarly, in mam-

---

[11] These results may represent a specific behavioral example of what is apparently a rather common theme in the evolution of ontogenetic processes. Kleinenberg (1886) has referred to this as the principle of substitution, in which it is held that the ontogeny of an evolutionary novelty is dependent on the presence of the ancestral structure (or function) for which the novelty is substituted. For example, gill slits or visceral pouches are retained in the embryos of birds and mammals not because they are adaptive per se but because they are necessary for the formation of novel structures such as the eustachian tubes, the tonsils, and the thymus gland.

Similarly, in lower chordates the notocord is retained in the adult as a structural support for the body, whereas in higher vertebrates it is only present in the embryo, where it (or its anlage) serves to induce the formation of the nervous system. By comparison, although embryonic motility may have served an adaptive role in the development of behavior of the ancestors of birds and mammals (e.g., as an indispensible antecedent of adult locomotion), its retention in the embryos of "modern" animals may only reflect the fact that the epigenetic mechanisms controlling the prenatal ontogeny of evolutionary novelties in the nervous system (or other systems) of birds and mammals utilize the same ancestral mechanisms, one aspect of which involves the generation of embryonic motility. Perhaps the demonstrated role of embryonic motility in the normal development of muscle, bone, and joints in present birds and mammals reflects such a process in which the adaptive role of a functional trait in the ontogeny of an ancestor (e.g., antecedents of postnatal behavior patterns) has been replaced or substituted in the ontogeny of the descendants (e.g., by the control of muscle development).

mals, postnatal suckling is a transient behavior pattern that is later superceded by a different mode of ingestive behavior. Prevention of suckling has little, if any, detrimental effect upon the later acquisition or perfection of adult ingestive behavior (Hall, 1975). In humans, fetal motor behavior may help attain an appropriate intrauterine position for birth (Langreder, 1949). (Regrettably, space does not permit a further delineation of additional examples, but see Oppenheim, 1981a.)

Provisional structures and functions are commonplace and most striking among animals with complex life histories (e.g., insects, amphibians) in which major metamorphic changes occur, often transforming animals so drastically from one stage to the next that they end up being mere shadows of their former selves. Destruction followed by dramatic reorganization or even the appearance of entirely new features are familiar themes of development in such forms, and the nervous system and behavior are no exception. Although I do not wish to offend my colleagues in developmental psychology by claiming that the ontogeny of the nervous system and behavior in "higher" vertebrates is metamorphic in nature, I would argue that even some of the regressions, losses, and other changes that occur during human development are only slightly less dramatic than the changes that amphibians undergo in their transformation from tadpoles into frogs.

In the case of animals that exhibit striking metamorphic changes during development, it is likely that the resulting structural–functional transformations represent the expression of "new" genes. In discussing the changes involved in the development of higher insects (i.e., from egg to larva, from larva to pupa, and from pupa to adult), C. M. Williams has said that, "We may think of the genome as being subdivided into three different gene sets corresponding to the successive chapters in the construction manual" (1969, p. 133). The embryologist N. J. Berrill made the same point regarding metamorphic life cycles and has added that, "Metamorphic change during the developmental cycle is an acceleration or condensation of essentially *the same* basic processes characteristic of most forms of development" (1971, p. 242, my italics). And in a discussion of mammalian development, E. F. Adolph has stated that, "As far as we know, each stage of development is functionally complete in its own right" (1957, p. 131).

Even for humans it is clear that during development new genes do not stop being expressed at birth. But with the exception of puberty, the time-dependent penetrance of certain heritable mental and neurological diseases, and perhaps certain aspects of senescence, we simply do not know the extent to which major developmental changes in behavior occurring postnatally reflect neurobiological events. And we have little understanding of how, or even whether, such changes depend upon differential gene expression (or repression) or whether epigenetic processes similar to those that characterize embryogenesis are involved.

What is clear, however, is that our previously held beliefs about the relative immutability of the nervous system following embryonic development are gradually giving way to a more dynamic perspective in which changes in neural structure and function are seen as continuing postnatally and perhaps even throughout the life cycle. Even during those stages that have been commonly accepted as representing the major periods of neural growth and differentiation, it is now clear that there is a greater degree of change and transformation than was generally suspected. Neuronal cell death (Oppenheim, 1981b), the formation and regression of transient neuronal structures and synaptic connections (Mark, 1980; Purves and Lichtman, 1980; Innocenti, 1982), transient neural and behavioral functions (Oppenheim, 1981a), and the loss of neural structures and functions associated with aging (Finch and Hayflick, 1977) are becoming widely recognized as normal, albeit enigmatic, features of development.

Thus, from a neuroembryological standpoint, it would come as no great surprise if, in the future, some of the most complex behavioral mechanisms of humans and other animals throughout the life cycle are found to be related to neurobiological changes that involve the transformation, reorganization, suppression, or regression of previously acquired structures and functions, and which depend, to one degree or another, upon differential gene activity and epigenetic mechanisms usually only associated with early developmental stages.[12]

As Ross Harrison once prophetically commented, "from the earliest stages of development the nervous system is highly regulable, a state

[12] Although it should be obvious that I am not arguing that the appearance of stage-specific (and species-typical) behavioral capacities are an inevitable result of gene-related neurogenetic mechanisms, past confusion over this pseudoissue dictates that I offer a brief explanation of what I do mean. All I wish to imply here is that epigenetic processes similar to those at work in the embryo may continue to a greater or lesser degree during infancy, childhood, the juvenile period, and perhaps even later (see Wilson, 1978; Scarr-Salapatek, 1976; Erikson, 1968; and Bretherton and Ainsworth, 1980, for similar views with regard to human behavioral development). As is always the case in the expression of a phenotypic character, environmental factors of one kind or another are certainly involved and indeed they may often be critical for regulating the expression of a capacity, as well as for influencing the time or extent of its expression and its subsequent maintenance. The only apparent novelty in my argument is the suggestion that, in some cases, the neurobiological mechanisms that do develop and mediate behavior at one stage may be transient ontogenetic adaptations and thus not merely direct antecedents or incomplete forerunners of the mechanisms or traits manifest in subsequent stages (Oppenheim, 1981a).

which is retained for a long period, possibly as long as the organism is capable of growing and learning" (1935, in Harrison 1969, p. 118). The embryo, fetus, and larva may represent only the most obvious examples of a general tendency in development for neural structures and functions to exhibit transient and provisional features.

Finally, I would be remiss if I did not at least mention the possibility—however unpalatable it may be—that much of what occurs behaviorally during prenatal development, may in the final analysis, represent little more than epiphenomena. It has long been recognized by evolutionary and developmental biologists that the earliest stages of ontogeny are frequently more conservative and stable relative to later stages simply because, in general, the earlier in development a perturbation occurs—be it genetic or epigenetic in nature—the more widespread and disruptive its effects will be. Thus, it is conceivable that in the course of evolution, certain traits that occur during the early ontogenetic stages of a descendant and that were adaptive in the embryos of the ancestors may be retained, not because they continue to serve a specific adaptive function themselves but, rather, because they may be inextricably linked with phylogenetically "new" structures and functions that simply cannot be easily altered or discarded without adversely affecting later ontogenetic events. For instance, the cellular mechanisms regulating normal muscular development in the embryo may be so closely linked genetically with events involved in the production of embryonic motility that it may not be possible for natural selection to act on one without altering the other (see footnote 12). (Viewed from this perspective, it is perhaps less surprising that the elimination or disruption of embryonic motility appears to have little, if any, lasting or significant effects on later behavior.)

Although space does not permit me to discuss all of the implications of this notion, I would be remiss if I did not at least mention that it serves to illustrate an important point: namely, that an organism's ancestry places strong constraints on its ontogeny and, by implication, on its potential for further evolution. The significance of this relationship between ontogeny and phylogeny was understood by an earlier generation of biologists (e.g., de Beer, Schmalhausen, Waddington) but was subsequently ignored and only recently shows signs of being resuscitated (e.g., see Gould, 1977).

VI. Concluding Remarks

In this chapter I have attempted to sketch in rather broad outline the current status and future prospects of several issues concerning the ontogenetic study of behavior from a neuroembryological perspective.

No attempt has been made to review any of these matters in detail and as a result many interesting facts and issues have been slighted or ignored. In fact, a comprehensive and definitive examination of this field has yet to be written. In lieu of that, the interested reader who desires additional information will find ample material in the list of references.

Much of the fascination for studying the neuroembryology of behavior is derived from the persistent hope that was so aptly expressed by the poet Samuel Coleridge, when he noted that, "the history of man for the nine months preceding his birth, would probably be far more interesting and contain events of greater moment, than all of the threescore and ten years that follow it" (1885, p. 301). The father of this field, W. Preyer, acknowledged the same hope when he suggested that "The fundamental activities of the mind, which are manifested only after birth, do not originate after birth" (1888, p. xii). As I have noted earlier, it was Preyer's belief that the "impulsive" or spontaneous movements of the embryo form the starting point for the development of the mind (i.e., of psychological processes). The belief that embryonic behavior may be shown to make an important contribution to the development of even the most complex postnatal behavior patterns has continued to serve as an inspiration to investigators in this field. For instance, in his 1963 review, which inaugurated the most recent resurgence of interest in this field, Hamburger expressed the belief that "this field . . . has significant implications for neurogenesis and neurology *as well as for ethology and psychology*" (p. 343, italics added).

In contrast, developmental and child psychologists (and child psychiatrists) have, by and large, tended to be unaware of, or to either ignore or dispute, the potential contributions of the prenatal period to later psychological development. Against this background, however, the independence of thought of a few child psychologists such as Gesell (1954), McGraw (1935), and, more recently, Piaget stands in bold relief. All of these investigators have recognized the value of a neuroembryological approach to behavioral development. Concerning neural development and behavior, Piaget, for instance, has expressed the hope that it may some day be possible "to demonstrate relationships between mental structures and stages of nervous development and thus to arrive at that general theory of structures to which my early studies constitute merely an introduction" (Piaget, 1952, p. 256). And along the same lines, he has argued that "future theories (*of behavior*) will be acceptable only if they succeed in integrating interpretations of embryogenesis, organic growth and mental development" (Piaget and Inhelder, 1969, pp. 153–154). Finally, concerning the specific contribution of embryonic behavior to mental development, Piaget has stated his belief that "The point of departure of development should not be sought in reflexes . . . but in the spontaneous and total activities of the organism. There is a continuous progression from spontaneous movements . . . to acquired habits and from the latter to intelligence" (Piaget and Inhelder, 1969, pp. 5–6).

The study of behavioral development from a neuroembryological perspective has proved to be a useful and advantageous approach, providing fresh insights, establishing new facts, and helping resolve old issues. Although there has been a marked increase in our knowledge since Coghill's time, the belief that the neuroembryological study of behavior can shed important light on the postnatal development of behavior still lies more in expectation that in present accomplishment, and more in faith than in established fact. Due to its complexity and diversity, its creations and revelations, its progressions and regressions, we can be certain that our present notions of the development of the nervous system and behavior represent only a vague, inchoate, and highly simplified approximation of reality. We fool ourselves if we believe that the currently fashionable ideas about these problems are anything more than a fraction of the truth yet to be discovered.

Although the development of new techniques and approaches from neurobiology and genetics, including the use of neurological mutants, offers the hope for fundamental breakthroughs in our understanding of neurogenesis and behavioral development and of their genetic and epigenetic control, it would be regrettable if these promising innovations were to lead to a neglect of the older, more established approaches to these problems. There is still much to be gained from careful and thoughtful behavioral analyses of prenatal development, from the tedious and often unexciting descriptions of neuroanatomical development (perfected to such a high degree by Ramón y Cajál, Coghill, Windle, and others) and from the patient and judicious application of techniques of classical experimental embryology involving surgical deletions, additions, and transplantations of parts of the embryo. So far the signs are encouraging that the new techniques, approaches, and concepts will be successfully utilized and integrated with those from the past.

Although the technical revolution in neurobiology in the past 20 years has served as an alluring impetus for investigators of the developing nervous system to focus their attention on ever more molecular levels of neurogenesis, it should never be forgotten that it is the entire developing organism and its behavior that is our major concern—and that has also been the major focus of natural selection. Reductionism in the neuroembryological study of behavior, as in other areas of psychobiology, if judiciously used (and if the results are thoughtfully interpreted), represents an essential and powerful approach to many of the unresolved questions in this field.

However, the behavior of a developing (or adult) organism will not be accounted for solely in terms of the properties of its units or in terms of the properties of more molecular levels of organization; indeed, one could argue that one must first understand the normal or naturally occurring behavior *qua* behavior before one can even begin to ask intelligent questions about the contribution of more molecular levels to the organization of the behavior. But by studying the units, we may eventually be in a better position to understand other, more complex properties (e.g., behavior) that the units exhibit when in combination with one another. Although each higher level of organization or complexity in the nervous system is dependent on the properties of the levels below, it is the more complex, emerging properties, especially behavior, that are the primary focus of the approach advocated here. Eventually the bits and pieces derived from the lower levels will have to be integrated in such a way that our understanding of the entire organism and its changing relationship to the environment at all stages of ontogeny becomes the major leitmotiv for a valid neuroembryology of behavior. Viewed from this perspective, it should be obvious that the problems involved in attaining an understanding of the neuroembryology of behavior will be incredibly difficult (in part because they are still only dimly conceived) and the path to their resolution is likely to be long and tortuous and fraught with distractions, false hopes, and unforeseen setbacks. Should these comments seem unduly pessimistic, however, I hasten to add that, in my opinion, the challenges posed by these difficulties will be far surpassed by the prizes to be gained from their resolution.

VII. Summary

The neuroembryological study of behavior represents the attempt to understand the ontogeny of behavior and its associated physiological and anatomical mechanisms, primarily, but not exclusively, during the embryonic, larval, or prenatal period. A major goal is to determine the precursors of later adaptive behavior in the activity patterns and neurobehavioral mechanisms of the embryo, larva, and fetus.

All embryos are behaviorally active to one degree or another, and it has been shown for some species that this behavior is spontaneous, in the sense that sensory input is not necessary for its generation or pattern. Although it is presently uncertain whether this spontaneous embryonic behavior is related to, or necessary for, later behavioral development, it has been shown that the abolition of spontaneous behavior by experimental means leads to deficits in muscle, joint, and bone development. All embryos can also respond to various kinds of sensory stimulation. The question of whether sensory experience prior to birth or hatching is important for later neural and behavioral development has long been an issue of great concern. With the exception of several interesting studies on the role of auditory experience in bird embryos, however, there is little concrete evidence at present to indicate that this is the case.

Active neural mechanisms mediating physiological inhibition appear to develop very early in at least some vertebrate embryos. It is conceivable that inhibition is involved in suppressing the overt response of embryos to many kinds of sensory stimulation. It is also possible that inhibitory mechanisms develop precociously in order to subserve some provisional neurogenetic trophic-like function in early neural development. Or, as is the case with so many other aspects of embryonic and prenatal development, the early appearance of inhibition may merely reflect the development of a mechanism that is not adaptively needed until after birth or hatching.

Some features of embryonic neural or behavioral function may represent ontogenetic adaptations (i.e., adaptations to the *in ovo* or *in utero* environment that are not necessarily antecedents of later, postnatal or adult, adaptations). Although higher insects and amphibians, as well as other forms that exhibit striking life-history changes, present the clearest examples of ontogenetic adaptations, there is no reason to believe that other species, including birds and mammals, are fundamentally different in this respect. Different environments require different neurobehavioral and other adaptations and this is as true for ontogeny as it is for phylogeny.

With the application of the myriad new neurobiological techniques developed in the last few years—and by the judicious use of older approaches, including behavioral, neuroanatomical, embryological, and genetic studies—there is the justifiable expectation that the neuroembryological study of behavior may eventually help to resolve one of the remaining great mysteries of biology; namely, whether (and if so, how) prenatal life events contribute in any direct and significant way to later behavioral development. To accomplish this, however, radical new ways of thinking about these problems may be required.

ACKNOWLEDGMENTS

I am indebted to a few friends for their comments and suggestions on an earlier draft of this chapter. I wish to thank George Barlow, Viktor Hamburger, and David Miller for their efforts. The suggestions and comments of Lanny Haverkamp were especially helpful. Peter Witt kindly translated the quotations from W. Preyer.

A part of the expenses incurred in the preparation of this chapter were paid for by the Zentrum für interdisziplinare Forshung of the University of Bielefeld, Federal Republic of Germany. A longer, and rather different, German version of this chapter is to appear in Oppenheim (1983).

REFERENCES

Adam, J., and Dimond, S. J. (1971). *Anim. Behav.* **19**, 51–54.

Adolph, E. F. (1957). *Q. Rev. Biol.* **32**, 89–137.

Anokhin, P. K. (1964). *Prog. Brain Res.* **9**, 54–86.

Armitage, S. E., Baldwin, B. A., and Vince, M. A. (1980). *Science* **208**, 1173–1174.

Bain, A. (1855). "The Senses and the Intellect." Parker & Son, London.

Barcroft, J. (1938). "The Brain and its Environment." Yale Univ. Press, New Haven, Connecticut.

Barcroft, J., and Barron, D. H. (1939). *J. Comp. Neurol.* **70**, 477–502.

Barlow, H. B. (1975). *Nature (London)* **258**, 199–204.

Bekoff, A. (1976). *Brain Res.* **106**, 271–291.

Bekoff, A. (1981). *In* "Studies in Developmental Neurobiology: Essays in Honor of Viktor Hamburger" (W. M. Cowan, ed.), pp. 134–170. Oxford Univ. Press, London and New York.

Berger, H. (1900). *Arch. Psychiatr. Nervenkr.* **33**, 521–567.

Bergey, G. K., MacDonald, R. L., and Nelson, P. G. (1978). *Neurosci. Abstr.* **4**, 601.

Berrill, N. J. (1971). "Developmental Biology." McGraw-Hill, New York.

Bird, M. M. (1980). *Cell Tissue Res.* **206**, 115–122.

Bjärvall, A. (1967). *Behaviour* **28**, 141–148.

Brenner, S. (1973). *Br. Med. Bull.* **29**, 269–271.

Bretherton, I., and Ainsworth, M. D. S. (1980). *In* "Species Identity and Attachment: A Phylogenetic Evaluation" (M. A. Roy, ed.), pp. 311–332. Garland, New York.

Bruce, I. C., and Tatton, W. B. (1980). *Exp. Brain Res.* **39**, 411–419.

Burchfiel, J. L., and Duffy, F. H. (1981). *Brain Res.* **106**, 479–484.

Campbell, S. K. (1982). *Adv. Behav. Pediatr.* **4**, in press.

Carmichael, L. (1926). *Psychol. Rev.* **33**, 51–68.

Carmichael, L. (1927). *Psychol. Rev.* **34**, 34–47.

Carmichael, L. (1941). *Psychol. Bull.* **38**, 1–28.

Carmichael, L. (1954). *In* "Manual of Child Psychology" (L. Carmichael, ed.), pp. 60–185. Wiley, New York.

Carpenter, W. B. (1874). "Principles of Mental Physiology." H.S. King, London.

Carr, V. M., and Simpson, S. B. (1978). *J. Comp. Neurol.* **182**, 727–740.

Chef, R., ed. (1979). "Ultrasound in Prenatal Medicine: Contributions to Gynecology and Obstetrics," Vol. 6. Karger, Basel.

Chronwall, B., and Wolff, J. R. (1980). *J. Comp. Neurol.* **190**, 187–208.

Coghill, G. E. (1929a). *Arch. Neurol. Psychiatry* **21**, 989–1009.

Coghill, G. E. (1929b). "Anatomy and the Problem of Behavior." Cambridge Univ. Press, London and New York.

Coghill, G. E. (1933). *Science* **78**, 131–138.

Coghill, G. E. (1934). *Anat. Rec.* **58**, 55–56.

Coleridge, S. T. (1885). "Miscellanies, Aesthetic and Literary." Bell & Sons, London.

Corner, M. A. (1977). *Prog. Neurobiol.* **8**, 279–295.

Crain, S. M. (1973). *In* "Studies on the Development of Behavior and the Nervous System" (G. Gottlieb, ed.), Vol. 2, pp. 69–114. Academic Press, New York.

Crain, S. M. (1976). "Neurophysiologic Studies in Tissue Culture." Raven, New York.
Crain, S. M. (1980). *In* "Tissue Culture in Neurobiology" (E. Giacobini, ed.), pp. 169–185. Raven, New York.
Crain, S. M., Bornstein, M. B., and Peterson, E. R. (1968). *Brain Res.* **8**, 363–372.
Cumming, I. (1955). "Helvetius, His Life and Place in the History of Educational Thought." Routledge & Kegan Paul, London.
Darwin, E. (1796). "Zoonomia: Or the Law of Organic Growth," Vol. 2. E. Earle, Philadelphia, Pennsylvania (4th American edition, 1818).
Dawes, G. S., Fox, H. E., LeDuc, B. M., Liggins, G. D., and Richards, R. T. (1972). *J. Physiol. (London)* **220**, 119–143.
DeCasper, A. H., and Fifer, W. P. (1980). *Science* **208**, 1174–1176.
Decker, J. D. (1970). *J. Exp. Zool.* **174**, 349–364.
Delcomyn, F. (1980). *Science* **210**, 492–498.
Detwiler, S. R. (1936). "Neuroembryology." Hafner reprint, New York (1964).
Detwiler, S. R. (1947). *J. Exp. Zool.* **106**, 299–312.
Dimond, S. J., and Adam, J. H. (1972). *Anim. Behav.* **20**, 413–420.
Douglas, J. W. B., and Gear, R. (1976). *Arch. Dis. Child.* **51**, 820–827.
Dreyfus-Brisac, C. (1975). *Biol. Psychiatry* **10**, 485–496.
Dubignon, J. M., Campbell, D., and Partington, M. W. (1969). *Biol. Neonate* **14**, 270–278.
Dubowitz, L., Dubowitz, V., Romero, A., and Verghote, M. (1980). *In* "Multidisciplinary Approach to Brain Development" (C. DiBenedetta, R. Balazs, G. Gombas, and G. Porcellati, eds). pp. 523–536. Elsevier, Amsterdam.
Duffy, F. W., Snodgrass, R. S., Burchfiel, J. L., and Conway, J. L. (1976). *Nature (London)* **260**, 256–257.
Dweck, H. W., Saxon, S. A., Benton, J. W., and Cassady, G. (1973). *Am. J. Dis. Child* **126**, 28–34.
Erikson, E. H. (1968). "Identity, Youth and Crisis." Norton, New York.
Etkin, W. A., and Gilbert, L. I. (1968). "Metamorphosis." Appleton, New York.
Evans, R. M. (1973). *Can. J. Zool.* **51**, 759–770.
Faber, R. (1956). *Arch. Neerl. Zool.* **11**, 498–517.
Fantz, R. L., and Fagan, J. F. (1975). *Child Devl.* **46**, 3–18.
Fein, K., Stokes, B. T., and Gonya, T. (1978). *Exp. Neurol.* **60**, 221–230.
Fentress, J. C., ed. (1976). "Simpler Networks and Behavior." Sinauer Assoc., Sunderland, Massachusetts.
Finch, C. E., and Hayflick, L., eds. (1977). "Handbook of the Biology of Aging." Van Nostrand-Reinhold, Princeton, New Jersey.
Fitzhardinge, P. M. (1975). *Pediatrics* **56**, 162–172.
Francis-Williams, J., and Davies, P. A. (1974). *Dev. Med. Child Neurol.* **17**, 709–728.
Fried, R., and Gluck, S. (1966). *Psychonom. Sci.* **6**, 319–320.
Fromme, A. (1941). *Genet. Psychol. Monogr.* **24**, 219–256.
Gardner, J., Lewkowicz, D., and Turkewitz, G. (1976). *Dev. Psychobiol.* **10**, 471–480.
Gates, E. (1895). *Monist* **5**, 574–597.
Gesell, A. L. (1928). "Infancy and Human Growth." Macmillan, New York.
Gesell, A. L. (1954). *In* "Manual of Child Psychology" (L. Carmichael, ed.), pp. 335–373. Wiley, New York.
Goldman, P. S., and Galkin, T. W. (1978). *Brain Res.* **152**, 451–485.
Goodlin, R. C. (1979). "Care of the Fetus." Masson Press, New York.
Gos, M. (1935). *Bull. Soc. R. Sci. Liege* **4**, 194–199, 246–250.
Gottlieb, G. (1970). *In* "Development and Evolution of Behavior" (L. R. Aronson, E. Tobach, D. S. Lerhman, and J. S. Rosenblatt, eds.), pp. 111–137. Freeman, San Francisco, California.

Gottlieb, G. (1971a). *In* "The Biopsychology of Development" (E. Tobach, L. R. Aronson, and E. Shaw, eds.), pp. 67–128. Academic Press, New York.

Gottlieb, G. (1971b). "Development of Species Identification in Birds: An Inquiry into the Prenatal Determinants of Perception." University of Chicago, Chicago, Illinois.

Gottlieb, G., ed. (1973). "Studies on the Development of Behavior and the Nervous System," Vol. 1. Academic Press, New York.

Gottlieb, G. (1976a). *In* "Studies on the Development of Behavior and the Nervous System" (G. Gottlieb, ed.), Vol. 3, pp. 25–54. Academic Press, New York.

Gottlieb, G. (1976b). *In* "Studies on the Development of Behavior and the Nervous System" (G. Gottlieb, ed.), Vol. 3, pp. 237–280. Academic Press, New York.

Gottlieb, G. (1978). *J. Comp. Physiol. Psychol.* **92,** 375–387.

Gottlieb, G. (1979). *J. Comp. Physiol. Psychol.* **93,** 831–854.

Gottlieb, G. (1980). *J. Comp. Physiol. Psychol.* **94,** 579–587.

Gould, S. (1977). "Ontogeny and Phylogeny." Belknap Press, Cambridge, Massachusetts.

Greenough, W. T. (1976). *In* "Neural Mechanisms of Learning and Memory" (M. R. Rosenzweig and E. L. Bennett, eds.), pp. 255–278. MIT Press, Cambridge, Massachusetts.

Grier, J. C., Counter, S. A., and Shearer, W. M. (1967). *Science* **155,** 1692–1693.

Grobstein, P., and Chow, K. L. (1975). *Science* **190,** 352–358.

Hall, W. G. (1975). *Science* **190,** 1313–1315.

Hamburger, V. (1948). *J. Comp. Neurol.* **88,** 221–224.

Hamburger, V. (1963). *Q. Rev. Biol.* **38,** 342–365.

Hamburger, V. (1973). *In* "Studies on the Development of Behavior and the Nervous System" (G. Gottlieb, ed.), pp. 52–76. Academic Press, New York.

Hamburger, V. (1975). *In* "The Mammalian Fetus" (E. S. E. Hafez, ed.), pp. 68–81. Thomas, Springfield, Illinois.

Hamburger, V., and Narayanan, C. H. (1969). *J. Exp. Zool.* **170,** 411–426.

Hamburger, V., and Oppenheim, R. W. (1967). *J. Exp. Zool.* **166,** 171–204.

Hamburger, V., Wenger, E., and Oppenheim, R. W. (1966). *J. Exp. Zool.* **162,** 133–160.

Harris, W. A. (1980). *J. Comp. Neurol.* **194,** 303–317.

Harris, W. A. (1981). *Annu. Rev. Physiol.* **43,** 689–710.

Harrison, R. G. (1904). *Am. J. Anat.* **3,** 197–220.

Harrison, R. G. (1969). "Organization and Development of the Embryo." Yale Univ. Press, New Haven, Connecticut.

Harth, M. S. (1974). *In* "Ontogenesis of the Brain" (L. Jilek and S. Trojan, eds.), pp. 57–65. University Press, Prague.

Haverkamp, L. J., and Oppenheim, R. W. (1981). *Neurosci. Abstr.* **7,** 181.

Herrick, C. J. (1949). "George Ellet Coghill, Naturalist and Philosopher." Univ. of Chicago Press, Chicago, Illinois.

Hess, E. H. (1973). "Imprinting: Early Experience and the Developmental Psychobiology of Attachment." Van Nostrand-Reinhold, Princeton, New Jersey.

Hinde, R. (1970). "Animal Behavior: A Synthesis of Ethology and Comparative Psychology." McGraw-Hill, New York.

His, W. (1874). "Unsere Körperform und das physiologische Problem ihrer Entstehung." Vogel, Leipzig.

Hollyday, M., and Hamburger, V. (1977). *Brain Res.* **132,** 197–203.

Holt, E. B. (1931). "Animal Drive and the Learning Process." Holt, New York.

Holtzer, H., and Kamrin, R. P. (1956). *J. Exp. Zool.* **132,** 391–407.

Hooker, D. (1944). "The Origin of Behavior." Univ. of Michigan Press, Ann Arbor.

Hooker, D. (1952). "The Prenatal Origin of Behavior." University of Kansas, Lawrence.

Hunt, E. L. (1949). *J. Comp. Psychol.* **42,** 107–117.

Impekoven, M. (1976). *Adv. Study Behav.* **7**, 201–253.

Innocenti, G. (1982). *In* "Recovery from Brain Damage" (M. W. Van Hof and J. Mohn, eds.). Elsevier, Amsterdam (in press).

Jacobson, M. (1978). "Developmental Neurobiology," 2nd ed. Plenum, New York.

Kaback, M. M., and Valenti, C. (1976). "Intrauterine Fetal Visualization." Am. Elsevier, New York.

Kammer, A. E., Dahlman, D. L., and Rosenthal, G. E. (1978). *J. Exp. Biol.* **75**, 123–132.

Kelsey, S. M., and Barrie-Blackley, S. (1976). *Dev. Med. Child Neurol.* **18**, 753–758.

Kleinenberg, M. (1886). *Z. Wiss. Zool.* **44**, 212–273.

Koppanyi, T., and Karczmar, A. G. (1947). *Fed. Proc., Fed. Am. Soc. Exp. Biol.* **6**, 346.

Kuo, Z.-Y. (1932). *J. Comp. Psychol.* **13**, 245–271.

Kuo, Z.-Y. (1939). *Psychol. Rev.* **46**, 93–122.

Kuo, Z.-Y. (1967). "The Dynamics of Behavior Development." Random House, New York.

Lamarck, J. B. (1809). "Zoological Philosophy" (English translation by H. Elliot, 1914). Hafner reprint, New York (1963).

Landmesser, L. (1978). *J. Physiol. (London)* **284**, 391–414.

Langreder, W. (1949). *Z. Geburtschilfe Gynaekol.* **131**, 237–245.

Lauder, J. M., and Krebs, H. (1978). *Dev. Neurosci.* **1**, 13–30.

Luys, J. (1882). "The Brain and its Function." Appleton, New York.

Macfarlane, A. (1977). "The Psychology of Childbirth." Open Books, London.

McGraw, M. (1935). "Growth: A Study of Johnny and Jimmy." Appleton, New York.

McGraw, M. (1942). "The Neuromuscular Maturation of the Human Infant." Hafner reprint New York (1963).

Macklin, M., and Wotjtkowski, W. (1973). *J. Comp. Physiol.* **84**, 41–58.

Mark, R. F. (1980). *Physiol. Rev.* **60**, 355–395.

Matthews, S. A., and Detwiler, S. R. (1926). *J. Exp. Zool.* **45**, 279–292.

Maudsley, H. (1876). "The Physiology of Mind." Macmillan, New York.

Miller, B. F., and Lund, R. D. (1975). *Brain Res.* **91**, 119–125.

Miller, D. B., and Gottlieb, G. (1978). *Anim. Behav.* **26**, 1178–1194.

Model, P., Bornstein, M. B., Crain, S. M., and Pappas, G. D. (1971). *J. Cell Biol.* **49**, 363–371.

Mueller, J. (1843). "Elements of Physiology." Lea & Blanchard, Philadelphia, Pennsylvania.

Murray, P. D. G., and Drachman, D. B. (1969). *J. Embryol. Exp. Morphol.* **22**, 339–371.

Narayanan, C. H., and Malloy, R. B. (1974a). *J. Exp. Zool.* **189**, 163–176.

Narayanan, C. H., and Malloy, R. B. (1974b). *J. Exp. Zool.* **189**, 177–188.

Narayanan, C. H., Fox, M. W., and Hamburger, V. (1971). *Behaviour* **40**, 100–134.

Nelson, P. G. (1975). *Physiol. Rev.* **55**, 1–61.

Nelson, P. G., Ranson, B. R., Henkart, M., and Bullock, P. N. (1977). *J. Neurophysiol.* **40**, 1178–1187.

Nornes, H. O., Hart, H., and Carry, M. (1980). *J. Comp. Neurol.* **192**, 133–141.

Obata, K. (1977). *Brain Res.* **119**, 141–153.

Okado, N., Kakimi, S., and Kojima, T. (1979). *J. Comp. Neurol.* **184**, 491–518.

Oppenheim, R. W. (1966). *Science* **152**, 528–529.

Oppenheim, R. W. (1972a). *Proc. Int. Congr. Ornithol, 15th, 1970* pp. 283–302.

Oppenheim, R. W. (1972b). *Dev. Psychobiol.* **5**, 71–91.

Oppenheim, R. W. (1973). *In* "Studies on the Development of Behavior and the Nervous System" (G. Gottlieb, ed.), Vol. 1, pp. 163–244. Academic Press, New York.

Oppenheim, R. W. (1974). *Adv. Study Behav.* **5**, 133–172.

Oppenheim, R. W. (1975). *J. Comp. Neurol.* **160**, 37–50.

Oppenheim, R. W. (1978). *Perspect. Biol. Med.* **22**, 44–64.

Oppenheim, R. W. (1979). *Dev. Psychobiol.* **12**, 533–537.

Oppenheim, R. W. (1981a). *In* "Maturation and Development: Biological and Psychological Perspectives" (K. Connolly and H. F. R. Prechtl, eds.), pp. 73–109. Lippincott, Philadelphia, Pennsylvania.

Oppenheim, R. W. (1981b). *In* "Studies in Developmental Neurobiology: Essays in Honor of Viktor Hamburger" (W. M. Cowan, ed.) pp. 74–133. Oxford Univ. Press, London and New York.

Oppenheim, R. W. (1982). *In* "Perspectives in Ethology," Vol. 5 (P. Bateson and P. Klopfer, eds.). Plenum, New York (in press).

Oppenheim, R. W. (1983). *In* "Vehaltensentwicklung bei Mensch und Tier" (K. Immelmann, G. Barlow, M. Main, and L. Petrinovich, eds.). Parey, Berlin (in press).

Oppenheim, R. W., and Reitzel, J. (1975). *Brain, Behav. Evol.* **11**, 130–159.

Oppenheim, R. W., Pittman, R., Gray, M., and Maderdrut, J. L. (1978). *J. Comp. Neurol.* **179**, 619–640.

Parmelee, A. H. (1975). *Biol. Psychiatr.* **10**, 501–512.

Parmelee, A. H., Stern, E., and Harris, M. A. (1971). *Neuropaediatrie* **3**, 294–304.

Perlow, M. J., Freed, W. J., Hoffer, B. J., Seiger, A., Olson, L., and Wyatt, R. J. (1979). *Science* **204**, 643–647.

Piaget, J. (1952). *In* "A History of Psychology in Autobiography" (E. G. Boring, ed.), Vol. 4, pp. 237–256. Clark Univ., Worcester, Massachusetts.

Piaget, J., and Inhelder, B. (1969). "The Psychology of the Child." Routledge & Kegan Paul, London.

Pittman, R., and Oppenheim, R. W. (1979). *J. Comp. Neurol.* **187**, 425–446.

Pittman, R., Oppenheim, R. W., and Ramakrishna, T. (1978). *J. Exp. Zool.* **204**, 95–112.

Planck, M. (1936). "The Philosophy of Physics." Norton, New York.

Plykkö, O. O., and Woodward, D. M. (1961). *J. Pharmacol. Exp. Ther.* **131**, 185–190.

Pollack, E. D., and Crain, S. M. (1972). *J. Neurobiol.* **3**, 381–385.

Pollack, E. D., and Kuwada, J. (1976). *Am. Zool.* **16**, 187.

Powell, L. F. (1974). *Child Dev.* **45**, 106–113.

Prechtl, H. F. R. (1965). *Adv. Study Behav.* **1**, 79–98.

Prechtl, H. F. R., Fargel, J. W., Weinmann, H. M., and Bakker, H. H. (1975). *In* "Aspects of Neural Plasticity" (F. Vital-Durrand and M. Jennerod, eds.), pp. 55–66. INSERM, Paris.

Preyer, W. (1885). "Specielle physiologie des Embryo." Grieben, Leipzig.

Preyer, W. (1888). "The Mind of the Child." Appleton, New York.

Provine, R. R. (1973). *In* "Studies on the Development of Behavior and the Nervous System" (G. Gottlieb, ed.), Vol. 1, pp. 77–102. Academic Press, New York.

Provine, R. R. (1976). *J. Insect Physiol.* **22**, 127–131.

Provine, R. R. (1979). *Behav. Neural Biol.* **27**, 233–237.

Provine, R. R. (1980). *Dev. Psychobiol.* **13**, 151–160.

Provine, R. R., and Rogers, L. (1977). *J. Neurobiol.* **8**, 217–228.

Purves, D., and Lichtman, J. W. (1980). *Science* **210**, 153–157.

Rajecki, D. W. (1974). *Behav. Biol.* **11**, 525–536.

Reitzel, J., and Oppenheim, R. W. (1980). *Dev. Psychobiol.* **13**, 455–461.

Reitzel, J., Maderdrut, J. L., and Oppenheim, R. W. (1979). *Brain Res.* **172**, 487–504.

Rice, R. D. (1977). *Dev. Psychol.* **13**, 69–76.

Riesen, A., ed. (1975). "The Developmental Neuropsychology of Sensory Deprivation." Academic Press, New York.

Roberts, E. (1976). *In* "GABA in Nervous System Function" (E. Roberts, T. H. Chase, and D. B. Tower, eds.), pp. 515–539. Raven, New York.

Rosen, M. G. (1978). *Harpers Mag.* April, pp. 46–47.

Rosner, B. S., and Doherty, N. E. (1979). *Dev. Med. Child Neurol.* **21,** 723–730.

Roux, W. (1881). "Der Kampf der Theile im Organismus." Engelmann, Leipzig.

Roy, M. A., ed. (1980). "Species Identity and Attachment: A Phylogenetic Evaluation." Garland, New York.

Saint-Anne Dargassies, S. (1977). "Neurological Development in the Full-Term and Premature Neonate." Excerpta Medica, Amsterdam.

Saito, K. (1979). *J. Physiol. (London)* **294,** 581–594.

Salk, L. (1962). *Trans. N. Y. Acad. Sci.* [2] **71,** 753–763.

Salk, L. (1973). *Sci. Am.* **228,** 24–29.

Scarr-Salapatek, S. (1976). *In* "Origins of Intelligence" (M. Lewis, ed.), pp. 165–197. Plenum, New York.

Sedlacek, J. (1964). *Physiol. Bohemoslov.* **13,** 411–420.

Sedlacek, J. (1976). *Physiol. Bohemoslov.* **25,** 505–509.

Sedlacek, J. (1977a). *Physiol. Bohemoslov.* **26,** 9–12.

Sedlacek, J. (1977b). *Physiol. Bohemoslov.* **26,** 311–314.

Sharma, S. C., Provine, R. R., Hamburger, V., and Sandel, T. T. (1970). *Proc. Natl. Acad. Sci. U.S.A.* **66,** 40–47.

Sigman, M., and Parmelee, A. H. (1974). *Child Dev.* **45,** 959–965.

Singer, H. S., Skoff, R. P., and Price, D. L. (1978). *Brain Res.* **141,** 197–209.

Slotkin, T. A., Smith, P. G., Lau, C., and Bareis, D. L. (1980). *In* "Biogenic Amines in Development" (H. Parvez and S. Parvez, eds.), pp. 29–48. Elsevier, Amsterdam.

Spelt, D. K. (1948). *J. Exp. Psychol.* **38,** 338–346.

Stein, B. E., Clamann, H. P., and Goldbert, S. J. (1980). *Science* **210,** 78–80.

Stehouwer, D. J., and Farel, P. B. (1980). *Brain Res.* **195,** 323–335.

Stokes, B. T., and Bignall, K. E. (1974). *Brain Res.* **77,** 231–242.

Taub, E. (1976). *In* "Neural Control of Locomotion" (R. Herman, S. Griller, P. S. G. Stein, and D. Stuart, eds.), pp. 675–705. Plenum, New York.

Taylor, A. C. (1944). *J. Exp. Zool.* **96,** 159–185.

Thibeault, D. W., Laul, V., and Gulak, H. (1975). *Pediatr. Res.* **9,** 107–110.

Tracy, H. C. (1926). *J. Comp. Neurol.* **40,** 253–369.

Truman, J. W. (1976). *J. Comp. Physiol.* **107,** 39–48.

Tschanz, B. (1968). *Z. Tierpsychol., Suppl.* **4,** 1–103.

Tschanz, B., and Hirsbrunner-Scharf, M. (1975). *In* "Evolution in Behavior" (G. P. Barends, C. G. Beer, and A. Manning, eds.), pp. 149–156. Oxford Univ. Press, London and New York.

Tsumoto, T., Eckart, W., and Creutzfeldt, O. D. (1979). *Exp. Brain Res.* **34,** 351–363.

Vince, M. A. (1973). *In* "Studies on the Development of Behavior and the Nervous System" (G. Gottlieb, ed.), Vol. 1, pp. 286–323. Academic Press, New York.

Vince, M. A. (1979). *Anim. Behav.* **27,** 908–918.

Vince, M. A., Reader, M., and Tolhurst, B. (1976). *J. Comp. Physiol. Psychol.* **90,** 221–230.

Visintini, F., and Levi-Montalcini, R. (1939). *Schweiz. Arch. Neurol. Psychiatr.* **43,** 1–45.

Waddington, C. H. (1957). "The Strategy of the Genes." Allen & Unwin, London.

Watanabe, K., Iwase, K., and Hara, K. (1972). *Dev. Med. Child Neurol.* **14,** 425–435.

Weismann, A. (1894). "The Effect of External Influences upon Development (The Romanes Lecture)." Oxford Univ. Press (Clarendon), London and New York.

Weiss, P. A. (1939). "Principles of Development: A Text in Experimental Embryology." Hafner reprint, New York (1969).

Weiss, P. A. (1941). *Comp. Psychol. Monogr.* **17,** 1–96.

Weiss, P. A. (1955). *In* "Analysis of Development" (B. H. Willier, P. A. Weiss, and V. Hamburger, eds.), pp. 346–401. Saunders, Philadelphia, Pennsylvania.

Weiss, P. A. (1970). *In* "The Neurosciences: Second Study Program" (F. O. Schmitt, ed.), pp. 53–61. Rockefeller Univ. Press, New York.

White, J. L., and Labarba, R. C. (1975). *Dev. Psychobiol.* **9**, 569–577.

Williams, C. M. (1969). *Dev. Biol., Suppl.* **3**, 133–150.

Wilson, R. S. (1978). *Science* **202**, 939–948.

Windle, W. F. (1940). "Physiology of the Fetus." Saunders, Philadelphia, Pennsylvania.

Windle, W. F., and Orr, D. W. (1934). *J. Comp. Neurol.* **60**, 287–307.

Wolff, P. H. (1966). *Psychol. Issues Monogr. Ser.* **5**, No. 17.

Wolff, P. H., and Ferber, R. (1979). *Annu. Rev. Neurosci.* **2**, 291–307.

Yntema, C. L. (1943). *J. Exp. Zool.* **94**, 310–349.

Young, D., ed. (1973). "Developmental Neurobiology of Arthropods." Cambridge Univ. Press, London and New York.

Ziskind-Conhaim, L. (1981). *Neurosci. Abstr.* **7**, 769.

INDEX

CONTENTS OF PREVIOUS VOLUMES

315